Drylands

T0199765

Many important areas of the world, including such key areas as the south-western United States, large parts of Israel, most of Australia and virtually all the Arab Gulf countries, together with much of northern Africa, can be classified as drylands or arid zones. Environmental management in these areas is extremely important in order to conserve and use resources in the best possible way. Misuse of resources, as has repeatedly happened in the past, can often have severe deleterious consequences with long-term implications. This book examines the drylands of the world and their management. It begins by describing drylands in a systematic way, by assessing the way human societies have evolved in response to the harsh and demanding nature of the physical landscape and by classifying the different human uses of drylands. It goes on to present detailed case studies of the human management of drylands, illustrating the different kinds of management adopted for different purposes, and contrasting the different approaches in different parts of the world. It includes case studies of dryland urban areas, a topic which has hitherto been much neglected and which is becoming increasingly important with the rapid growth of many dryland urban areas. The book concludes by assessing the likely future of drylands.

Peter Beaumont is Professor and Head of the Department of Geography, University of Wales, Lampeter, Dyfed, Wales, UK.

Routledge Natural Environment:
Problems and Management Series
Edited by Chris Park
University of Lancaster

The series offers a contemporary treatment of critical environmental topics. Adopting an interdisciplinary, international approach, it will be an important source for the academic, the practitioner, and the student of environmental affairs.

The Roots of Modern Environmentalism
David Pepper

Environmental Policies: An International Review
Chris C. Park

The Permafrost Environment
Stuart A. Harris

The Conservation of Ecosystems and Species
G. E. Jones

Environmental Management and Development in Drylands
Peter Beaumont

Chernobyl: The Long Shadow
Chris C. Park

Nuclear Decommissioning and Society: Public Links to a New Technology
Edited by Martin J. Pasqualetti

Green Development: Environment and Sustainability in the Third World
W. M. Adams

Environmental Policy and Impact Assessment in Japan
B. Barrett and R. Therivel

Radioactive Waste: Politics and Technology
Frans Berkhout

The Diversion of Land: Conservation in a Period of Farming Contraction
Clive Potter, Paul Burnham, Angela Edwards, Ruth Gasson and Bryn Green

Waste Location: Spatial Aspects of Waste Management, Hazards and Disposal
Edited by Michael Clark, Denis Smith and Andrew Blowers

Land, Water and Development: River Basin Systems and their Sustainable Management
Malcolm Newson

Environmental Problems in Eastern Europe
F. W. Carter and David Turnock

Drylands

Environmental Management and Development

Peter Beaumont

London and New York

First published 1989
by Routledge
2 Park Square, Milton Park, Abingdon, Oxon, OX14 4RN
29 West 35th Street, New York NY 10001

New in paperback 1993

Transferred to Digital Priting 2004

British Library Cataloguing in Publication Data

A catalogue record is available for this book
from the British Library.

ISBN 0–415–00457–8
     0–415–09663–4 (pbk)

*Library of Congress Cataloging in Publication Data
has been applied for.*

ISBN 0–415–00457–8
     0–415–09663–4 (pbk)

To Margaret, Mark,
and Adam

Contents

List of figures x

List of tables xv

List of plates vxii

Preface xviii

Part One: *Systematic Study of Drylands*
1. Nature of the Dryland Environment 3
    1.1 Definitions of Drylands 3
    1.2 Climatic Characteristics 8
        1.2.1 Precipitation 8
        1.2.2 Temperature 13
        1.2.3 Relative humidity 15
        1.2.4 Wind 16
    1.3 Evapotranspiration and Water Balance 17
    1.4 River Runoff 21
    1.5 Groundwater 28
    1.6 The Nature of Arid Landforms 33
    1.7 Arid Zone Soils 36
    1.8 Vegetation 40
    1.9 Fauna 43
    1.10 Dryland Ecosystems 46
    1.11 Climatic Change 47
2. Human Use of Drylands 57
    2.1 Traditional Societies 57
        2.1.1 Introduction 57
        2.1.2 Evolution of traditional societies 58
        2.1.3 Later developments 67
        2.1.4 Traditional societies at the present day 71
        2.1.5 Aborigines 74
    2.2 Modern Societies 82
        2.2.1 Introduction 82
        2.2.2 Development planning 85
        2.2.3 Population growth 89
        2.2.4 Agriculture and land reform 94
        2.2.5 The future 96
3. Intensive Use of the Dryland Environment 97
    3.1 Urbanisation 97
        3.1.1 Introduction 97
        3.1.2 Growth rates of cities 98

3.1.3 Urban management problems 99
3.1.4 Geomorphological hazards 109
3.1.5 Constructional limitations 114
3.1.6 Tucson — water for a desert oasis 116
3.2 Irrigation 125
3.2.1 Introduction 125
3.2.2 Irrigation water sources 128
3.2.3 Irrigation systems 130
3.2.4 Irrigation water quality 133
3.2.5 Distribution of irrigated agriculture 135
3.2.6 Irrigation development in the USA 138
3.2.6.1 The early years 138
3.2.6.2 The Bureau of Reclamation 141
3.2.6.3 Irrigation in the USA today 145
3.2.7 Conclusion 150
3.3 Mineral exploitation 152
3.3.1 Introduction 152
3.3.2 Mining activity and environmental problems 154
3.3.3 Patterns of mineral exploitation 156
3.3.4 Case study — mining in Western Australia 158
3.3.5 Conclusion 161
4. Extensive Use of the Dryland Environment 163
4.1 Rain-fed Agriculture 163
4.1.1 Introduction 163
4.1.2 Case study — northern Jordan 168
4.1.3 Case study — Australian wheat production 175
4.2 Pastoral systems 179
4.2.1 Introduction 179
4.2.2 Traditional pastoralism 179
4.2.3 Commercial pastoralism 182
4.2.4 Case study — South Africa 186
4.2.5 Conclusion 190

Part Two: *Regional Resource Management — Case Studies*
5. The Sahara and Central Australia: Pastoralism Under Different Management Systems 195
5.1 The Sahara 195
5.1.1 Introduction 195
5.1.2 The environment 196
5.1.3 Traditional land use 198

5.1.4 Case study — Niger 202
    5.1.4.1 Introduction 202
    5.1.4.2 Land use evolution 203
    5.1.4.3 The environment and its resources 204
    5.1.4.4 Modernisation of land use 207
    5.1.4.5 The Sahel drought, 1968–73, and its aftermath 209
5.1.5 Conclusion 213
5.2 Central Australia 215
    5.2.1 Historical development 215
    5.2.2 Pastoral practices 217
    5.2.3 Case Study — Gascoyne basin 221
        5.2.3.1 The environment 221
        5.2.3.2 Pastoral development 222
    5.2.4 Case Study — Alice Springs 227
        5.2.4.1 Historical development 227
        5.2.4.2 The current situation 231
    5.2.5 Conclusion 235
6. River Basin Development: the Nile and the Colorado 238
6.1 The Nile Basin 238
    6.1.1 Introduction 238
    6.1.2 The basin 239
    6.1.3 Hydrology 243
    6.1.4 Basin irrigation 246
    6.1.5 Perennial irrigation 248
    6.1.6 The Aswan High Dam 251
    6.1.7 Current position 255
    6.1.8 Future 260
6.2 The Colorado Basin 263
    6.2.1 Introduction 263
    6.2.2 The basin 263
    6.2.3 Water resources 265
    6.2.4 Historical development 266
    6.2.5 Major projects 267
    6.2.6 Future water supply and demand 272
7. Oases: Isfahan and Salt Lake City 277
7.1 Introduction 277
7.2 The Isfahan Oasis 278
    7.2.1 Introduction 278
    7.2.2 Environment 283
    7.2.3 Land use 285
    7.2.4 River management 288

7.3   Salt Lake City and the Wasatch Oasis   293
 7.3.1 Introduction   293
 7.3.2 The new settlements   296
 7.3.3 Irrigation   299
 7.3.4 Escalente — a pioneering Mormon settlement 301
 7.3.5 The Future   305

8. The Great Plains of the USA:
 Changing Patterns of Exploitation   307
 8.1   Introduction   307
 8.2   The Environment   309
 8.3   The History of Settlement   310
 8.4   Drought and Risk   311
 8.5   The Irrigation Movement — the High Plains
   of Texas   314
 8.6   Importation of water   325

9. Rapid Economic Development in the Gulf:
 The Impact of Oil Revenues   332
 9.1   Historical   332
 9.2   Oil and Oil Revenues   334
 9.3   The Urbanisation Process   338
 9.4   Environmental Impact   345
 9.5   Saudi Arabia — Eastern Province   347
 9.6   Water   350
 9.7   The Future   354

10. Israel: Integrated Water Development   357
 10.1   Introduction   357
 10.2   Climate   358
 10.3   Hydrology   360
 10.4   Water Resource Development   364
 10.5   National Water Carrier   367
 10.6   Pollution and Environmental Issues   371
 10.7   Re-use of Waste Waters   373
 10.8   Water Legislation and Policy   376
 10.9   Water Use   379
 10.10 The Future   382

11. Soviet Central Asia:
 Water Transfer and Irrigation Development   386
 11.1   Introduction   386
 11.2   Irrigation Development   387
 11.3   Water Transfer Schemes   393
 11.4   Impact of the Water Transfers   399

12. The Los Angeles Conurbation: Problems of
    Environmental Management and Resource Provision   402
    12.1 Introduction   402
    12.2 The Environment   403
    12.3 Water Supply   408
    12.4 Sewage Collection, Treatment and Disposal   416
        12.4.1 Introduction   416
        12.4.2 Sewage treatment techniques   417
        12.4.3 The sewage systems   419
    12.5 Environmental Problems Caused by Too
         Much Water   425
    12.6 Air Pollution   432
        12.6.1 Introduction   432
        12.6.2 Nature of the smog problem   433
        12.6.3 Management of air pollution effects   439
        12.6.4 Temporal trends   443
        12.6.5 Conclusions   447

Part Three: *Prospect*
13. The Future of Drylands   451

Bibliography   475
Index   512

Figures

1.1 The arid regions of the world 7
1.2 Annual rainfall of Australia showing the 10 percentile and 90 percentile distribution 9
1.3 Annual number of raindays in the south-west of the USA 11
1.4 Variations in precipitation along a transect from the Mediterranean Sea to the Jordanian desert 12
1.5 Average annual evaporation in Australia from a Class A pan 18
1.6 Water balance diagram for Berkeley, California 20
1.7 Mean annual water surplus in the Middle East (after Thornthwaite, Mather and Carter, 1958) 22
1.8 River discharge variations along the Zayandeh River in the Isfahan Basin, Iran 23
1.9 The number of large dams inaugurated each year throughout the world since the mid nineteenth century 24
1.10 Annual flows on the River Colorado, USA, below all major diversions, 1910–80 25
1.11 Flow of the River Nile at Aswan before and after the construction of the Aswan High Dam 27
1.12 Directions of water movement in the Nubian Sandstone aquifer system 31
1.13 Major aquifer systems of Saudi Arabia (after Burdon 1973) 32
1.14 Basin model of arid zone landforms 34
1.15 Soil moisture dynamics at different depths in desert soils 42
1.16 Diurnal temperature variations for different habitats in the Egyptian desert 45
1.17 Long-term variations of astronomical parameters defining insolation climates and insolation for various latitudes 48
1.18 Oxygen 18 versus depth in the Vostòk ice core, Antarctica, with the definition of the successive stages and indication of the ages corresponding to the limits between these stages 49
1.19 Histograms of lake-level status for 1000 year time periods 50

1.20   Average annual precipitation for 18 California
       stations   52
1.21   Rainfall variations in southern Africa   54
1.22   Annual average surface air temperature changes   55

2.1   Distribution of the wild ancestors of wheat and
       barley in the Middle East   59
2.2   Map of selected ancient sites in the Middle East   63
2.3   Traditional surface irrigation canals on the Varamin
       Plain, Iran   73
2.4   Tribal divisions of the Walbiri tribe, Australia   76
2.5   The demographic transition model   90
2.6   Population growth in selected arid countries and
       states in the USA from the mid-nineteenth century
       to the present day   91

3.1   Urbanisation of the Colma Creek watershed,
       California   112
3.2   Groundwater quality variation at Suez, Egypt   115
3.3   Per capita water use in Tucson, Arizona   120
3.4   Monthly water use by single family dwelling in
       Tucson, Arizona in relation to evapotranspiration
       minus precipitation (ccf = 100 ft^3)   122
3.5   Water use in the Casa del Agua, Tucson, Arizona   124
3.6   A qanat: cross-section and plan   129
3.7   Wetting stages of a field during surface irrigation 131
3.8   Diagram for the classification and use of irrigation
       waters   134
3.9   Irrigation water withdrawals from the major
       watershed units of the USA   147

4.1   The northern highlands of Jordan   169
4.2   Relationships between precipitation and crop yields
       at Dair Abu Said in the northern highlands of
       Jordan   171
4.3   Slope gradients in the Wadi Ziqlab, northern
       highlands of Jordan   172
4.4   Land capability classes in the Wadi Ziqlab in the
       northern highlands of Jordan   174
4.5   Expansion of the dry-farming grain belt of southern
       Australia   176
4.6   South Africa: relief and rainfall   187

5.1	The Sahara desert: mean annual precipitation and seasonal variations	197
5.2	The Eghazer and Azawak regions of Niger	203
5.3	Boreholes, wells and the northern limit of rainfed agriculture in Niger	208
5.4	Relationships between the growth of cattle numbers and annual precipitation in Niger	210
5.5	Niger: areas susceptible to desertification	212
5.6	Major land use areas in Australia	216
5.7	Gascoyne Basin, Western Australia: location and station boundaries	222
5.8	Relationships between sheep numbers and annual precipitation in the Gascoyne Basin, Australia	224
5.9	Erosion vulnerability in the Gascoyne Basin, Australia	225
5.10	Alice Springs district, Australia	228
5.11	Relationships between cattle numbers and annual precipitation in the Alice Springs district, Australia	234
6.1	The Nile basin and its major hydraulic works	242
6.2	Annual variability of flow of the River Nile at Aswan, 1871–1965	245
6.3	Relationships between cropped area and population growth in Egypt from the early nineteenth century to the present day	257
6.4	The Colorado River Basin showing the major hydraulic facilities	264
6.5	Colorado River Storage Project (CRSP). The location of storage capacity and participating projects	269
6.6	Central Arizona Project	271
6.7	Welton–Mohawk irrigation project	273
7.1	The Isfahan Oasis and the Zayandeh River basin, Iran	279
7.2	Flow of the Zayandeh River at Pol-e-Zamankhan and Pol-e-Varzaneh (1964–5)	285
7.3	Cultivated area and districts of the Isfahan Oasis	287
7.4	Regulators and main canals of the modern irrigation network, Isfahan Oasis	289
7.5	Isfahan Master Plan (1968)	291
7.6	Major new projects in the Isfahan basin	292
7.7	Wasatch Oasis and Salt Lake City, Utah	294

7.8	Major sites of Mormon settlement 1847–57	297
7.9	The site of Escalante, Utah	302
8.1	Location and extent of the Great Plains	308
8.2	High Plains of Texas	315
8.3	Growth of the irrigated area on the High Plains of Texas	316
8.4	Changes in the area of harvested cropland and the percentage of the cropped area which is irrigated, High Plains of Texas	319
8.5	Texas Water Plan: the main features	321
8.6	Outcrop of the Ogallala aquifer	324
8.7	The NAWAPA (North American Water and Power Alliance) Project	326
8.8	The Lewis Smith water diversion scheme	328
8.9	The Corps of Engineers plan for diverting water onto the Great Plains	329
9.1	The Gulf: major physical and economic features	333
9.2	Oil production from the major oil producing states in the Gulf	335
9.3	Oil production and oil revenues for Saudi Arabia	337
9.4	Urban centres along the Gulf	339
9.5	Electrical power consumption in megawatts for the State of Qatar (1983 and 1984)	343
9.6	Proposed Gulf power grid	344
9.7	LANDSAT images showing changes and developments in the Al Khobar-Dhahran-Dammam-Al Qatif agglomeration between 1972 and 1981	346
9.8	The new town and industrial site of Jubail, Saudi Arabia	348
10.1	Average annual precipitation for Israel (1931–60)	359
10.2	Water balance diagram for the Jordan River	361
10.3	National Water Carrier and related water networks in Israel	368
10.4	Dan Region Sewage Reclamation project (after Tahal 1977)	374
10.5	Water use in Israel	381
10.6	Irrigation water use and irrigated land in Israel	384
11.1	Arid and semi-arid lands of the USSR	386

11.2	Major river basins in the USSR and average flow values	388
11.3	Irrigation projects in the Ukraine, the Lower Volga and Central Asia	390
11.4	Planned river diversions in the European part of the USSR	394
11.5	Planned river diversions in the Siberian part of the USSR	397
12.1	Site of Los Angeles	405
12.2	Annual precipitation variations in Los Angeles	407
12.3	Water diversion from the Owens Valley and the Colorado River to Los Angeles	409
12.4	California State Water Project	412
12.5	Water costs along the California Aqueduct	413
12.6	Sources of water for the Los Angeles Municipal Water District under dry year conditions	414
12.7	Hyperion sewage works and collector systems, Los Angeles	415
12.8	Thermal processing of sludge, Hyperion sewage works, Los Angeles	420
12.9	Flood control systems for Los Angeles	425
12.10	Storm damaged areas in Los Angeles in 1983	429
12.11	A. Los-Angeles: Carbon monoxide 1984. Number of days on which Federal standard was exceeded (8-HR CO > 9.3 ppm) B. Los-Angeles: Total suspended particulate 1984. Annual Geometric Mean, $\mu g/m^3$	431
12.12	Los Angeles: Nitrogen dioxide 1984. Annual average values, pphm	436
12.13	Los Angeles: Ozone 1984. Number of days on which Federal Standard was exceeded (1-HR 0_3 > 12 pphm)	437
12.14	Air management areas, Los Angeles	438
12.15	Trends in ozone levels over Los Angeles 1975–85	441
12.16	Trends in carbon monoxide levels over Los Angeles 1975–85	444
12.17	Trends in lead levels over Los Angeles 1975–85	445
13.1	Recreational and military lands in the arid southwest of the USA	465
13.2	Rainfall for Gao, Mali	470

Tables

1.1   Arid homoclimates after Peveril Meigs   5
1.2   Effects of altitude and inland location on
temperature — Bushehr and Kerman, Iran, 1964   14
1.3   Distribution of different soil types in the arid zone   37

2.1   Food sources of the Walbiri — fauna   78
2.2   Food sources of the Walbiri — flora   81
2.3   Arid nations of the world   84
2.4   Sectoral distribution of investment in the First
Oman Development Plan 1976–80   88

3.1   Number of dryland cities of more than 100,000
population in different states   97
3.2   Major urban dryland centres with more than
2 million people   98
3.3   Urban populations and growth rates in the Middle
East, 1960–80   99
3.4   Magnitude of city metabolism for a city of one
million population in the USA   107
3.5   Geomorphological hazards in drylands   110
3.6   Land use changes in the Colma Creek basin,
California   113
3.7   Erosion rates from areas of different land use in
Colma Creek, California, 1969–70   113
3.8   Consumptive water use patterns, Tucson Active
Management Area, 1980   123
3.9   Irrigated area by continent, 1984   136
3.10   Irrigated area by country, 1984   137
3.11   Irrigation water use by regions, USA, 1980   146
3.12   Water use by agriculture in the 17 western US states
by the year 2000 assuming different water prices   149

5.1   Mean monthly rainfall and number of rain-days,
Tahoua and Agadez, 1921–54   205
5.2   Estimates of livestock numbers and losses in Tchin
Tabaraden and Agadez, 1968–74   211
5.3   Amounts of land applied for in the Alice Springs
district, 1872–85   229

6.1	Monthly average discharge for the major rivers in the Nile basin	244
6.2	Estimates of water supply and demand in Egypt, 1990	261
6.3	Virgin flow of the Colorado River at Lees Ferry, Arizona	265
7.1	Annual precipitation at four stations in the Isfahan oasis, Iran	284
7.2	Population growth in the state of Utah, USA	301
8.1	Range of water deliveries to the High Plains in the Corps of Engineers plan	330
9.1	Installed electricity generating capacity of the Gulf States, 1986–2000	342
9.2	Wheat production in Saudi Arabia	349
9.3	Water budget for Eastern Province, Saudi Arabia	353
10.1	Renewable non-saline water resources of Israel	379
10.2	Major storage reservoirs in Israel	380
10.3	Israel — water balance, 1977–2000	385
11.1	Water resources of the Syrdarya and Amudarya Rivers	396
11.2	Predicted decrease in flow of surface waters to the Caspian and Aral Seas as a result of industrial and agricultural activity	398
12.1	Air pollution sources in the Los Angeles basin	434
12.2	Air Pollution Episode criteria — Los Angeles	440

Plates

2.1 Traditional village in semi arid zone of northern
     Tunisia   60
2.2 Traditional irrigation canals and planned irrigation
     development on the Varamin Plain, Iran   61

3.1 Modern road development on an alluvial fan in
     Las Vegas, Nevada   100
3.2 Traditional irrigation canal on the Varamin Plain, Iran 101
3.3 Traditional well, Battina coast, Oman   102
3.4 Traditional means of diverting water into irrigation
     canals along major rivers in Iran   103
3.5 Groundwater extraction by deep well in the
     Isfahan Oasis   104
3.6 Karaj Dam, Iran   105
3.7 Pre-formed concrete canalets carrying irrigation water
     to an agricultural project in western Turkey   106

4.1 Gully plugs to reduce downstream movement of
     sediment in the Wadi Ziqlab in the northern highlands
     of Jordan   166
4.2 The effects of soil erosion in the northern highlands
     of Jordan   167

6.1 Canals taking water from the Colorado River at the
     Imperial Dam   240
6.2 Gillespie Dam on the Salt River near Phoenix,
     Arizona   241

7.1 The Shah Mosque in the centre of the city of Isfahan 280
7.2 The Isfahan Oasis from the air   281
7.3 The Isfahan Oasis from LANDSAT   282

12.1 Major flood channel in Los Angeles   404

Preface

The aim of this book is to provide the reader with an understanding of the ways in which drylands have been managed and developed. It sets out to do this by introducing a variety of examples differing widely in both time and space which illustrate alternative ways of resource use. By so doing it is hoped to show that there is no single or indeed optimum method or model for dryland management or development. Two of the most crucial factors are what are the objectives of the decision makers who are carrying out the management or development policy and what are the technology levels accessible to them. Equally significant is the political importance of the decision makers and the energy sources to which they have access. Even in ancient Egypt kings could build pyramids by the use of slave labour.

In the twentieth century two major changes have had a growing impact on drylands. First, there is the rapid rise in population, especially since 1950. This has created tremendous pressure on many dryland environments, sometimes resulting in the breakdown of a fragile ecosystem and growing desertification. Secondly, there is the impact of technology. With the internal combustion engine man can now command tremendous amounts of energy which are capable of reshaping parts of the natural environment. Equally important was the development of reinforced concrete in the late nineteenth century. This, through the construction of dams has permitted rivers to be controlled and rapid access communications networks to be built into the remotest parts of the arid zone.

Today, even the most isolated communities are in touch with modern societies through the ubiquitous portable radio and more recently television. This information transfer, however, tends to be uni-directional. The remote areas learn about the capital city and the rest of the world through electronic media, but in many cases little knowledge of these remote areas gets back to the capital, except in times of disaster.

To many countries, particularly those with a variety of environments, drylands have often been considered as somewhat marginal zones, with little economic value. As a result capital investment in them is restricted and development slow even today. In contrast in parts of the Arabian/Persian Gulf and the 'Sun-Belt' of the south-west of the USA massive investment has taken place over the past two or three decades to produce some of the most affluent communities of the world, yet with very different cultural backgrounds.

Finally, it is hoped that the book will stimulate interest in the economies and ecosystems of the world's drylands.

Part One

Systematic Study
of Drylands

1

Nature of the Dryland Environment

1.1 DEFINITIONS OF DRYLANDS

The term drylands is one which is easy to understand in a general sense, but is difficult to define in precise terms. The same is also true for concepts such as aridity and drought. With all of these terms it is water shortage which is the critical variable, though other factors are also important. In this study the term 'drylands' will be used to distinguish all those areas which experience regular water shortage on a seasonal or longer-term basis. This obviously includes all the arid lands of the world, defined by their extremely low precipitation totals, but will also encompass regions which may record significant rainfall for at least part of the year. In this context, the term 'drylands' involves a much broader interpretation than is the case with the more commonly used term, 'arid lands'.

Many workers have attempted to define precisely the term 'aridity' so that areas could be cartographically delimited for planning purposes. Any definition must, therefore, be quantifiable and at the same time produce a result which ideally has some physical meaning on the ground, either in terms of vegetation patterns or soil formation processes. Water is, of course, the key to understanding aridity, but absolute amounts, in isolation from other factors, tell us little of the dynamic processes which are operating in any particular region. Precipitation is the main input of water into any drainage basin and the two major outputs are evapotranspiration and streamflow. Of these two, evapotranspiration is the independent variable and streamflow the dependent one. Hence, in any consideration of aridity, the focus of any study must be on the relationships between precipitation and evapotranspiration.

Precipitation is an environmental attribute which is relatively easy to measure, and records, often over many years, are available from many dryland regions. In contrast, data on actual evapotranspiration — that is, the water loss from plant and soil surfaces — are extremely scarce for almost all parts of the world, even at the present day. This has meant that workers in the arid zone have therefore tended to overestimate the significance of absolute precipitation amounts, and been forced to utilise what are often crude measures of temperature or energy input as surrogates for evapotranspiration data. Not surprisingly the formulae which have been produced have lacked the degree of precision which is necessary for accurate planning purposes.

Over the years, many definitions of aridity in the meteorological/ climatological sense have been put forward. In a review paper, Wallen (1967) identifies three bases for definition, which he defines as the classical, index and water balance approaches. The classical approach focuses on climatic elements and their relationships to vegetation belts and agricultural zones. Initially, simple assessments of precipitation amounts were used, though these were later amplified by studies of variability and intensity. The effects of temperature were originally confined to mean or extreme values, but later, more sophisticated methods were employed, making use of heat sums, limiting temperatures and annual fluctuations.

The index approach seeks to delimit regions with differing degrees of aridity by the application of standard formula. The best known example of this approach is the climatic classification of Koppen (1931), which defined the boundaries of the arid zone in terms of annual precipitation and temperature indices. Koppen identified two arid climates, deserts (Bw climates) and steppes and semi-deserts (Bs climates), which together accounted for slightly more than 26 per cent of the Earth's land surface. A similar index of aridity developed by Martonne (1926) has also been widely used.

To provide a more scientific approach to the subject and in particular to explore the relationships between precipitation and evapotranspiration, the water balance concept was developed. Initially, the two main workers in the field were Thornthwaite and Penman, who both developed the concept of potential evapotranspiration independently in 1948 (Penman, 1948; Thornthwaite, 1948). Of the two, Penman devised the more sophisticated formula by making use of both the turbulent transfer and energy balance approaches, whereas Thornthwaite's work relied upon the energy balance approach alone. In experimental work, Penman's formula has been shown to produce results which are closely comparable with direct measurements obtained from evaporating pans and lysimeters. As a result it has been widely adopted throughout the world for the estimation of potential evapotranspiration. Many other workers have developed formulae for specific needs. Perhaps the best known is that of Blaney and Criddle (1950), which has been widely employed to calculate irrigation water requirements in the USA and elsewhere.

It is, however, the work of Meigs during the 1950s which today forms the basis of most definitions of arid conditions (Meigs, 1953). He aimed to produce a world map, which would reflect plant-growing conditions and, therefore, be valuable for agricultural planning purposes. Meigs employed Thornthwaite's method to calculate potential transpiration to produce a moisture index

Table 1.1: Arid homoclimates by Peveril Meigs

Moisture: The extremely arid classification is based on rainfall records which show at least one year without rain. The arid and semi-arid classifications are based on the deficit of precipitation in relation to potential evapotranspiration using the index described by Thornthwaite (1948).

    E Extremely arid
    A Arid
    S Semi-arid

Season of precipitation:

    a No distinct season
    b Summer precipitation
    c Winter precipitation

Temperature: In the climatic classification symbol used on the map, e.g. *Sb*24, the first digit represents the coldest month and the second digit the warmest month based on mean monthly temperatures as follows:

    0 — below 0°C
    1 — 0–10°C
    2 — 10–20°C
    3 — 20–30°C
    4 — above 30°C

Hot	(24, 33, 34)
Mild	(22, 23)
Cool winter	(12, 13, 14)
Cold winter	(02, 03, 04)

Source: Meigs, 1953.

which illustrated the relationships between precipitation and evapotranspiration. This was then used to produce a threefold division of the arid zone into extremely arid, arid, and semi-arid (Table 1.1). The map so produced also provided information on the season of precipitation by the use of three categories — no distinct season, summer precipitation, and winter precipitation. Temperature information was included on the mean monthly temperatures of the coldest and warmest months respectively in 10°C groupings.

The most recent definitive map of the arid zones of the world has been prepared by the Man and Biosphere (MAB) programme of UNESCO and was published in 1979 (UNESCO, 1979). It represents an updating of the well-known Meigs map of 1952, which was also produced for UNESCO. As yet the new map has not become widely known and many workers still continue to use the Meigs map. The main divisions of the UNESCO map, representing differing degrees of bioclimatic aridity, are delimited using values of the ratio P/ETP (Figure 1.1). P is the mean annual precipitation and ETP is the mean potential evapotranspiration calculated using the Penman formula. Four categories are recognised: the hyper-arid has a P/ETP ratio of less than 0·03; the arid from 0·03 to 0·20; the semi-arid from 0·20 to 0·50; and the sub-humid from 0·50 to 0·75. The sub-humid classification is a new one not used in the earlier Meigs map.

The hyper-arid zone consists of true desert climates where precipitation is extremely low and irregular in occurrence. Perennial vegetation is almost totally absent and neither pastoral nor arable farming is possible using naturally occurring rainfall. The arid zone receives annual precipitation totals of between 80 and 350 millimetres (mm), with inter-annual rainfall variability of between 50 and 100 per cent. Some scattered vegetation permits low-intensity grazing, but no rain-fed agriculture can be maintained on a continuing basis. The semi-arid zone is characterised by precipitation totals of between 200 and 700 mm. Grassland and scrub vegetation predominate, providing high-quality grazing. Rain-fed agriculture can easily be maintained in this zone, but yields show marked variations from year to year as inter-annual variability is between 25 and 50 per cent. The sub-humid zone contains a variety of vegetation types, from savannahs to broken woodlands. Inter-annual rainfall variability is less than 25 per cent and productive arable farming is the primary human land use.

Each of these categories is further sub-divided on the basis of

Figure 1.1 The arid regions of the world

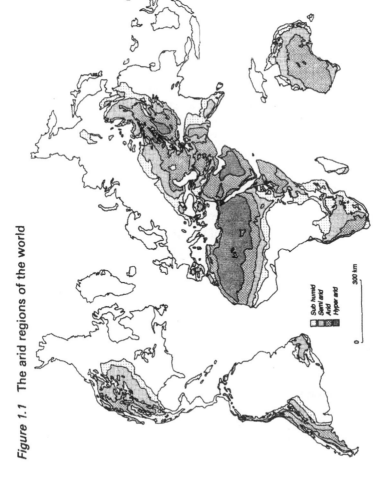

Sub humid
Semi arid
Arid
Hyper arid

0   300 km

Source: Redrawn from the *Map of the World Distribution of Arid Regions.* © Unesco 1977.
Reproduced by permission of Unesco.

temperature. The temperature of the coldest month in degrees centigrade is used to define four classes. These are:

(a) warm winters with coldest-month temperatures between 20 and 30°C;
(b) mild winters with coldest months from 10 to 20°C;
(c) cool winters with coldest months from 0 to 10°C; and
(d) cold winters where the temperature of the coldest month is less than 0°C.

In turn these temperature classes are further sub-divided in terms of the mean temperature of the hottest month of the year using values of 10, 20 and 30°C as the boundaries between the categories. Finally, the map also shows the lengths of periods of drought in relation to the seasons and precipitation regimes.

Although the UNESCO (1979) map is currently the definitive work on the arid zone, it has to be realised that the data on which the map is based are still very restricted for many parts of the world. This means that the lines drawn on the map often have a spurious accuracy, particularly in the more isolated parts of the arid zone. Having defined the arid lands it is important next to consider their distribution over the surface of the Earth. In examining the UNESCO map (Figure 1.1) the first point which strikes one is the concentration of the world's drylands in a belt stretching across north Africa and into south-west and central Asia. Indeed, so important is this zone that it accounts for about two-thirds of all arid lands. Elsewhere, a much less continuous pattern is found. Australia is undoubtedly the driest continent, but only accounts for 13 per cent of the arid zone. In the New World, pockets of aridity are found, which are locally important, though nowhere do they match the intensity and scale of Africa, Asia or Australia. In total these New World drylands account for 14 per cent of the arid zone.

1.2 CLIMATIC CHARACTERISTICS

1.2.1 Precipitation

Of all the climatic parameters it is precipitation which plays the dominant role in drylands. By definition all arid lands receive low

Figure 1.2 Annual rainfall of Australia showing the 10 percentile and 90 percentile distribution

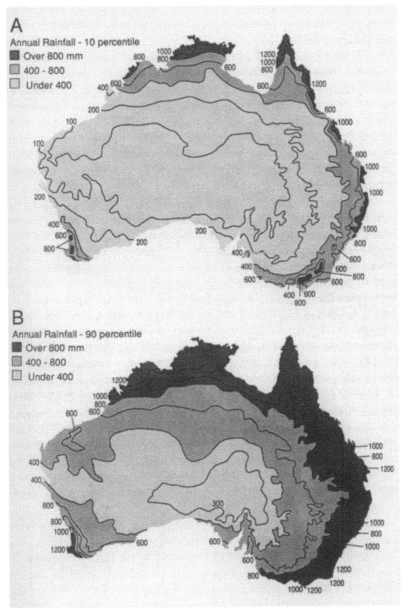

Source: Redrawn from a map from the Rainfall volume of the *Climatic Atlas of Australia* supplied by courtesy of the Bureau of Meteorology, Australia.

annual precipitation totals, with stations like Aswan in Upper Egypt often experiencing many years with no rainfall whatsoever. An equally important characteristic of drylands is their high rainfall variability. Indeed, as precipitation totals decrease, annual rainfall variability increases. This means that the reliability of precipitation at any station will decrease as precipitation totals decline. Variability, though, can provide either much more or much less water than mean annual totals would suggest. Indeed, to talk of mean annual precipitation within the arid zone conveys a false impression of what conditions are actually like. In many cases a truer reflection is conveyed by the inter-quartile range (which brackets 50 per cent of available observations) or the 10 percentile to 90 percentile range (which brackets 80 per cent of available observations) (Figure 1.2).

The form of precipitation in the arid zone depends largely on temperature conditions. Throughout many of the drylands, snowfall is not unknown, and in some highlands it is the chief form of precipitation in the winter months. A feature of many drylands is that high intensity precipitation can occur locally, which in turn produces flash-flood conditions. Regrettably there are still few continuously recording rain gauges in the arid zone and so much of our knowledge of rainfall intensities comes from a few countries, such as the USA, Australia and Israel, with high-quality meteorological networks. Although heavy rainfall can occur even in the driest areas, it is not usually common. Indeed, in most parts of the USA the vast majority of rain-days record 24-hour falls of less than 6·35 mm (0·25 inches). Nowhere in the USA, west of the New Mexico–Colorado line, expects on average even one heavy rainfall or more than 50·8 mm (2 inches) per day on an annual basis (Visher, 1966). At Tehran, Iran, analysis of daily precipitation data going back to 1898 has revealed that a fall of 35 mm would be expected to occur on average once in every ten years, whilst a fall of 75 mm would be expected only once in a thousand years (Gordon and Lockwood, 1970).

Despite precipitation being low in drylands, it is often surprising to discover on how many days rainfall occurs. In the USA, for example, most of the arid and semi-arid lands record between 20 and 60 days with precipitation falls of more than 0·01 inches (Figure 1.3). Another feature of drylands in the USA is the relatively high proportion of the annual precipitation which falls as snow. This is partly a reflection of the high altitude of the western USA. In most of Nevada almost one-third of the total annual

Figure 1.3 Annual number of raindays in the south-west of the USA

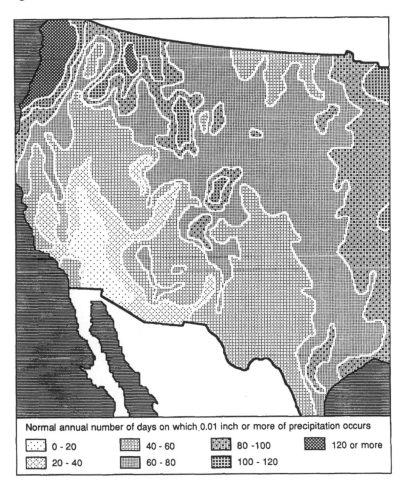

Source: Redrawn from Visher S.S., 1966 *Climatic Atlas of the United States*.
Reproduced by permission of Harvard University Press.

precipitation is accounted for by snowfall, while in eastern
Arizona and New Mexico average values fall between 10 and 20
per cent.

A feature of dryland precipitation in some areas is its 'patchy'
nature (Sharon, 1972). What this means is that precipitation can
occur locally from an isolated storm system, yet all around it the
land remains dry. Over time the total rainfall received will average

11

Figure 1.4 Variations in precipitation along a transect from the Mediterranean Sea to the Jordanian desert

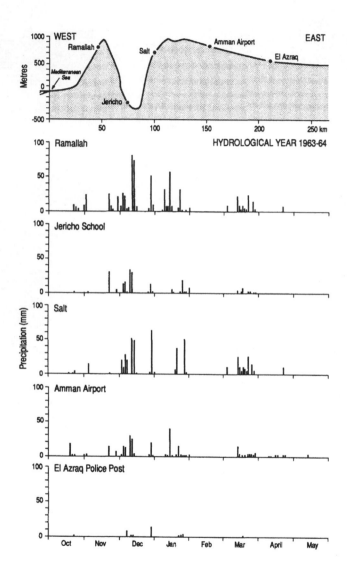

Source: Redrawn from Beaumont P., Blake G.H. and Wagstaff J.M., 1988, *The Middle East — A Geographical Study.* Reproduced by permission of David Fulton Publishers, London.

out to similar long-term values as individual storms will follow different routes.

In the arid zone there can be marked changes in precipitation over short distances in response to topography and other factors. In the eastern Mediterranean, for example, a west to east transect from the coast across Israel and Jordan reveals an increase in precipitation over the highlands and a decrease as the desert is approached (Figure 1.4). The precipitation records show rain-day sequences of two to five days duration associated with the passage of depressions and their associated frontal systems. At the drier stations the number of rain-days is fewer and the daily amounts of precipitation less than at the highland stations.

1.2.2 Temperature

It is much more difficult to generalise about temperature in the arid zone than it is about precipitation, because marked altitudinal variations mean that temperatures can vary widely, even over relatively short distances. On the whole, altitudinal and locational differences impose the greatest effect on diurnal and seasonal variations, by reducing minimum temperatures. Maximum temperatures tend to show surprisingly little variation. The net result is that annual temperature ranges increase with altitude and also with increasing distance from the sea.

An example of this can be shown by figures for two stations in southern Iran with similar latitudinal locations (Table 1.2). Bushehr is a port at sea level on the Persian Gulf and Kerman a town at almost 1750 metres (m) on the Iranian plateau. In terms of mean temperatures there is a major difference between the two stations, though it should be noted that the summer temperatures show the smallest variation. It is, however, when absolute temperature ranges (both annual and monthly) are considered that major differences appear. At Bushehr the annual range is 40·6°C, while at Kerman it is a massive 64·2°C. With monthly temperatures both stations reveal minimum ranges during the summer months, but the Kerman station always records significantly higher values than Bushehr. What is particularly interesting is that during the summer months the absolute maximum temperatures at the two stations are very similar. Indeed, the monthly maximum temperatures never deviate by more than about 7°C, whereas minima can differ by more than 20°C.

13

Table 1.2: Effects of altitude and inland location on temperature —
Bushehr and Kerman, Iran, 1964

	Bushehr *El*, 4m Lat. 28° 59′ N			Kerman *El*, 1749m Lat. 30° 15′ N		
	Mean	Highest max.	Lowest min.	Mean	Highest max.	Lowest min.
January	9·3	18·8	−1·0	−3·1	12·2	−24·4
February	15·0	24·8	3·4	8·0	20·0	−6·4
March	20·8	33·0	8·6	13·4	26·6	−1·5
April	22·6	36·4	12·0	16·7	29·5	4·0
May	27·2	39·6	14·4	21·3	34·6	5·6
June	31·2	39·4	23·8	22·0	39·8	10·2
July	31·9	38·6	23·0	27·8	39·6	14·4
August	31·7	38·0	22·0	25·8	38·8	10·8
September	28·9	37·4	19·0	19·2	35·0	1·2
October	23·2	33·0	11·6	11·9	31·8	− 5·8
November	19·6	29·4	7·8	8·4	24·5	−12·4
December	13·8	25·0	2·0	1·3	19·0	−15·4
Year (Mean, max. and min.)	22·9	39·6	−1·0	14·4	39·8	−24·4
Max. range (between highest max. and lowest min.)		40·6			64·2	

Source: *Climatic yearbook of Iran 1964.*

In the largest desert area, the Sahara, maximum average daily temperatures of more than 45°C occur in the interior of the western parts (Thompson, 1965). Indeed, everywhere outside the highlands in the Sahara experiences average daily temperatures above 37·5°C in July. During January the effects of interior heating are less marked and there is a general increase in average daily maximum temperatures from around 15–17°C on the Mediterranean coast to 35°C on the southern margin of the arid belt. Average daily minima in July are highest in the western Sahara at more than 30°C. Elsewhere, minima at this time are usually between 20 and 25°C. In January the average daily minimum temperatures fall below 5°C in a wide belt across the northern part of the Sahara. Along the coast they are ameliorated slightly by marine influences, whilst to the south they rise gradually to around 15°C.

In Australia average maximum temperatures rise above 39°C in the north-west (Department of Science, Bureau of Meteorology, 1975a). Throughout most of the interior, values above 36°C are common in summer. In winter a pronounced south to north gradient is seen, which is similar though reversed in direction to that observed in the Sahara. Along the Nularbor Plain, average maximum temperatures are slightly above 15°C, while in the extreme north of the arid belt averages rise to over 30°C. Average minimum temperatures in January vary from 15 to 27°C. A marked gradient is seen away from the southern coastal belt, but over most of the interior and north temperatures are above 21°C. In July a very different pattern is observed with this time the greatest temperature gradient along the northern coast. Over most of the southern part of the interior, average minimum temperatures are between 3 and 6°C.

Within the arid and semi-arid area of the USA, it is much more difficult to generalise about temperature conditions owing to the considerable altitudinal variations which occur over short distances. In July the highest daily maximum temperatures of in excess of 38°C occur in south-east California and along the western border of Arizona (Visher, 1966). Elsewhere, temperatures of between 27 and 32°C predominate. In January the effects of altitude are reduced and a more general north to south gradient prevails. Daily maximum temperatures along the Mexican border reach above 16°C, while in the north temperatures drop below −7°C. Daily minimum temperatures in July are much more difficult to generalise about as they are greatly influenced by relief. In the extreme south, minima do not fall below 21°C but in the higher mountains values of less than 5°C are recorded.

1.2.3 Relative Humidity

Relative humidity plays a vital role in human comfort in the arid zone. At first it might seem rather surprising that high humidities can occur in dryland areas where water shortages are the norm. However, many arid zones are close to the coast and the resulting high humidities reflect this. The Persian Gulf, for example, is bordered by some of the driest areas in the world, yet places such as Bahrain Island rarely have average monthly humidities of less than 60 per cent, and during the winter months these values can rise to close to 80 per cent. Relative humidities do exhibit marked diurnal and seasonal variations. During the daily cycle the highest relative humidities are recorded during the early hours of the morning, while minimum values occur between 1400 and 1500 hours. On a seasonal basis, highest humidities in the arid zone are recorded during the winter months, especially in those areas characterised by winter precipitation maxima.

In Australia the lowest relative humidities on a continental basis are recorded during October (Department of Science, Bureau of Meteorology, 1978). This is because during the summer period, the northern part of the country comes under the influence of the summer monsoon. At this time of year well over half the continent experiences relative humidities of less than 20 per cent during the early afternoon. In contrast, during the winter when eastward-moving depressions are crossing the southern part of the continent, the belt of highest humidities is squeezed northwards. What is clear in both cases, but particularly during October, is the very sharp relative humidity gradient which can occur in coastal localities. In parts of western Australia, for example, coastal relative humidities can be more than 60 per cent, while less than 150 kilometres (km) inland they are less than 20 per cent. Over North Africa an east–west belt stretching from almost the Atlantic to the Red Sea records mid-day relative humidities of less than 20 per cent throughout the year, though the position of the belt does show seasonal movements. This belt represents the largest continuous area of low humidities anywhere in the world.

High humidities associated with high temperatures produce considerable discomfort for human beings and it is not surprising, therefore, that many coastal desert locations have not proved attractive for extensive human settlement until the advent of air conditioning technology from the 1950s onwards (Mather, 1974).

1.2.4 Wind

In many parts of the arid zone the wind has an important effect on human perception of the dryland environment. One of the reasons for this is that many dryland areas take the form of extensive plains with little vegetation cover over which the wind can blow with unchecked force. In the settlement of the Great Plains of North America it was the wind (intensity and constancy) rather than any other climatic element which had the greatest impact on the settlers. In its most devastating manifestation during the droughts of the 1930s, it produced the Dust Bowl, eroding the soil and lifting clouds of dust thousands of metres into the air.

In all arid climates the prevailing low-moisture contents mean that surface soils are dry and, therefore, prone to disturbance by wind. A feature of dryland climates is haziness due to the presence of dust in the atmosphere. When strong winds are generated, usually in association with frontal systems, dust storms which severely reduce visibility for many days at a time are common. When prevailing winds from a given direction are frequent and when fine particulate material is plentiful, drifting sand dunes may develop. These are often small in size, but they can render cultivation impossible as well as disrupting communication systems.

1.3 EVAPOTRANSPIRATION AND WATER BALANCE

Evaporation and evapotranspiration form the major losses of water from the land surface in drylands. It is therefore essential to be able to measure or estimate the amount of evaporation and evapotranspiration in different areas. Both are difficult parameters to measure, however, and our knowledge of them is much less comprehensive than with precipitation and temperature conditions.

The instrument which is most commonly used for evaporation measurement is the US Weather Bureau Class A pan. This is a circular pan, 1.22 m in diameter and 254 mm in depth, which is filled with water and supported by a timber framework that stands on the ground. The rate of evaporation is measured by the drop in water level through time. In Australia comprehensive evaporation data have been compiled using the Class A pan as the main data source. Records show that within the confines of arid and semi-arid Australia, annual evaporation totals are in excess of 3 m and in parts of northern Western Australia and Northern Territories they reach in excess of 4.4 m (Figure 1.5). Considerable variations occur on a monthly basis. In January (the hottest month) almost all the interior experiences evaporation totals above 400 mm, with peaks of over 600 mm. By way of contrast July reveals a south to north gradient across the interior from less than 100 mm on the Nularbor Plain to over 250 mm along the shores of the Gulf of Carpentaria.

Detailed evaporation pan data are also available for the drylands of the USA. These reveal highest rates in excess of 3,048 mm over southern Nevada and eastern California, with much of

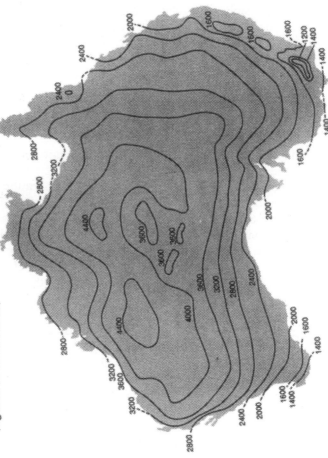

Figure 1.5 Average annual evaporation in Australia from a Class A pan (figures in millimetres)

Source: Redrawn from a map from the Evaporation volume of the *Climatic Atlas of Australia* supplied by courtesy of the Bureau of Meteorology, Australia.

the south-west recording values in excess of 2,032 mm per annum. Information on annual evaporation from lakes and reservoirs in the USA reveals a similar pattern to that of the pan data, but the very high peak recordings are not duplicated by this data set.

In many cases it is often easier to estimate evapotranspiration rather than to measure it, and over the years a number of formulae have been devised to estimate potential evapotranspiration. Potential evapotranspiration is defined as the maximum quantity of water capable of being lost as water vapour in a given climate by a continuous cover of vegetation when the soil is kept saturated. It is obviously a function of climatic conditions and in particular of the amount of incoming solar radiation. Actual evapotranspiration is the amount which occurs in a given locality, and this is less than or equal to potential evapotranspiration owing to soil moisture deficiencies which develop at certain times of the year. Whereas potential evapotranspiration is relatively easy to estimate using empirical formulae, actual evapotranspiration is very difficult to measure without complicated and expensive equipment.

Most of the formulae for estimating potential evapotranspiration make use of one or both of two approaches, the turbulent-transfer and energy-balance methods (World Meteorological Organisation, 1966). The turbulent-transfer approach is based on the idea that it is theoretically possible to determine the rate of evaporation from any surface by sole reference to the moisture content of the atmosphere and the wind speed at two points close to the ground. The basic problem with the method is that measured differences are usually small and so observations must be accurate and frequent. The energy balance method is simpler and based on the premiss that to change water from a liquid to a gas, a certain amount of energy (600 calories per gram at 0°C) is required. If the total amount of incoming solar radiation can be determined and if the proportion of this radiation which is used for evaporation can be evaluated, the amount of water loss can be calculated. A number of formulae have been devised to estimate potential evapotranspiration, including well-known ones by Blaney and Criddle (1950), Penman (1948, 1963), and Thornthwaite and Mather (1957). Of these the Penman formula, which uses both the turbulent transfer and energy balance approaches, appears to provide the best estimate available for potential evapotranspiration.

In an arid area the availability of water is critical to all forms of life and determines the likelihood of groundwater recharge and the volume of runoff in rivers and streams (Rodier, 1985). Evapo-

Figure 1.6 Water balance diagram for Berkeley, California

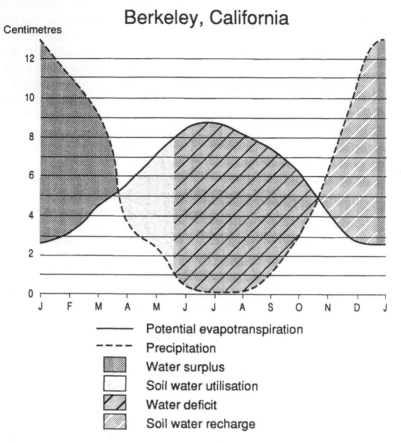

Berkeley, California

---- Potential evapotranspiration

- - - - Precipitation

Water surplus

Soil water utilisation

Water deficit

Soil water recharge

Source: Redrawn from *Water — Yearbook of Agriculture 1955*, United States Department of Agriculture.

transpiration is often so great that water surpluses, representing precipitation minus evapotranspiration for some recording interval, are non-existent. Under these conditions vegetation can only flourish by obtaining water from soil moisture storage until this source is in turn exhausted. Water balance diagrams can be constructed to reveal these changes (Figure 1.6). Berkeley in California experiences a Mediterranean climate with a dry season from April to October during which evapotranspiration exceeds precipitation. During April and May potential evapotranspiration can be maintained by water being withdrawn from soil moisture storage. By June, though, this water is exhausted and a water deficit situation then occurs until November when once again rainfall exceeds evapotranspiration. During this period the water deficit has to be met by irrigation for plant growth to be possible.

1.4 RIVER RUNOFF

Using the water balance approach it is possible to construct maps of water surplus (precipitation minus evapotranspiration) for arid regions (Figure 1.7). In the Middle East it is clear that water surpluses are confined to the highland areas, yet obviously human settlements flourish in areas where no water surpluses are indicated. The reason for this is that these water surpluses are transported from the highland to lowland area by rivers and aquifers. Perhaps the best example in south-west Asia is the Tigris–Euphrates river system which carries the water surpluses of central Turkey to the dry plains of Iraq. In the case of the Euphrates River almost 90 per cent of the total discharge is generated in Turkey and nothing at all is produced within Iraq (Beaumont, 1978a). In contrast, water use for irrigation within the basin has over the years been concentrated in Iraq.

It is interesting that in arid areas such as the USA and the Middle East, where high mountains are located close to arid lowlands, river regimes are dominated by snowmelt peaks. In the uplands most of the precipitation, which falls dominantly during winter, occurs as snow. This means that the water is locked in the solid form as snow pack and so does not begin to melt until spring and early summer. As a consequence major rivers like the Colorado and the Euphrates reveal maximum flows in spring and early summer. A very similar picture is seen on the central plateau of Iran where many small rivers flow from the Elburz and Zagros Mountains into an area of inland drainage. All of these reveal the same snowmelt peak in early summer when the month of maximum flow can record more than 20 times the volume of the month of lowest discharge (Beaumont, 1973). This pattern is also repeated in the western USA.

Even in the most arid regions it is still possible for perennially flowing streams to occur. Perhaps the best example here is the Nile, one of the major rivers of the world, which flows through Egypt where for most of its journey precipitation totals average

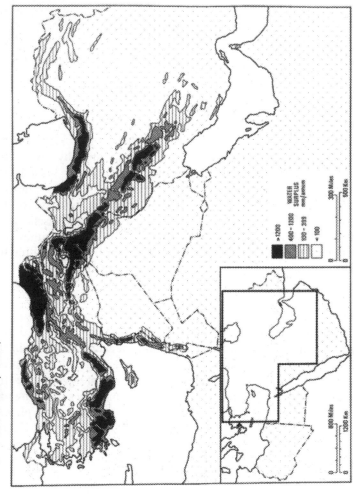

Figure 1.7 Mean annual water surplus in the Middle East (after Thornthwaite, Mather and Carter, 1958)

WATER
SURPLUS
mm/year

> 1200
400 – 1200
100 – 399
< 100

300 Miles
500 Km

600 Miles
1200 Km

Source: Redrawn from Beaumont P., Blake G.H. and Wagstaff J.M., 1988, *The Middle East — A Geographical Study*. Reproduced by permission of David Fulton Publishers, London.

less than 50 mm per annum. This is only possible because water surpluses are being moved from one area to another — in this case from the plateau of east Africa and the highlands of Ethopia. It is, however, only the largest streams which are perennial in the arid zone; with most of the smaller rivers, flow only occurs when water surpluses are generated following individual storm events. Such flash floods can, however, produce extremely large discharge peaks.

A feature of rivers in arid regions which is often overlooked is that once outside the highland areas, where the water surpluses are generated, the rivers often reveal a decrease in discharge in a downstream direction. This is partly the result of naturally occurring phenomena, such as evaporation and percolation of water into the bed of the stream, but more importantly it is the result of the diversion of water for irrigation purposes. Such a

Figure 1.8 River discharge variations along the Zayandeh River in the Isfahan Basin, Iran. (Pol-e-Zamankhan is the gauging station furthest upstream and Pol-e-Varzaneh furthest downstream)

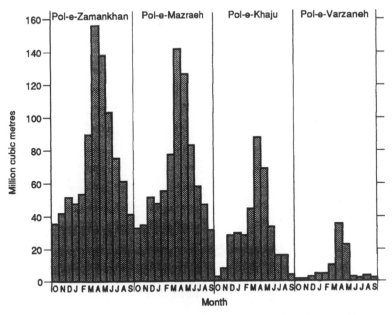

Source: Redrawn from Clarke J.I. and Bowen-Jones H. (eds.), 1981, *Change and Development in the Middle East*. Reprinted by permission of J.I. Clarke & H. Bowen-Jones.

situation is clearly seen in the case of the Zayandeh River, which provides water for the Isfahan Oasis, Iran (Figure 1.8).

In many parts of the arid zone it is no longer possible to talk about 'natural' river regimes, for even if humans have not constructed major engineering works to regulate river flow, the land use changes which have been made over hundreds, if not thousands, of years, have through vegetation removal, soil erosion and the construction of urban centres significantly altered the runoff patterns. The net effect of this activity has normally been to reduce soil moisture storage capacities and to increase the rapidity of runoff. This makes rivers have a more flashy flow regime, as well as greater flood volumes. Unless this runoff can be stored in reservoirs, it does mean that the quantity of water which is available for beneficial uses has been effectively reduced.

Figure 1.9 The number of large dams inaugurated each year throughout the world since the mid nineteenth century

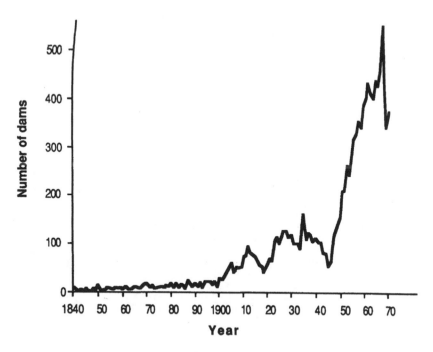

Source: Redrawn from Beaumont P., 1978, 'Man's impact on river systems — a world wide view', *Area*, v. 10, pp. 38–41. Reprinted by permission of the Institute of British Geographers.

In more recent times river regimes have been substantially changed as a result of human activities. This has been especially the case in the period since the Second World War, which has seen a major surge of dam-building activity (Beaumont, 1978b) (Figure 1.9). The impact of this has been greatly to increase water storage capacities by the construction of reservoirs. This has permitted the diversion of large quantities of water for irrigation and urban usage, but it has also meant that downstream flows of water below the reservoirs have been substantially reduced. This is seen along the Colorado River, where irrigation use began early in the present century and where major water storage projects date from the 1930s. The decline in the residual flow of the river is clearly seen (Figure 1.10). The very low flow rates between 1935 and 1939 were the result of the filling of Lake Mead behind the Hoover Dam (Bredehoeft, 1984). It is seen that by 1960 all the water of the Colorado River was being used, residual flows being extremely low. In 1979 and 1980 exceptional floods along the lower Colorado occurred below the main reservoirs and resulted in higher than normal residual flows. Such events are, however, unusual.

Figure 1.10 Annual flows on the River Colorado, USA, below all major diversions, 1910–80

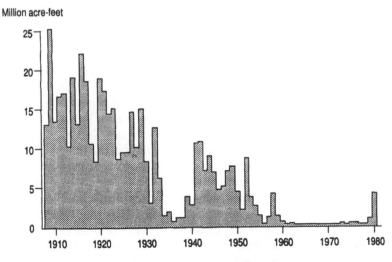

Source: Redrawn from Engelbert G.A. and Scheuring A.F., 1984, *Water Scarcity — Impacts on Western Agriculture*. University of California Press. Reprinted by permission of the University of California Press.

What has already happened along the Colorado looks as if it might occur on the River Euphrates, which is also an international river, with its watershed located in Turkey, Syria, Saudi Arabia and Iraq. Some 88 per cent of the river flow is generated in Turkey and a further 12 per cent in Syria. No significant quantities of water are produced in either Iraq or Saudi Arabia. Until the mid-twentieth century almost all the water used in the basin was consumed by Iraq for irrigation. This use in the Tigris–Euphrates lowlands goes back at least 6,000 years to some of the earliest irrigation systems in the world and gave rise to civilisations like those of Akkad and Sumer. Since about 1950 this position has been changing rapidly as both Syria and Turkey developed major irrigation systems using water from the Euphrates (Beaumont, 1978a). Already both countries have constructed, or are in the process of building, major dams and associated irrigation projects and these should be in full production by the beginning of the twenty-first century. The result will be that Turkey and Syria will begin to make large demands on the waters of the Euphrates, when previously such needs were extremely small, or in the case of Turkey virtually non-existent.

When the new projects are in full operation it seems possible that the consumptive use of water within the Euphrates basin may well be in excess of the average flow of the river at Hit, Iraq (Beaumont, 1978a). This will cause particular problems during periods of dry years which are relatively common in the Euphrates basin: for example, between 1955 and 1962 the mean annual flow was only 76 per cent of the long-term average. Under these conditions it is inevitable that all the water demands will not be met, and it is unclear how this situation will be faced. Already Turkey, Syria and Iraq claim to have reached general agreement about future use of the waters of the Euphrates, but in a time of acute water shortage it might well be that both Turkey and Syria view their own water needs as more important than those of Iraq. Such actions would have an extremely serious impact on Iraq's irrigated lands. Under such conditions the potential for conflict is very great and in the past, hostilities over water have only just been averted. In 1975, when Syria was filling the Tabqa Dam, relations between Syria and Iraq became very strained as the result of a disagreement about water use; and for a time an armed conflict between the two countries seemed a distinct possibility (Arab Report and Record, 1975). The situation is made even more intriguing by the fact that although Iraq's claim to a major share of

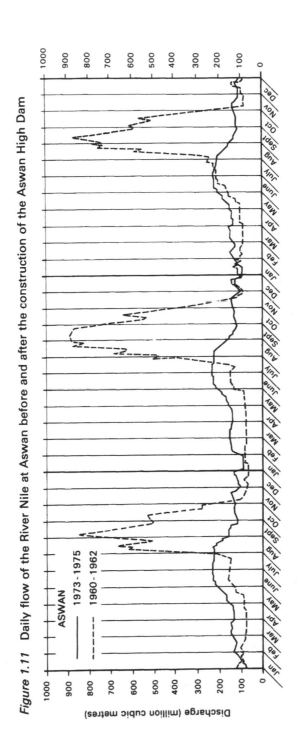

Figure 1.11 Daily flow of the River Nile at Aswan before and after the construction of the Aswan High Dam

Source: Redrawn from Clarke J.I. & Bowen-Jones H. (eds.), 1981, *Change and Development in the Middle East.* Reprinted by permission of J.I. Clarke and H. Bowen-Jones.

the waters of the Euphrates is undoubtedly strong, all the river flow is generated within the borders of Turkey and to a lesser extent Syria. These countries might claim that these resources belonged to them.

The taming of another great river of the world is seen in the case of the Nile. Here the construction of the Aswan High Dam in the 1960s has meant that the Nile flood, on which the traditional basin irrigation system of Egypt used to depend, has been eliminated. In its place is a more controlled release of the waters of Lake Nasser, which is more in keeping with the actual pattern of modern perennial irrigation demand (Figure 1.11).

1.5 GROUNDWATER

Groundwater accumulates as a result of water surpluses percolating into the ground. Rocks and sediments which are capable of storing and transmitting significant quantities of groundwater are called aquifers. The total volume of water which can be stored in an aquifer depends on the porosity (the volume of pore spaces filled with air) of the rock or sediment. In general, unconsolidated materials have higher porosity than rocks which have been consolidated and had their constituent particles cemented together. However, porosity is not necessarily a good measure of the likely value of an aquifer because not all the water which is contained within it can be extracted for human use. The smaller the pore spaces, the more tightly the water is held by suction forces within the aquifer. The important quality of an aquifer is its specific yield, which effectively means the amount of water which can be drained from a unit volume of rock. The water which remains is defined as specific retention. Specific yield plus specific retention gives porosity. Clay has a very high porosity, but its specific yield is extremely low. In contrast, gravel has a high porosity, but its specific yield is also high. Consolidated rocks, such as sandstone and shale, both have lower porosities and specific yields than the unconsolidated material from which they were formed. The other important characteristic of an aquifer is the speed with which water moves through it. This is called transmissibility. In general the most important factor here is the size of the individual pore spaces. When these are large, water movement is more rapid than when they are small. Gravel and coarse sands are the materials which possess the highest transmission rates.

Two major types of aquifer are recognised. Unconfined aquifers are those in which the water table forms the upper surface of the zone of saturation in the rock, and an impermeable layer the lower boundary. With this type of aquifer, fluctuations of the water table correspond with changes in the volume of water being stored within it. A confined aquifer is one in which groundwater is under a pressure greater than that of the atmosphere as the result of impermeable beds overlying as well as underlying the water-bearing rocks. In confined aquifers the water volume remains relatively constant and they function mainly as conduits transporting water from recharge areas to natural or artificial discharge points.

Alluvial aquifers are found along river valleys and beneath alluvial fans. These are usually shallow, operate as unconfined systems and are composed of unconsolidated sands and gravels. They are small in area and possess water tables which respond rapidly to local precipitation conditions. Recharge of these aquifers takes place along broad gravel-floored valleys in uplands or on the upper parts of alluvial fans where the river leaves its highland course. In these areas water percolation into the underlying gravels can be very rapid and it is not unusual for all the waters of a river or stream to disappear while crossing a large alluvial fan. Rates of water yield from alluvial aquifers are high, though total storage volumes are limited. While the bed of a river above an alluvial aquifer can be dry for many months of the year, water will still be flowing through the aquifer, although at only very limited flow rates. However, it is not uncommon for an impermeable rock layer to cross a valley floor and so cause the water in an aquifer to rise to the ground surface and form river flow for a short section before percolating underground once again. Some form of settlement is often located where these perennial flow sections occur: for example, the original site of Los Angeles is located at such a point (Nelson, 1983).

Deep rock aquifers normally occur in sandstone and limestone formations. They often function as confined systems and can cover enormous areas. Two examples are the Ogallala aquifer of the Great Plains of the USA and the Nubian Sandstone aquifer of north Africa. Much of the water in these large rock aquifer systems seems to be of a fossil nature, having been emplaced during conditions of greater water availability many thousands of years ago. Natural recharge of these aquifers occurs in upland or foothill zones where surface outcrops of the aquifer are found. The

29

actual mode of recharge is uncertain, but it seems likely that it is achieved by concentration of runoff into river channels, followed by percolation through the bed into the underlying aquifer. In this way water draining from a considerable area can be recharged into an aquifer which may have only a limited surface outcrop. Surprisingly little is yet known of the recharge rates of these large aquifers at the present day, though actual rates are believed to be low. In the case of the Ogallala aquifer, for example, recharge is believed to be about only 0·5 centimetres (cm) per year (Klemt, 1981). Other work has suggested that contemporary recharge in very arid areas may be greater than previously thought. In the Dahna sand dunes of Saudi Arabia it has been shown that even when precipitation totals are as low as 80 mm/year, about one-quarter of this amount can percolate downwards and recharge groundwater systems (Dincer *et al.*, 1974).

Water surpluses can be transported large distances in aquifer systems, with two good examples being provided from the Middle East. In north Africa the Nubian Sandstone aquifer, consisting of sands, sandstones, clays and shales, attains thicknesses in excess of 3,500 m. Although a number of different aquifer units are present within it, it functions as a single multi-layered artesian system covering about 2·5 million square km (Hammad, 1970; Himida, 1970; Shata, 1982). Water moves over 1,000 km in a north-easterly direction from recharge areas in northern Chad to discharge into a series of depressions in Egypt. The largest of these is the Qattara depression, but there are also the Siwa, Farafra, Bahariya and Dakhla oases (Figure 1.12)

In the Arabian Peninsula eleven major aquifers have been identified, ranging in age from the Cambrian to the Quaternary. These have been grouped into four aquifer systems (Burdon, 1973, 1982) (Figure 1.13). The two oldest — the Palaeozoic and the Triassic — contain mainly sandstone aquifers, with the Basement Complex forming a major aquiclude (impermeable formation which may contain water but is incapable of transmitting water in significant quantities) beneath them. Both seem to function as closed and confined aquifer systems, suggesting that natural discharge from them is minimal. The Wasia and Biyadh formations are the main water bearing strata of the Cretaceous aquifer system. This also functions mainly as a confined system, though some water transfer to overlying aquifers may take place. The Eocene aquifer system, which is largely made up of carbonate units, is also a confined system. Discharge from it takes place

Figure 1.12 Directions of water movement in the Nubian
Sandstone aquifer system

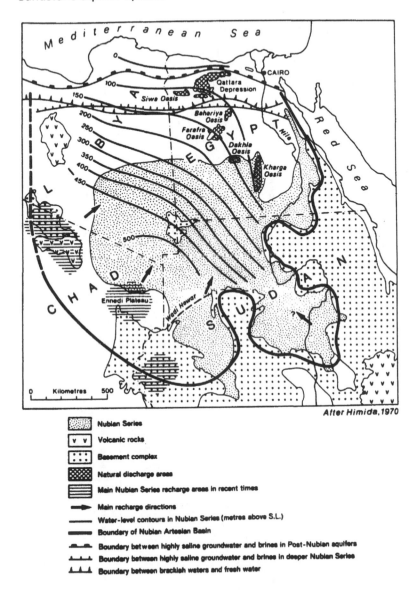

After Himida,1970

Nubian Series

Volcanic rocks

Basement complex

Natural discharge areas

Main Nubian Series recharge areas in recent times

Main recharge directions

Water-level contours in Nubian Series (metres above S.L.)

Boundary of Nubian Artesian Basin

Boundary between highly saline groundwater and brines in Post-Nubian aquifers

Boundary between highly saline groundwater and brines in deeper Nubian Series

Boundary between brackish waters and fresh water

Source: Redrawn from Himida I.H., 1970, 'The Nubian Artesian Basin, its
regional hydrological aspects and palaeohydrological reconstruction',
Journal of Hydrology (New Zealand), v. 9, pp. 88–119. Reprinted by
permission of the New Zealand Hydrological Society.

Figure 1.13 Major aquifer systems of Saudi Arabia (after Burdon 1973)

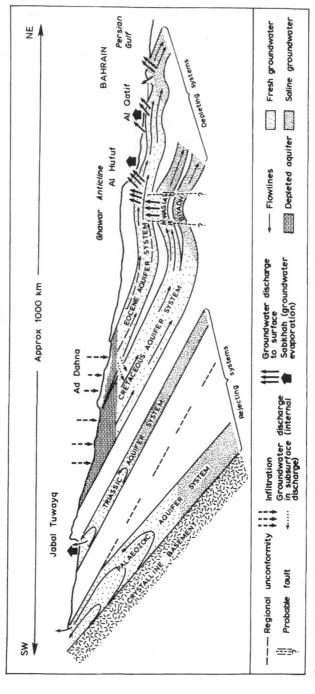

Source: Redrawn from Beaumont P., 1977, 'Water and development in Saudi Arabia', *Geographical Journal*, v. 143, pp. 42–60. Reprinted by permission of the Royal Geographical Society.

through terrestrial and submarine springs, as well as by evaporation from coastal sabkhas (marshy areas where groundwater rises to the surface) along the Gulf coast (Pike, 1970, 1983). Together, these aquifer systems transport water surpluses from western Arabia some 500 to 600 km eastwards to the extremely arid areas fringing the Persian Gulf. It appears that much of the water in these aquifers is fossil, with major recharge taking place between about 16,000 and 35,000 years ago (Bakiewicz et al., 1982; Thatcher et al., 1961).

1.6 THE NATURE OF ARID LANDFORMS

There are neither landforms nor processes unique to the arid zone, but there are certain conditions which prevail in these areas and which do have an important effect on the way that landforms develop (Cooke and Warren, 1973). The arid nature of the terrain means that vegetation cover is not continuous in the driest parts and, as a result, large areas of bare ground are found. This bare ground is susceptible to both wind and water erosion to a much greater extent than in more humid regions.

The complex nature of landforms in the arid zone makes it extremely difficult to generalise about them. It is helpful, however, to make use of a simple basin model to provide a framework with which to examine both the human and physical environment. In its simplest form the model is divided into four zones: uplands, alluvial fan, alluvial plain and salt lakes or salt desert (Figure 1.14). In any particular part of the arid zone, not all of these zones will be found, though in places like the south-west USA, Iran and central Asia, the model can be applied exactly.

The uplands are characterised by resistant rock types and steep slopes. The amplitude of relief may be hundreds, or possibly thousands of metres. Climatic conditions may often be harsh, with freeze–thaw action common in winter. As a result soil profiles are poorly developed and large areas of bare rock occur. Coarse scree deposits mantle the lower slopes. Major river valleys dissect the uplands and these tend to be flat-floored owing to the deposition of material from the adjacent slopes. The uplands are zones of maximum erosive activity as a result of the severe weather conditions, mass movement on the steep slopes and the considerable fluvial activity which occurs along the main valley lines. Fault lines commonly separate the uplands from adjacent lowlands.

Figure 1.14 Basic model of arid zone landforms

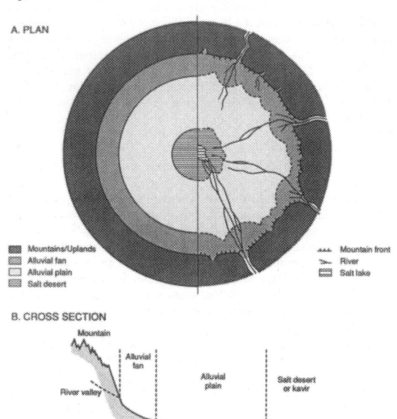

A. PLAN

Mountains/Uplands
Alluvial fan
Alluvial plain
Salt desert

▲▲▲ Mountain front
≥ River
▭ Salt lake

B. CROSS SECTION

Mountain

Alluvial fan

Alluvial plain

Salt desert or kavir

River valley

Source: Redrawn from Beaumont P., Blake G.H. and Wagstaff J.M., 1988, *The Middle East — A Geographical Study.* Reproduced by permission of David Fulton Publishers, London.

The other three zones are depositional in nature. The alluvial fan zone is situated at the margin of the uplands and is characterised by slopes with angles between about 3° and 15°. Amplitude of relief is very low (normally only a few metres), though sometimes the major streams can be incised into the upper parts of the fan. Coarse sedimentary material makes up most of the alluvial fan zone, with deposition from both fluvial and mudflow activity (Bull, 1977). Mudflows from the uplands usually only reach the upper part of the fans, while fluvial activity continues on the two other zones as well.

The alluvial plain and salt desert zones are continuations of the alluvial fan zone. The alluvial plain possesses lower slope-angles than the fans and in many parts of the arid zone forms the dominant morphological feature. It is formed by the depositional activity of running water and consists of fine gravels, sands and silts. The finer material is commonly removed by the wind leaving a stone or desert pavement behind. The sediment eroded by the wind is subsequently deposited as loess or as sand dunes in the lower parts of this zone or on the margins of the adjacent salt desert.

The salt lake or desert zone only occurs in basins with no drainage outlet. Material in this zone is usually fine grained, with the coarser material having been deposited in its passage over the alluvial plain. Salt crusts are often found on the surface and saline deposits occur in the soil profile. Slopes are very gentle (usually less than one degree) and as a consequence any water bodies are shallow and cover extensive areas. Water movement into this zone only occurs during the season of maximum precipitation and in some years no water at all will accumulate. By late summer all water will have been evaporated away and any dissolved material deposited as salt crusts.

Landform changes are slow over most of the arid zone. Indeed, many of the features of central Australia have remained essentially unchanged for millions of years. The only major exceptions to this are river channels and sand dune areas where considerable movement over a few years can occur, often with very great significance for human settlement. With rivers, both lateral shifts in position and changes in base-level can occur. In the arid southwest of the USA there is considerable evidence for the cutting of steep-sided channels, known as arroyos, into the flat alluvial floodplains of many river systems. This arroyo cutting occurred mainly between 1865 and 1915, with a burst of activity being witnessed in many areas in the 1880s (Cooke and Reeves, 1976). Much less incision has taken place since 1915. The effect of the arroyo cutting was to make communications more difficult, to increase sediment production and to alter the hydrology and vegetation of the former floodplain areas. There is still no complete agreement as to what caused this period of arroyo cutting and it would seem that no single cause is responsible. Key factors appear to have been changes in the valley floor vegetation associated with the advent of the white race and its cattle herds, together with the impact of features which concentrated runoff and hence increased erosion (such as embankments, roads and irrigation canals).

It seems likely that quite spectacular changes in river courses occurred across alluvial plains before modern river engineering techniques were introduced in the late nineteenth century. Unfortunately, few detailed accounts of such changes exist. An exception is the Los Angeles River in California. Before 1825 this river entered the Pacific Ocean through Ballona Creek into Santa Monica Bay. Then, following extensive flooding in the same year, its course changed southwards around the Palos Verdes Hills to flow into San Pedro Bay (Nelson, 1983, p. 88). In total this represented a southwards shift of its mouth of more than 10 km. Sand dune movements are not quite as spectacular in their speed of occurrence but equally devastating in their impact on settlements. In the Middle East there are a number of accounts of oases being threatened by moving sand over a time-scale of only a few years (Hidore and Albokhair, 1982).

1.7 ARID ZONE SOILS

Dryland soils exhibit certain characteristics which distinguish them from the soils of other regions, as the result of prevailing climatic conditions (Dan, 1973; Kovda et al., 1979). They tend to be shallow, coarse-textured and possess weak horizon development. The low moisture content limits biological activity, while organic matter contents are low owing to the high summer temperatures. The dominant direction of water movement in the soil profile is upwards and so calcium carbonate and other salt accumulations are commonly found in the top metre or so of the profile. The deposition of these salts means that dryland soils are alkaline in reaction. The infrequent though often heavy rainstorms ensure that fine material is removed from the surface to produce superficial stone layers, commonly known as desert pavements.

Soils in floodplain, terrace and former lake environments are finer-grained and often lack any stones whatsoever. Here the water table is normally close to the surface and the upward movement of water in the soil profile can lead to the formation of salt crusts, which greatly restrict vegetation growth and development. The better-drained floodplain soils usually provide the most fertile agricultural soils in arid areas.

Although many countries have devised soil classification schemes, the one most commonly used in the west is a United States system known as the 'Seventh Approximation' (Soil Survey Staff, 1960). The system was designed with soil surveying in view, and it concentrates on identifying soil properties which affect soil use as well as being indicative of soil-forming processes. Two of the most important ones are soil moisture and temperature regimes. It is a taxonomic system with the soils being classified by the properties they possess, rather than on the basis of genetic factors (as with many of the earlier soil classifications).

Ten major categories are recognized, though only five occur widely in the arid zone (Dregne, 1976). These are Alfisols, Aridosols, Entisols, Mollisols and Vertisols. Although soil-forming factors do not play a determining role in the Seventh Approximation, some of the major categories do reflect climatic and vegetation conditions. In drylands, for example, the Aridosols are associated with desert scrublands; the Mollisols with semi-arid and sub-humid grasslands; and Alfisols with cool forests and hot

Table 1.3: Distribution of different soil types in the arid zone

	Alfisol	Aridosol	Entisol	Mollisol	Vertisol	Total
		(Thousands of square km)				
Africa	2,085	4,890	10,320	120	245	17,660
% of arid region	11·8	27·7	58·4	0·7	1·4	100·0
Asia	–	5,920	4,855	2,850	780	14,405
% of arid region	–	41·1	33·7	19·8	5·4	100·0
Australia	440	2,765	2,285	–	760	6,250
% of arid region	7·0	44·2	36·6	–	12·2	100·0
North America	165	1,950	350	1,790	100	4,355
% of arid region	3·8	44·8	8·0	41·1	2·3	100·0
South America	380	790	1,175	490	–	2,835
% of arid region	13·4	27·9	41·4	17·3	–	100·0
Spain	–	185	69	–	–	254
% of arid region	–	72·8	27·2	–	–	100·0
World	3,070	16,500	19,054	5,250	1,885	45,759
% of arid region	6·71	36·09	41·64	11·47	4·12	100·0

Source: Dregne, 1976.

savannahs. Entisols and Vertisols are not restricted to particular climatic and vegetation conditions. Each of these main categories is capable of being further divided into smaller groups on the basis of specific soil properties. Two of the orders, the Entisols and the Aridosols, dominate the soils of the drylands and account for about 77 per cent of the total area (Table 1.3).

The most widely occurring soils in drylands are the Entisols. These are mineral soils with little or no horizon development. They are young soils and are characteristic of areas being eroded or subjected to deposition. A number of variants occur. Aquents are wet Entisols found mainly on floodplains, deltas or coastal marshes, where the water table is high. Fluvents develop on alluvial sediments, such as fans and flood plains, but here the water table is at depth and drainage through the soil is good. Although commonly flooded, water logging is not a problem with them. Othents are poorly developed soils found most often on rocky hillslopes subjected to high erosion rates. Psamments are Entisols developed on shifting or stabilised dunes; they are especially prone to erosion.

Aridisols are mineral soils with weak horizon development. Salt contents are often high and the profiles are well drained. These soils are exceptionally dry and soil moisture for plant growth is absent for many months at a time. Stone armouring to form desert pavements is a common characteristic on fans and plains.

Alfisols possess a light-coloured surface layer and an argillic (clay-sized particles) horizon below. Pans can also develop. They experience sufficient precipitation to keep soil moisture levels above wilting point for at least three months during the growing season. Ustalfs are Alfisols formed under hot summer conditions. They are well drained and reddish in colour and are the typical soils of the savannahs. In contrast, Xeralfs occur in Mediterranean climatic regimes with winter rainfall maxima.

Mollisols are found in the damper semi-arid and sub-humid grasslands. They are dark in colour with a high base status, and they usually make very fertile agricultural soils. Vertisols are deep soils with high clay contents which experience deep cracking as drying takes place. They normally occur on gently sloping or level areas. In Australia, surface-cracking soils cause enormous financial losses. Each year about 50,000 houses are damaged by cracking and claims for repairs account for 80 per cent of all housing insurance claims (Considine, 1984).

A feature of many dryland soils is their susceptibility to erosion by both wind and water. Wind erosion is prevalent where surface soil horizons experience long periods of dryness and where vegetation cover is minimal. Soil particles of about 0·1 mm in diameter (fine sand size) are the most susceptible to wind erosion. In general, erobility increases as sand content increases, with the most critical particle size range being between 0·05 and 0·42 mm in diameter. Certain fine-grained soils have a resistance to wind erosion due to the formation of thin surface crusts following rainfall. These crusts bind the particles together and make erosion more difficult. Anything which increases resistance to the wind at low levels will reduce erosion potential: for example, desert pavements (where stones are concentrated on the surface) increase friction and slow down the wind, and a similar effect is provided by vegetation. Indeed, the removal of vegetation, through ploughing or overgrazing, is often one of the main causes of wind erosion. In Australia monitoring of soil erosion is now being attempted by the use of satellite systems (Bell, 1987).

Despite low precipitation totals the water erosion of soil predominates over that of wind in most dryland environments. The initiation of erosion is often caused by raindrop impact and splash which breaks up soil crumbs and dislodges particles in a chiefly downslope direction. Particles hit by raindrops have been observed to have moved laterally by as much as 2 m. Wischmeier and Mannering (1969), working on soil properties affecting erosion, showed that the most erodible soils were those which were rich in silt and low in clay and organic matter content. Of all the factors they examined, silt content seemed to have the most important effect on erodibility. In the arid zone it is the soils derived from loess which have the highest silt contents.

Water erosion in drylands manifests itself as either sheet or gully erosion. Sheet erosion predominates in upland areas where relatively thin soils occur over bedrock, whereas gully erosion is confined to areas of weakly consolidated rocks and thick soils. In many parts of the arid zone soil aggregates are unstable when water is added, so that surface pores and cracks are quickly filled with fine-grained particles during a rain storm. This greatly reduces the infiltration capacity of the soil and promotes overland flow, which in turn increases erosion rates. Other factors influencing erosion rates are the steepness and the length of slope segment.

All these factors and others are incorporated in the 'Universal

39

Soil Loss Equation' of Wischmeier and Smith (1965). The prediction equation is of the form:

$$A = RKLSCP$$

where A is the estimated soil loss in tons/acre; R is a rainfall factor dependent on intensity and duration; K is soil erodibility; L is the length of slope factor; S is a slope gradient factor; C is a cropping management factor; and P is the type of erosion control being practised. The rainfall and soil erodibility factors are essentially fixed by nature, but most of the others can be changed to varying degrees by human activity.

1.8 VEGETATION

The transition from forest vegetation to desert scrub is a gradual and almost imperceptible one in lowland areas (Crawford and Gosz, 1982; Furley and Newey, 1983). However, in many regions it is complicated by the presence of uplands, which, with their own microclimates, are usually damper than the surrounding lowlands and also experience lower temperatures throughout the year. In these uplands the vegetation cover can often be quite dense, so making the distinction between uplands and arid lowlands a remarkably sharp one.

As more arid conditions are approached, woodland becomes discontinuous and is eventually reduced to individual or small clumps of trees. These trees are usually smaller than similar species in damper conditions and often take on a gnarled appearance. Grassland takes over as the dominant vegetation, though in turn this degenerates into desert scrub, comprising small, wide-spreading thorny bushes and related species with increasing amounts of bare ground between plants. Under extreme aridity virtually no vegetation at all may be present, except along wadis where plants with deep root systems are able to tap groundwater. This gradient from humid to arid conditions is paralleled by a marked reduction in biomass (amount of living organic material) per unit area. In scrub deserts the biomass is less than a twentieth of that occurring in semi-arid savannahs. What is also quite remarkable in the semi-arid and arid regions are the variations in biomass from year to year in response to changes in moisture conditions.

In the arid and semi-arid zones, plants of the three basic types (the hydrophytes, the mesophytes and the xerophytes) are all found with micro-environmental conditions playing a vital role in their distribution (Cloudsley-Thompson, 1977). All plants require moisture in the soil to survive, but the different basic types can withstand widely varying conditions.

Hydrophytes can survive in soils which are completely saturated with water. Such conditions are, of course, limited in drylands, but they can occur along perennial streams and in marshland environments. Mesophytes and xerophytes need air to be present in the soil and cannot survive when the moisture content in the soil reaches field capacity or above for any length of time. Mesophytes will die when the soil moisture content falls below the wilting point, but xerophytes become dormant at this moisture level and can survive moisture depletion to as low as 5 per cent.

During dry periods, plants have to maintain their moisture content at acceptable levels and at the same time keep an active gas exchange with the atmosphere. Under these conditions water loss is minimised, but this can only be achieved by the cessation of growth. The ephemerals effectively escape from drought by a rapid growth cycle during a short period of favourable soil moisture contents. During very dry conditions they survive as seeds or in storage roots. Plants such as succulents are able to resist drought by possessing water storage cells which can be recharged under moist conditions. Some plants have succulent leaves, some have succulent stems, while others have specialised storage roots or tubers. During drought many of these plants reduce water loss to a minimum, but their growth rates suffer accordingly.

The true drought endurers are the xerophytes, which can continue to grow provided that moisture is present in small but fairly continuous amounts. The xerophytes are drought-adapted plants and can survive waterless periods and still remain active. They grow in such a way as to maximise possible water supplies, usually by an extensive lateral root system and a high root-to-leaf ratio. Their water loss is minimised by partial closure of the stomata, by possession of a small number of deeply set stomata, by having hairy or leathery leaves, or by a small area of transpiring surface. Other plants have deep root systems. An advantage of possessing deep roots is that the deeper and moister layers of the soil can be tapped for water (Noy-Meir, 1973). Figure 1.15 shows that although close to the surface there are considerable variations in soil moisture content, at depths of more than 60 cm,

Figure 1.15 Soil moisture dynamics at different depths in desert soils

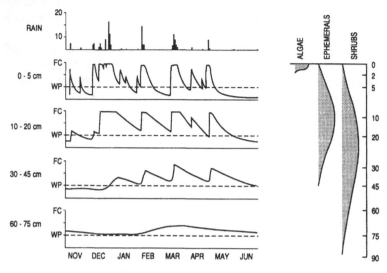

Note: W.P. = wilting point; F.C. = field capacity. Right:- vertical distribution of activity of plant types (scale distorted to fit left part).
Source: Redrawn from Nov Meir S. 1973, 'Desert ecosystems — environment and producers', *Annual Review of Ecology and Systematics*, v. 4, pp. 25–51. Reproduced, with permission, from the Annual Review of Ecology and Systematics by Annual Reviews, Inc.

moisture contents remain above the wilting point all the year round.

Halophytes, which are adapted to saline conditions, are also found in the drylands. Their distribution is usually limited to halomorphic soils where continuous upward movement of moisture followed by evaporation leads to high salt concentrations in the upper soil horizons. Such conditions can occur naturally in closed desert basins, but are more common today on irrigation projects where overwatering has taken place. The true halophytes, as opposed to those plants which possess a high salt tolerance, are able to store appreciable quantities of salt without damage and some are even able to secrete salt through specialised glands.

Interestingly, many plants have developed mechanisms to ensure that they obtain as large a proportion of available water resources as possible. One of the best examples is provided by the creosote bush, which by exuding chemicals prevents other plants from growing close to it. Its root system can, therefore, tap the maximum amount of soil moisture possible without competition.

The low biomass per unit area of arid and semi-arid vegetation has meant that humans have been able to change greatly the vegetation pattern over the years, often unintentionally. By grazing in the more arid parts and by removal of forest in the wetter areas, the vegetation has been so modified that it is doubtful whether any truly natural vegetation exists today. In the more favourable environments, from the point of view of soils and slopes, the natural vegetation has been replaced by cultivated fields. In the semi-arid regions where precipitation exceeds 250 mm, rain-fed agriculture predominates, while below this figure irrigation takes over.

1.9 FAUNA

For animals the arid and semi-arid areas (with their high summer temperatures and at least seasonal aridity) provide a very harsh environment (Cloudsley-Thompson, 1977). These difficulties are compounded by low biomass production, which in turn means that carrying capacities are low (Crawford and Gosz, 1982). While smaller animals can often shelter from the extremes of heat, the larger animals have developed various adaptations so that they can respire and excrete with minimum loss of water and make use of cooling mechanisms which do not depend on the evaporation of water. It should be remembered that the really extreme conditions persist in most environments for only a few months and that at other times of the year much more amenable conditions prevail. Some animals minimise water loss by producing highly concentrated urine and relatively dry faeces, while others have no sweat glands at all. Indeed, only the largest mammals such as the camel, donkey, oryx and gazelle have a surface-to-volume relationship which is small enough to permit them to use the evaporation of sweat as a cooling mechanism.

The key to animal survival in desert conditions is the uptake of water when it is available. A camel can lose over a quarter of its body weight when denied access to water, but is then able to consume a comparable amount of water to that lost in a very short period of time when water does become available. To minimise water loss the camel does not begin to sweat until its body

43

temperature reaches 40·7°C. It also seems to possess the ability to store heat by day and lose it at night when water losses are smaller.

Despite the low productivity of desert ecosystems a wide range of animal life is present in drylands, including insects, arthropods, snakes, rodents, birds and large mammals. Arthropods possess large surface areas in relation to their mass and would therefore not be able to survive high rates of water loss. Animals such as woodlice, centipedes and millipedes overcome this difficulty by spending daylight hours in moist micro-environments and only coming out to feed at night. On the other hand, arachnids and most insects have an impervious cuticle which protects them from water loss, but despite this, most arachnids and insects become active under nocturnal conditions when temperatures are low and relative humidity values reach their maximum. Even scorpions, with their impervious skins, are most active by night, though they are able to tolerate high temperatures and marked desiccation. Desert snakes tend to be nocturnal too.

Only a few insects, such as grasshoppers, beetles and spiders, are active under the hottest desert conditions and these usually have long legs to keep them away from the hottest surfaces. Although lizards are often active during the day, they do seek shelter during the hottest conditions by keeping in the shade or burying in the sand.

Small animals, such as the kangaroo rat, jerboa, gerbils and other rodents, survive in deserts by making use of favourable micro-environmental conditions, such as burrows and holes, where the climatic stresses are less (Figure 1.16). Many have also developed behavioural patterns which mean that most of their activities are carried out at night. Rodents, birds and insects often obtain their water intake through the food they eat rather than from free water. Birds and the larger mammals are active during the day, though these also seek shade during the hottest periods. The American jackrabbit and the kangaroo rat, for example, evidently never take in liquid water.

In most drylands there is one time of the year when damper conditions prevail. At this time there is a surge in plant growth, which is accompanied by a marked increase in animal activity. Caterpillars and grasshoppers consume the new vegetation, while butterflies, bees and wasps collect nectar from flowers and ants harvest grass seeds. Predators, such as scorpions and spiders, also find an abundance of food at this time. Migratory birds return and raise their young, and most other reptiles, birds and mammals give

Figure 1.16 Diurnal temperature variations for different habitats in the Egyptian desert

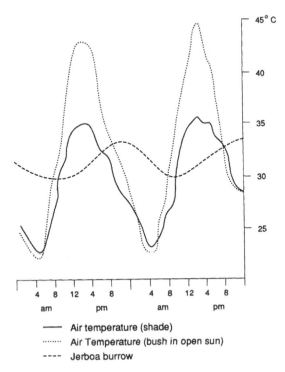

———— Air temperature (shade)
·········· Air Temperature (bush in open sun)
– – – – Jerboa burrow

Source: Redrawn from Cloudsley-Thompson J.L., 1974, *Desert Life*, Aldus Books, London. Reprinted by permission of J.L. Cloudsley-Thompson.

birth at this time. With gazelles the young are born about one month after the rainy season has begun, when plenty of grazing is available. The camel's mating season coincides with the time of rainfall and the gestation period is twelve months, so that the young are born at the next annual cycle of plant growth.

In the deserts of the Old World the main large herbivores were the antelopes, gazelles and wild asses, while in North America they were the pronghorn and the mule-deer. In Australia the kangaroos and the wallabies filled the same ecological niche. All of these animals have low water requirements and are extremely mobile, which means that they are capable of travelling large distances in search of water.

Today most of the carnivores in drylands are small and include

45

lynxes, hyaenas, foxes, coyotes and dingoes. In earlier days larger ones were present, including the lion in the Kalahari and Sahara deserts and the puma on the margins of the North American deserts. Lions were reputedly common in Algeria until the time of the French occupation in the early nineteenth century (Cloudsley-Thompson, 1977, p. 109).

In the drylands of the Old World many of the predators and larger herbivores have been hunted to extinction. In the Arabian Peninsula the Arabian oryx has been eliminated in the wild and species such as the ostrich, which until the beginning of the twentieth century were found in the region, have now disappeared completely. At the same time as the carnivores and herbivores were being hunted and killed, humans were introducing domestic animals of their own, particularly cattle, sheep and goats, to compete with native herbivores for scarce resources. This meant that the native animals were pushed further and further into marginal areas, but even here they were not safe as the range of domesticated animals increased. The result has been that throughout the north African and Asian deserts the natural fauna has been reduced to only very small numbers and extinction of further species seems likely. Perhaps the best example comes from North America where the bison of the Great Plains were virtually eliminated by overhunting in a 30-year period towards the end of the nineteenth century.

Within the desert areas today certain animals still play an important role in transport. In traditional societies the camel, together with the donkey, is widely used for transporting goods. The Bactrian camel can carry loads of up to 450 kilograms (kg), while the Arabian camel or dromedary can only manage lighter loads of up to 270 kg, but can transport such loads for up to 50 km each day (ibid., p. 101). Today it is the domesticated grazing species (cattle, sheep and goats) which dominate animal numbers on the margins of all the deserts of the world. To all intents and purposes these have replaced the natural fauna and it is their grazing activities which are having such a large impact on vegetation and its regeneration.

1.10 DRYLAND ECOSYSTEMS

Up to now attention has been focused on the systematic aspects of drylands. It is important to realise, however, that all these attributes form part of an ecosystem with complex interdependent interactions. Dryland ecosystems are extremely heterogeneous with wide variations in topography, climatic conditions and soils. The unifying factor is low annual amounts of precipitation. Even here, though, there are variations as in some environments, precipitation is relatively reliable on a seasonal basis, while in others rain may never occur for years at a time. Given this fact drylands have low biological productivity with values ranging from 2·6 to 816 grams (g) dry matter per square metre per year (Hadley and Szarek, 1981).

Noy-Meir (1973) suggested that it was pulses of precipitation which triggered off and controlled processes such as production, consumption and decomposition, and in turn, these affected energy flow and nutrient cycling. A number of workers have attempted to relate various ecosystem parameters, such as primary productivity, plant litter production and decomposition rates, with climatic variables like actual evapotranspiration. While these produce good correlations in many environments, the results are not as convincing in dryland areas (Crawford and Gosz, 1982).

Work on dryland ecosystems has suggested that while water availability is the key factor controlling biological productivity, deficiencies of nitrogen and phosphorus can also limit primary production (West, 1981). The nature of the dryland environment also makes it difficult to calculate budgets for different elements as considerable spatial variations occur. It is well known, for example, that nitrogen builds up in areas beneath desert shrubs, whereas only a few metres away very low values can be recorded. Average values for such environments can, therefore, sometimes give a false impression of what is actually going on.

1.11 CLIMATIC CHANGE

In the arid zone, climate is the major control of land use as many activities are carried out close to their minimal water requirements. Hence, even small climatic fluctuations can produce major effects on crop growth and animal survival. One of the chief issues facing all managers of the arid environment is that of assessing the importance of climatic change, particularly over short time periods (Nemec, 1985). During the Quaternary era there have been marked natural climatic oscillations which have produced major ice advances in the northern hemisphere on a number of occasions. There is still controversy as to the causes of these climatic fluctuations, though nowadays most of the evidence would suggest variations in the amount of solar radiation reaching the surface of the Earth, as a result of variations in the orbital eccentricity of the Earth, in the longitude of perihelion and in the Earth's obliquity or tilt. These views, which were first put forward in a coherent way by Milankovitch, suggest that there are regular variations in energy levels at different latitudes, and that when these coincide they can lead to warm or cold intervals occurring

Figure 1.17 Long-term variations of astronomical parameters defining insolation climates and insolation for various latitudes

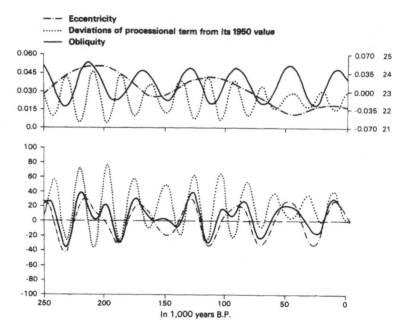

Note: The upper part shows eccentricity e (dashed-dot line), the deviations of processional parameter Δ e (sin $\overline{w}$) from its 1950 AD value (dotted line), and obliquity (full line). The left-hand scale is related to e; the right-hand scales, respectively, to Δ (e sin $\overline{w}$) and obliquity. The lower part shows deviations of solar radiation (cal cm^{-2} day^{-1}) from their 1950 AD values for the calorific northern hemisphere summer half-year at 80°N (dash-dot line); 65°N (solid line), and 10°N (dotted line).
Source: Redrawn from Berger A.L., 1978, 'Long-term variations of calorific insulation from the earth's orbital elements', *Quaternary Research*, v. 9, pp. 139–167. Reproduced by permission of Quaternary Research.

(Lockwood, 1980; Milankovitch, 1930) (Figure 1.17). The last major cold phase occurred between 25,000 and 15,000 years ago. The approach of Milankovitch is theoretical, but over the last two decades increasing empirical evidence has been obtained which confirms the basic tenets of the theory. The first of this work dealt with cores taken from deep ocean basins (Emiliani and Shackleton, 1974), while more recently, cores from the Antarctic ice sheet have revealed a similar pattern of climatic variation over the last 100,000 years or so (Figure 1.18) (Lorius *et al.*, 1985).

Figure 1.18 Oxygen 18 versus depth in the Vostòk ice core, Antarctica, with the definition of the successive stages and indication of the ages corresponding to the limits between these stages

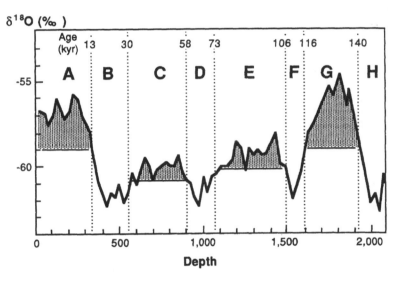

Source: Redrawn from Lorius *et al.*, 1985, 'A 150,000 year climatic record from Antarctic ice', *Nature*, v. 316, pp. 591–596. Reprinted by permission from Nature, Copyright © Macmillan Magazines Limited.

In terms of arid-zone management, what is needed is more specific spatial and temporal data recording climatic changes over the individual continental masses. As yet relatively little definitive information exists, though that which does accords well with the deep sea core and ice core evidence. Perhaps the most interesting data to date have been obtained from the study of the levels of

Figure 1.19 Histograms of lake-level status for 1000 year time periods from 30,000 years BP to the present for (a) southwestern United States, (b) intertropical Africa, and (c) Australia

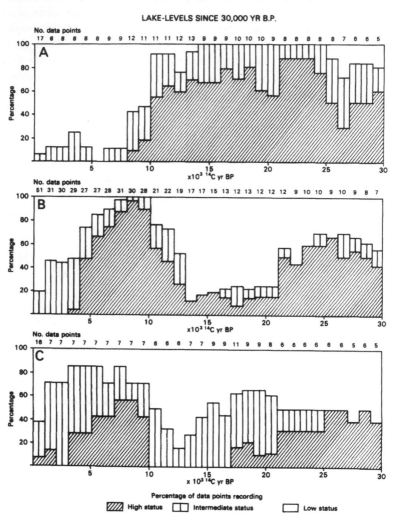

Source: Redrawn from Street F.A. & Grove A.T., 1979, 'Global maps of lake-level fluctuations since 30,000 years BP', *Quaternary Research*, v. 12, pp. 83–118. Reproduced by permission of Academic Press.

inland drainage lakes situated in the arid zones of south-west USA, Africa and Australia (Street and Grove, 1979). Lake levels are particularly good indicators of climatic conditions as they integrate precipitation and evapotranspiration values to produce a given runoff figure. In the south-west USA high lake levels seem to have existed from 30,000 to 10,000 years ago, to be followed by much lower values (Figure 1.19). A more varied picture is seen in inter-tropical Africa, which broadly conforms with the evidence from Australia. High lake levels prevailed from 30,000 to 21,000 years ago, but much lower levels occurred between 21,000 and 12,000 years BP. A second period of high lake levels was recognised from about 10,000 to 5,000 years ago, and then over the last 5,000 years a marked decline in levels has been recorded. These lake levels reflect changes in the position of the main atmospheric boundary contact zones, such as the Inter-tropical Covergence Zone. Major differences do seem to occur in lake levels between the tropical and extra-tropical zones: for example, in the tropical zone, high lake levels correlate with interglacial or interstadial conditions, whereas in more temperate areas, such as the south-west USA, high lake levels are associated with glacial episodes. One point does seem to be clear — namely, that throughout the tropics there was a substantial increase in surface water beginning about 10,000 years ago. It would also appear that the last 1,000 years or so have been relatively dry compared with earlier times.

Our knowledge of climatic fluctuations over the last few hundreds of years is very restricted. However, tree-ring chronologies for parts of the south-west of the USA have permitted the exploration of precipitation changes over time. Estimates for parts of California have revealed that the period from 1901–61 experienced a higher mean precipitation total than at any time since 1670–1720 (Fritts and Gordon, 1980) (Figure 1.20). This clearly illustrates the danger in using mean values of either precipitation or river discharges for planning purposes, particularly when the most recent period of record appears to have been considerably wetter than the long-term average.

Our knowledge of climatic fluctuations over periods of hundreds of years is very restricted. River discharges do have the advantage of integrating a whole range of climatic effects within a drainage basin into a single value of water volume, but unfortunately only few long-term records exist. The River Nile provides an obvious exception, though even here the accuracy and continuity of the record leaves much to be desired (Bell, 1970).

To obtain a higher resolution of temporal climatic fluctuations it is essential to study the period of instrumental record covering the

51

Figure 1.20 Average annual precipitation for 18 California stations reconstructed from 52 western (USA) tree ring chronologies

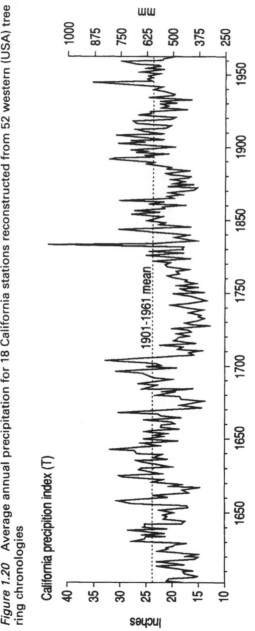

Note: Dots represent eight-year weighted averages used to smooth out the annual values. The horizontal line corresponds to the 1901–1961 mean value.

Source: Redrawn from Friths H.C. & Gordon G.A., 1980, 'Reconstructed annual precipitation for California', in Hughes M.K. *et al.*, *Climate from tree-rings*, Cambridge University Press. Reproduced by permission of Cambridge University Press.

last 150 years. Unfortunately most of the arid parts of the world do not have meteorological data going back this far and so our ability to register climatic trends, periodicites or merely random variations is restricted. However, work in South Africa has suggested that from the early part of the twentieth century to the present day, there has been a sequence of wet and dry periods, which implies a periodicity in the causative factors (Figure 1.21) (Tyson, 1980; Tyson et al., 1975).

Almost everywhere else in the arid zone, and particularly in Third World countries, there are insufficient data to identify the nature of climatic fluctuations which may be occurring. This is especially true in the Sahel region of Africa where instrumental records prior to the Second World War are very scarce. Here, major droughts have occurred between 1968–73 and also in the early 1980s, which appear to have recurrence levels of at least 50 years. Certainly the last drought of comparable magnitude seems to have taken place between 1910 and 1915.

Besides these naturally occurring changes it is important to consider the possibility that human activity may be affecting climatic conditions. The most likely cause would seem to be the increase in carbon dioxide levels as the result of burning fossil fuels over the last 150 years. There is now no doubt that carbon dioxide levels in the atmosphere are increasing, but the effect that this is likely to have on climatic conditions is still uncertain. Recent reports originating within the USA suggest that the most likely effect by the latter part of the twenty-first century will be a marked increase in average temperatures (Rind, 1984). Over the major arid parts of the world this warming trend may be as much as 4°C (Figure 1.22). What impact this will have on precipitation distribution is uncertain. However, the increased temperatures will produce higher evapotranspiration rates and so many areas may, from the environmental standpoint, become more arid in nature. Where these human-made trends are paralleling natural climatic changes, the movement towards aridity may be particularly rapid. Elsewhere it is possible that the two trends could cancel each other out, at least over short time periods. An interesting assessment of the likely effect on the High Plains of the USA is given by Glantz and Ausubel (1984).

The important point for the environmental manager is that everyone must accept that climatic change is a reality (Beaumont, 1988). In the past most managers have assumed that there was no climatic change because it could not be proved with the data which

Figure 1.21 Rainfall variations in southern Africa

The smoothed regional summer rainfall series for 1910–1967 (shaded) and the fitted curve extrapolated to 1977 (solid line). The comparison between predicted and observed rainfall for the period 1968–77 is shown (dotted).

Mean, Fitted and Predicted Rainfall

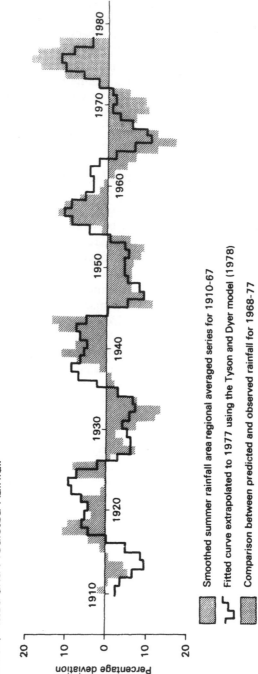

Smoothed summer rainfall area regional averaged series for 1910-67

Fitted curve extrapolated to 1977 using the Tyson and Dyer model (1978)

Comparison between predicted and observed rainfall for 1968-77

Source: Redrawn from Tyson P.D. & Dyer T.D., 1978, 'The predicted above normal rainfall of the seventies and the likelihood of droughts in the eighties in South Africa', *South African Journal of Science*, v. 74, pp. 372–77. Reproduced by permission of the South African Journal of Science.

Figure 1.22 Annual average surface air temperature changes (degrees C) due to doubled CO_2 as determined by the GISS model

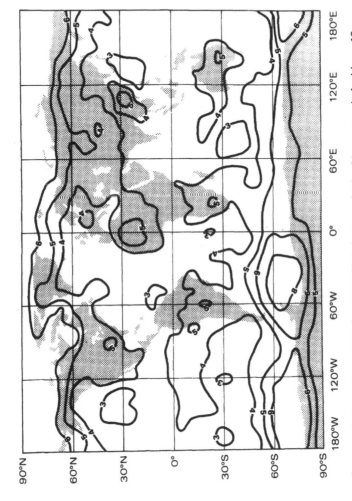

Source: Redrawn from Rind D., 1984, 'Global climate in the 21st century', *Ambio,* v. 13, pp. 148–151. Reproduced by permission of Ambio.

were available to them. In future we must accept that climatic change is taking place, although our instrumental records are probably not complete enough for us to recognise it easily. Such a change in attitude should enable us to react more quickly to changes in the next few decades.

2

Human Use of Drylands

2.1 TRADITIONAL SOCIETIES

2.1.1 Introduction

The societies which are found in drylands in the twentieth century are the result of evolution over time. With some groups these changes have been very slow over hundreds of years and have resulted in a society which is in equilibrium with its environment. Such societies can have different forms, with economies based on hunting and gathering, pastoral farming or sedentary agriculture. In other cases, changes have been so rapid in modern times that entirely new economies have been produced which are dependent not on local agricultural produce (as with the case of traditional societies), but on fossil-fuel subsidies instead.

Until the twentieth century a number of traditional societies lived in the arid and semi-arid regions of the world by employing a range of solutions to the problem of supplying food and other materials necessary for human survival. At the simplest level were the hunter–gatherer communities which harvested the resources that nature had made available. Of necessity their way of life required considerable movement and population numbers were not large in any one area. The nomads were groups of people who obtained their existence from the husbandry of animals, and the movement of these animals along well-defined routes to harvest the available plant resources. Some groups exploited considerable altitudinal ranges in search of pastures, while others covered large horizontal distances. Almost all nomads, though, require contact with sedentary farming populations where animal products can be

57

exchanged for foodstuffs such as cereals. Sedentary settlements can be divided into those which grow crops under rain-fed agriculture and those in drier climates where irrigation has to be practised. In general the irrigated systems require a more complex social organisation, as the allocation of water resources can be a complex and difficult matter.

2.1.2 Evolution of traditional societies

The stage when humans lived by hunting and gathering was by far the longest stage in human evolution, lasting from the origin of *homo sapiens* as a separate species until the twentieth century in certain parts of the world. During this time important changes took place. People learnt to control and use fire and to fashion ever more sophisticated and efficient stone tools through the Palaeolithic and Mesolithic. One of the most important changes in human lifestyle was the development of agriculture, i.e. the domestication of plants and animals (Bender, 1975; Flannery, 1973; Harlan, 1978; Higgs and Jarman, 1972; Isaac, 1970; Reed, 1977; Ucko and Dimbleby, 1969). In tracing these developments it is important to distinguish between seed agriculture and vegeculture.

Vegeculture plants, such as manioc and yams, are cultivated by vegetative reproduction. This type of agriculture probably developed in the tropics on the boundary between forest and grassland. It is well known from south-east Asia, where a series of plants (including taro, yams, breadfruit, sago palm, coconuts and bananas) were indigenous. Evidence of vegeculture dates back at least 11,000 years. Vegeculture provided a number of advantages: there was no need to clear all the vegetation before planting and roots could be harvested individually over a long period, which meant that less storage was needed. The tropical roots were also usually associated with other plants, such as bananas and coconuts, which provided an alternative food source. As a result food supply was relatively easy to secure and, given the warm climate, food plants were available all the year round.

In contrast, seed cultivation seems to have developed under semi-arid conditions in south-west Asia. South-west Asia includes a range of differing environments from more than 300 m below to more than 4,000 m above sea level. What is unusual about the region is that environmental changes from plains to high moun-

Figure 2.1 Distribution of the wild ancestors of wheat and barley in the Middle East

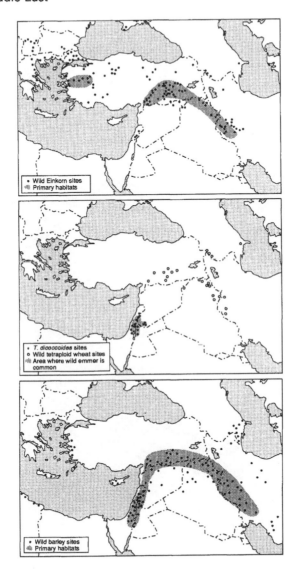

Note: wild einkorn (Triticum boeticum); wild emmer (Triticum dicocciodes), wild barley (Hordeum spontaneum).
Source: Redrawn from Zohary M., 1969, 'The progenitors of wheat and barley', in Ucko P.J. & Dimbleby G.W. (eds) *The domestication of plants and animals*, Duckworth & Co. Reproduced by permission of Duckworth & Co.

Plate 2.1 Traditional village in semi arid zone of northern Tunisia.
Note the internal courtyards and the high walls along the streets.

Plate 2.2 Traditional irrigation canals and planned irrigation development on the Varamin Plain, Iran.

tains can occur over very short distances. These topographical changes are mirrored by climatic variations from hot desert to arctic-alpine conditions over a few kilometres. In turn this has produced a rich variety of flora and fauna occupying a range of different yet spatially close ecosystems.

The earliest evidence for cereal domestication is found in a broad belt stretching from Turkey in the west, southwards along the highlands of the eastern Mediterranean as well as south-westwards along the foothills of the Zagros Mountains of Iran and Iraq. In this region there occurred the wild plants from which wheat and barley, the main cereals, were domesticated (Figure 2.1). Also in this region the wild progenitors of sheep, goats, cattle and pigs were found together. Compared with vegeculture, seed agriculture was much more demanding in the conditions it required: for example, the ground had to be cleared of all vegetation before sowing could occur. As the cereals are annuals, all the seed must be harvested during a relatively short period of time. In turn there is a requirement that the seed must be carefully stored for use during the rest of the year. All this requires a degree of organisation of society which is greater than that needed for vegeculture, together with a degree of permanence for the individual settlements.

The actual process by which agriculture began is unknown, but by about 12,000 years ago hunters and gatherers were reaping wild wheat and barley, grinding grain and storing it in pits (Mellaart, 1975). By 8,000 years ago a system of village communities dependent on cereals (wheat and barley) and the tending of sheep and goats existed throughout the region from Greece eastwards to the interior plateau of Iran. The crucial period in terms of agricultural development, therefore, was between 12,000 and 8,000 years ago, as this period saw the replacement of hunter–gatherer communities by sedentary systems. Over the years new plants were added to the range of those domesticated, including by about 6,000 years ago, the olive, vine and fig. Cattle and pigs were domesticated later than sheep and goats, with cattle being used for draught purposes as well as for meat.

Although the details of the transition from hunter–gatherer to agriculturalist are not known in detail, enough information is available to sketch out a likely sequence of events. About 20,000 years ago, it would seem that humans began to change from hunting large game almost exclusively, to a lifestyle which made use of a broader range of available foodstuffs, including small

Figure 2.2 Map of selected ancient sites in the Middle East

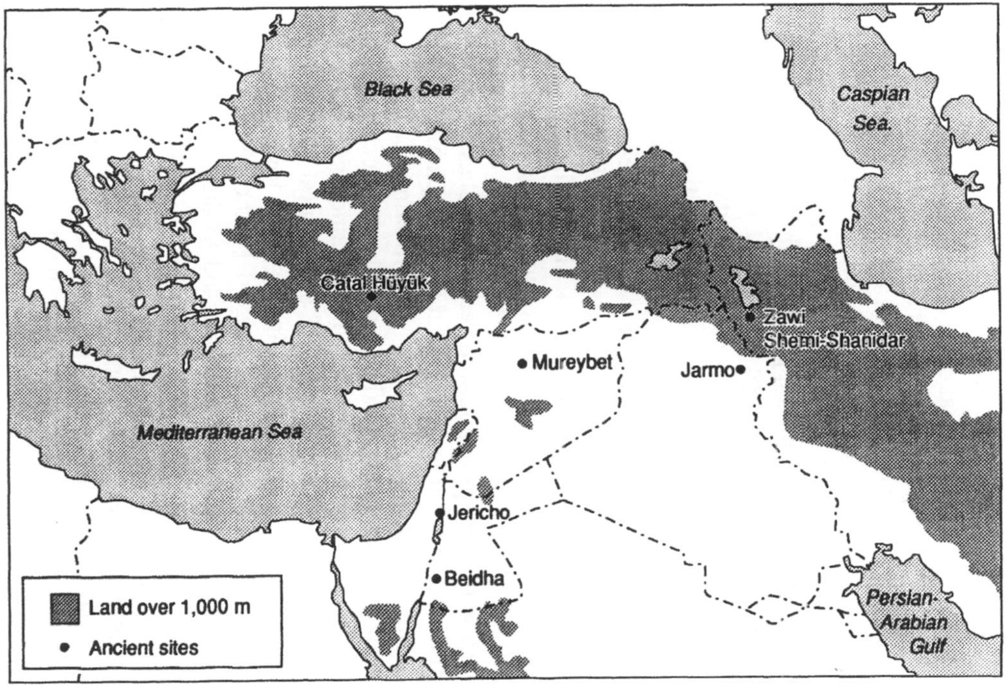

game, fish, molluscs, birds, as well as fruits, nuts and other plant foods. In the semi-arid regions of the Middle East these plant foods included grass seeds from the species which were the wild ancestors of wheat and barley.

The basic problem for a primitive society in using wild grass seeds as a food source was that large quantities of the seeds had to be obtained during a harvest period of perhaps three or four weeks, and that these seeds then had to be stored in a dry condition until they were used. This was by no means the end of the problem, for the seeds had to be heated to remove the husks. Following separation the seed needed to be ground down to flour using grinding stones. These demanding conditions meant that if these people were to make use of the wild grass seeds, they had to establish semi-permanent dwellings in the growing areas, where the seeds could be processed and stored. It would seem, therefore, that the establishment of settlements, albeit only small ones, was a prerequisite before the domestication of grass species could take place. Indeed, it would appear that the domesticated varieties of the cereals did not emerge until about a thousand years after the first permanent settlements had been established.

One of the earliest of these sites where there is evidence for the collection, roasting and storage of grains is at Zawi Chemi-Shanidar in Iraq (Solecki, R.L., 1964; Solecki, R.S., 1963, 1964, 1971) (Figure 2.2). Here there are two related sites some 4 km apart. Zawi Chemi is an open-air site, while 400 m higher is the cave of Shanidar. Earliest occupation of the sites is from 8920 BC (10,870 BP) and many occupation levels are found covering a period of perhaps a thousand years. Actual use of these sites is not known, though it is suggested that the open-air site was occupied during the summer and the cave in winter. At Shanidar bone remains indicate that wild goat was the most common meat source, but many other animals were hunted and molluscs collected. The cave also reveals considerable agricultural activity with storage pits, querns, mortars and grinding slabs for processing the wild grains. By about 8000 BC (9950 BP) the herding of sheep had become important and this later led to their domestication.

An even more intriguing site is found at Mureybet in north Syria on the east bank of the Euphrates (Cauvin, 1972, 1973). This was a settlement of Natufian hunters and fishermen, with at least four occupation levels indicating some degree of permanency. The earliest site was destroyed by fire about 8640 BC (10,590 BP). The faunal remains include gazelle, aurochs, and wild ass, as well as bird and fish bones and freshwater molluscs. The remains of plants including wild barley have been found, together with what have been interpreted as pits for roasting the wild grains and perhaps meat and fish. The next level, which contained about 200 houses, appears to have ended by 8142 BC (10,092 BP). At this stage the food resources of the Euphrates were still widely used and some hunting occurred. There is, however, no evidence of animals being herded. Particularly significant about this level is the fact that it reveals a wide range of charred grains and seeds from a very early period, including wild einkorn, wild barley, lentils and bitter vetch. What is intriguing here is that wild einkorn and wild barley are not found naturally in this low environment. This suggests that seeds from elsewhere must have been transported into this new environment for cultivation. As such it would seem a step beyond the mere collection of naturally growing wild species. The successful nature of the food production system in the settlement is attested by the fact that the site was occupied for most of the eighth millennium BC.

It would seem, therefore, that pre-domestication sedentary villages developed in response to the necessity to store and process the wild grain seeds. The fact that an infrastructure had to be built with houses, storage pits, ovens and grinding facilities meant that the people were not quick to abandon them. The success of the early sites was also assured by the presence in the same area of other food resources such as wild goats at Shanidar and aquatic resources at Mureybet. Over the years the wild grains changed to develop the characteristic feature of the domesticated varieties, which is the tough and non-brittle rachis. This meant that the seeds remained in the ear when harvested and had to be removed by the threshing process. Changes also occurred with the husks, which became easier to remove, without the need for roasting, which had been essential with the wild varieties.

The most remarkable early settlement yet discovered in the Middle East, where domestic cereals were grown, is at Jericho in Jordan (Kenyon, 1969–70). This is a settlement fed by springs at the foot of the rift valley escarpment. It was used by Natufian hunter–gatherers over a long period, but took on a new significance during the Pre-Pottery Neolithic between 8350–7350 BC (10,300–9,300 BP). The site, which only receives about 100 mm of precipitation annually, is spread over 4 hectares (ha) and probably had a population of about 2,000 people. It is without doubt the earliest known town in the world and has probably the oldest irrigation system. The earliest levels had no defensive walls, but thereafter fortifications became an important part of the settlement. No domestic animals were found though gazelle, which was the chief meat source, was possibly herded as well as hunted. The major food source of the settlement, though, was domesticated wheat and barley (in the form of emmer wheat and two-row hulled barley) grown under irrigated conditions. Where and when these cereals were domesticated before being brought to Jericho still remains unknown. Legumes were also cultivated. With this food supply Jericho proved a stable and successful settlement which lasted for a thousand years before being mysteriously abandoned in 7350 BC (9300 BP). The chief wealth of the site was the cereals which the spring waters allowed to be produced in large quantities. The fortifications suggest that this wealth was desired by other groups.

After slow beginnings the domesticated cereals spread rapidly through the Middle East and settlements from a wide variety of environments are known. At Beidha in southern Jordan, an early Natufian open-air site was reoccupied at the beginning of the seventh millenium BC (Kirkbride, 1968). Originally it seems to have been a temporary or possibly seasonal settlement, but round houses were built later. This level was destroyed by fire in 6650 BC (8600 BP). Thereafter development of the site took place over a long period, with changes from round to rectangular structures. Food for the site was provided by the cultivation of domesticated wheat and barley. Legumes were also grown and pistachio nuts and acorns collected. Domesticated goats were herded, as possibly were gazelles too. Other animals including aurochs, ibex and wild boar were hunted.

A similar though larger site is known from Jarmo, on the Chemchemal Plain of Iraq (Braidwood *et al.*, 1960). This site covered 1·5 to 2 ha and was located in oak–pistachio woodlands. It contained 20–30 houses and lasted for approximately 500 years between 6500 and 6000 years BC (8450–7950 BP). It was a permanent village which primarily depended on the production of domesticated emmer wheat, though two–row hulled barley and einkorn were also present. Legumes were cultivated and pistachio nuts and acorns collected. Domesticated goats and to a lesser extent sheep were herded, and other animals hunted.

The largest known Neolithic agricultural site in the Middle East, covering about 12 ha, is at Catal Hüyük on the Konya Plain on central Anatolia (Mellaart, 1967). Excavations provide occupancy dates from 6250 to 5400 BC (8200–7400 BP), though deeper layers are certainly present. At its zenith it is thought that Catal Hüyük had about a thousand houses, with a total population of between 5,000 and 6,000. The economy of the settlement was based on irrigated agriculture and cattle breeding, trade and industry. Domestic emmer, einkorn, breadwheat and six-rowed barley were grown, together with peas, bitter vetch and vetchling. The wild ancestors of the cereals were not indigenous to the plain, suggesting that (as at Jericho) people must have imported these varieties from other areas where domestication originally occurred. Acorns, pistachios and almonds, together with fruits such as crabapple, juniper berries, hackberry and capers were also present. It would seem that domestic cattle provided the inhabitants with more than 90 per cent of their meat, as well as transport. Wild sheep were abundant, but goats were rare. Hunting of wild animals also seems to have been important.

The village communities which had been established up to 8,000 years ago were generally located in foothill locations, where precipitation was sufficient to permit rain-fed agriculture. Individual villages were generally not large, though many seem to have been occupied for long periods. Crop yields would have been low and unreliable as they are today in similar regions. At first a shifting type of cultivation around the villages was probably practised, with livestock grazing on the stubble which was left behind as well as on adjacent rangelands. The basis of the settlement was, therefore, a true mixed farming type of economy, with hunting and gathering still very important. The fields of growing grain would have proved attractive to sheep, goats, gazelle and other herbivores and would undoubtedly have made it easier for people to herd and later domesticate these animals.

2.1.3 Later developments

Between 8,000 and 6,000 years ago there occurred a marked change in the areas which were opened up for settlement with a movement down into the floodplains of the Tigris–Euphrates lowlands. In contrast, the Nile floodplain was not settled until much later at about 5,500 years ago. In the Tigris–Euphrates lowlands, and indeed in the Nile valley, the low annual precipitation totals meant that the cultivation of cereals was only possible by the use of irrigation. The Tigris–Euphrates lowlands (i.e. Mesopotamia) proved to be a very demanding environment in which to grow crops. The main flood along the river took place in late spring and early summer as a result of snowmelt in the high mountains of eastern Turkey. This meant that the crops were growing in the fields when the flood arrived. Serious damage occurred if it could not be contained. Soil salinity always presented a problem as well (Jacobsen and Adams, 1958).

To harness the water resources of the region in an efficient manner it was essential that large-scale hydraulic works were constructed. However, such works could only be built during periods of political stability, and they required considerable societal organisation to provide the necessary labour force (Adams, 1965, 1972, 1978). This meant that the early irrigation schemes were relatively small and local in extent. Indeed, it was not until the development of the Sumer empire in the second millenium BC (4000–3000 years BP) that the first major irrigation works were established. In these floodplain environments new cereal varieties were introduced, although it is not certain whether the new varieties were a response to the different environment, or whether new varieties which had been developed elsewhere permitted the floodplain environment to be opened up for the first time. Whatever the reason, six-row barley replaced two-row barley and breadwheat joined emmer and einkorn. Of the two latter varieties, einkorn became the dominant one in the Tigris–Euphrates lowlands, as it proved much more tolerant to the saline soils. Another important change was the growing need for animal labour to work the heavy floodplain soils, leading in turn to the development of the ox-drawn plough.

Along the Nile, irrigation development proved much easier (Butzer, 1976, 1984). The Nile flood, which occurs in late summer, watered the fields and built up soil-moisture levels. It also deposited silt with many plant nutrients and washed away any salt accumulations. Once the flood had receded, the cereals were planted and then grew without any further water being added. The crops were harvested in April and May and then the land was left fallow until after the next flood. This long dry period proved invaluable in minimising the number of pests which would damage the crop. Flood control measures were in operation in Egypt by about 5,000 years ago. The nature of the environment did not demand the complex hydraulic works so necessary in Mesopotamia, permitting available labour to be used for other purposes.

Thus, by about 5,000 years ago, it was possible to distinguish two types of agriculture in south-west Asia. The first, and older type, is a farming system based largely on rain-fed agriculture in the foothills throughout the region. Wheat and barley were the chief crops and sheep, goats, cattle and pigs were all kept. Shifting cultivation was probably practised, though in the more favourable sites long-term continuity of settlement seems to have occurred. In the drier areas small-scale irrigation systems were established using spring or river waters, but all of these were restricted in extent. The second type was large-scale irrigated agriculture on major floodplains where precipitation totals were low. These systems made use of similar crops and animals to those of rain-fed agriculture and were best developed along the Tigris–Euphrates and Nile floodplains. Both were associated with major urban developments.

The reason why plants and animals were domesticated just when they were still remains a mystery, though many theories have been put forward. Some argue that domestication took place as a result of growing population pressures in the region. With this theory it is postulated that increasing pressures on resources necessitated greater food production and that domestication of plants and animals was the consequence. It does seem plausible that as stone tools became more sophisticated it would have been easier for humans to hunt animals and, therefore, provide an assured food supply, which in turn may have led to increased population numbers (Cohen, 1975; Hassan, 1973; Sengel, 1973). Some have claimed that it was deteriorating environmental conditions which caused humans to adopt a strategy of widely differing food sources and that this in itself is an indicator of the growing pressures which were being put on the available resource base (Harris, 1980). Others have argued that domestication of plants and animals was inevitable as a result of the societal evolution. This may be so, but it does not explain why it should occur between 12,000 and 8,000 years ago, following a period in excess of two million years when humans had existed as hunter–gatherers.

Yet another theory postulates that domestication of plants and animals was associated in some way with important climatic changes at the end of the Devensian glacial period. Over north-west Europe the last major ice advance took place between about 25,000 and 15,000 years ago. Our knowledge of conditions in the Middle East still remains fragmentary, but it is known that in the highlands of Turkey and Iran at this time, cold conditions prevailed with a snowline perhaps 1,000 m lower than at present.

Relatively little palynological evidence is available, but a diagram from Lake Zeribar in the Zagros Mountains indicates that a cool steppe climate persisted until about 11,000 years BP (Zeist and Wright, 1963). Then between 11,000 and 5,500 years ago, a rapid climatic amelioration took place, with trees such as oak returning, to produce a climate that was probably warmer and drier than that of today. Finally, over the last 5,500 years conditions similar to those currently experienced have prevailed. The evidence would therefore seem to suggest at least a coincidence between the time of a significant climatic change in the region and the beginnings of agriculture. The intriguing question is whether this is in some way a cause and effect relationship.

Whatever the reason for the domestication of plants and animals the fact that it occurred had important effects on the human utilisation of the environment. Two land use management aspects are of great significance. The first is that settlement forms, such as villages and towns, now take on a degree of permanence which they previously lacked. Seed agriculture can only be practised successfully by settled communities and so a new type of society had to come into existence, which in the case particularly of the irrigation civilisations had to be able to organise an efficient division of labour.

Perhaps even more important, though, was the effect on land carrying capacities as far as humans were concerned. By the introduction of agriculture, humans could substitute themselves for animals at the end of a food chain. This meant that the amount of energy which humans were able to obtain direct from a crop from a given unit area of land increased considerably compared with that which could have been gained by eating a herbivore grazing the same area. Some workers have suggested that, as a result of domestication, the world population grew rapidly by at least 25 times between 12,000 and 6,000 years ago. It has been estimated that the population of the Middle East in 10,000 BP was 100,000, but that by 6,000 BP it had risen to 3·2 million (Carneiro and Hilse, 1966; Harris, 1980).

What there can be no doubt about was the rapid spread of agriculture throughout the Middle East and then, from about 6,000 years ago, its rapid diffusion westwards into Europe and eastwards into Asia. When this occurred it meant that there was a spread of the relatively high population densities associated with agriculture into areas which had only been sparsely populated previously by hunter–gatherers. By about 4,000 years ago most of

Eurasia and north Africa had been colonised by agriculturalists, leaving any hunter–gatherers in the lands beyond this ever-advancing agricultural frontier.

Within the arid zone experimentation with agriculture continued, leading to interesting solutions to the problem of dealing with water scarcity. An interesting example is provided by the Nabateans of the Negev, who lived some 2,000 years ago (Evanari et al., 1971; Hillel, 1982). These people perfected an efficient method for rainwater harvesting, permitting them to grow crops in areas where the average rainfall was less than 100 mm. The idea behind the system was to encourage runoff on hillslopes by smoothing the slopes and producing a complex series of conduits and channels. As a result infiltration into the soil was minimised. The water which occurred as runoff was collected in the various channels and led towards low-lying regions, usually floodplains, where check dams would halt water movement and encourage infiltration to take place. Having had their soil-moisture levels recharged, these lowland areas could then be sown with cereals. Any excess water from the highest field would be led on to the next one downstream. The channel systems in the lowlands easily permitted the water flow to be diverted from one field to another without difficulty.

The old runoff farms of the Nabateans usually had several cultivated fields watered by watersheds of 10 to 50 ha. Each of these watersheds was divided into catchments of 1–3 ha in size, designed in such a way as to prevent a major flood wave being generated at the farm itself. Instead, the water was fed in a controlled manner down to the farm so as to minimise erosion and water damage. At Avdat in the Negev, it has been calculated that each hectare of cultivated land received runoff from about 20 ha of adjacent slopes, as well as any precipitation which might fall directly upon it. This meant that the cultivated areas would normally expect to receive water depths equivalent to between 300 and 500 mm, in regions where average annual precipitation totals were only 100 mm per annum.

2.1.4 Traditional societies at the present day

The traditional societies which had evolved in the Neolithic period persisted in many parts of the world right through to the twentieth century with remarkably few changes. Hunter–gatherers such as

71

the Australian aborigines and the Kalahari bushmen still followed in the early years of the present century a way of life that dated back to times even before the Neolithic. In parts of the Middle East and Africa pastoral nomads have traversed the same routes for centuries, with the only major changes being variations in population and animal numbers in response to political conditions, disease, and climatic fluctuations.

On the margins of deserts throughout Africa and Asia agricultural villages and small market towns have followed dryland farming procedures, revealing little change over the last 6,000 years. Cereals have been the staple crops, together with vegetables, while the grazing of animals, such as sheep and goats, has taken place on adjacent poor-quality pastures. Draught animals provided the main motive force and the communities in each area were largely self-sufficient with regard to food, fibre for clothing and building materials. There was a small though important trade in luxury goods such as certain culinary utensils which could not be fashioned locally.

In the driest areas cultivation has only been possible by irrigation. Here, one can see enormous differences in scale. On major river systems, such as the Nile, Tigris–Euphrates and Indus, large-scale irrigation systems were still in operation, though in many cases the complexity and sophistication of the individual systems was not as great as it had been in earlier times. Certainly in this respect there was a deterioration in activity in many river basins from a heyday many hundreds of years earlier. With the Tigris–Euphrates system, for example, from the twelfth century onwards irrigation took place on a local non-co-ordinated basis, until the late nineteenth century, when the old canals were cleaned out.

On alluvial fans on the margins of the deserts, small but complex irrigation systems developed using both surface and groundwater sources. Some of these were extremely complex and permitted the irrigation of extensive areas. On the Varamin Plain, Iran, more than 40 hand-dug unlined irrigation canals provided water for an irrigation network supporting 100,000 people in more than 250 villages (Figure 2.3). This network was so complex and efficient that very little flow of the River Jaj actually crossed the alluvial fan and flowed out into the desert unused. While it is true that percolation losses through the unlined canals were high, these losses on the Varamin Plain were more apparent than real. The canals were, in effect, acting as groundwater recharge systems and

Figure 2.3 Traditional surface irrigation canals on the Varamin Plain, Iran

Source: Redrawn from Beaumont P., 1974, 'Water resource development in Iran', *Geographical Journal*, v. 140, pp. 418–431. Reproduced by permission of the Royal Geographical Society.

the water was then abstracted by a complex qanat system further down the alluvial fan (Beaumont, 1968).

2.1.5 Aborigines

One of the most primitive groups making use of arid and semi-arid environments was the aborigines of Australia. At the time of European occupation of the continent it is estimated that there were probably 300,000 in all, with most located on the northern and eastern peripheries of the country (Heathcote, 1975). They lived by hunting, gathering and fishing and followed a nomadic way of life. Their material possessions and tools were extremely simple, but their environmental knowledge and skills were of a high order. They did not, however, practise any form of cultivation or herding of animals.

Given their nomadic way of life, population densities were low, averaging one person per 26 square km over the continent as a whole. The distribution of aborigines before white settlement seems to have been concentrated mainly around the northern, eastern and southern coasts, with relatively few people living in the dry inland locations. The actual peopling of the continent is believed to have occurred during periods of low sea-level associated with glacial conditions, when land routes connected Australia with New Guinea and the islands to the north (Tindale, 1959). Two indigenous cultures have been recognised, though others may well have been present as well. The oldest (Tasmanoid) was mainly confined to Tasmania. These people possessed more negroid features and were shorter than the later (Australoid) group. The Tasmanoid culture also possessed simpler weapons and less complex hunter–gatherer strategies, and so are believed to be the older of the two. The Australoid group which peopled the main continental area was divided on tribal grounds into more than 500 units, with between 150 and 1,500 people in each.

All of these tribes were nomadic and never occupied permanent settlements. The essentials for life were weapons, fire and a few light containers of bark, woven grass or bone for food and water. Everything had to be carried by the people when they moved, as they lacked any domestic animal; and consequently, material possessions were kept to a minimum.

Given the range of environmental conditions prevailing in Australia it is not surprising that four different types of economy have been recognised amongst the aboriginal populations (McCarthy, 1959). In the sub-tropical forests of the north and north-east coasts, a hunting, fishing and gathering economy is recognised. On the east coast and on Tasmania, a fishing and shell collecting economy occurs, while inland are gatherers of the mountains and forests. Finally, there are the hunters and collectors of the dry arid zone.

The tribes of the interior provide the best example of primitive human adaptation to an arid environment. One of these is the Walbiri tribe which occupied the land of western Northern Territory prior to the advent of the white race. They occupied about 105,000 square km of land of varied terrain. Most of it receives between 12·5 and 25 cm of rainfall each year, though in the northern fringe it reaches 37·5 cm. Summer temperatures are very high, going well over 40°C in the period before the rainy season begins. In winter, temperatures drop to between 10 and 30°C and nights can be cold. Strong easterly winds at this time raise dust storms and make living conditions difficult. Throughout the region there is a lack of surface waters, with seasonally filled billabongs and soaks along the major creeks being the only water sources.

Four major divisions within the tribal area were recognised before the 1930s when the breakdown of the traditional ways of life occurred. These regions roughly coincide with the north, east, south and west parts of the tribal area (Figure 2.4). In the north the Waneiga country covered a large area, but was lacking in both variety and quality of vegetable and animal life. As a consequence population numbers were always low. To the east was the Yalpari (Lander) region. This was rich in vegetation resources and game, and groundwater was available along the Lander Creek all the year round. Population densities were always relatively high. Adjoining this region to the south was the Ngalia country, with plenty of vegetable and animal resources and adequate water supplies. Finally, in the west the Walmalla territory was poorly endowed with food resources and meagre water supplies were only available in the wet season. Reptiles provided the only assured animal food. It is estimated that prior to the coming of the Europeans there were between 1,000 and 12,000 members of the Walbiri tribe, which provided an overall carrying capacity of about one person for each 92 square km (Meggitt, 1962).

In each of these regions the tribal groups would be self-sufficient in terms of food and water resources in all but the most extreme

Figure 2.4 Tribal divisions of the Walbiri tribe, Australia

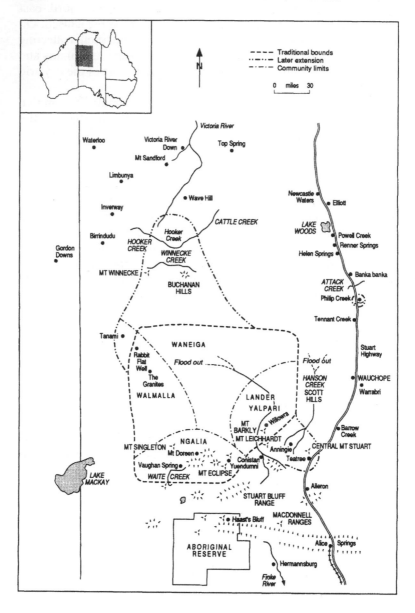

Source: Redrawn from Meggitt M.J., 1971, *Desert people — a study of the Walbiri Aborigines of Central Australia*, University of Chicago Press. Reproduced by permission of the University of Chicago Press.

conditions. During the year the Walbiri would follow routes while hunting and gathering, which would efficiently harvest the available plant and animal resources. The time of plenty was the autumn and winter when the water holes would be full and vegetable foods and game abundant. At this time the people would congregate into large groups and move from one water hole to another in search of plants and animals. As the food and water supplies dwindled, the large gatherings would begin to break up into smaller and smaller groups, until by late spring the travelling and gathering unit would be reduced to a single male, his wives, children and perhaps aged relatives. At this time food resources would be very restricted and life difficult. In general a strict division of labour was observed. Men would hunt game, while the women would gather vegetable foods and smaller wild life. The women would also prepare food, collect water and firewood and bring up the young children.

The key to the tribe's survival under these harsh environmental conditions was a deep understanding of desert conditions and the resources which were available. This knowledge was passed on from one generation to another by word of mouth and by the acquisition of skills when the children were growing up. The food resources which the Walbiri used in their desert environment were quite remarkable in terms of their variety, with more than 40 different plants and a hundred different animals (ibid.) (Tables 2.1 and 2.2). All of these were gathered or hunted. In times of plenty the more desirable foods would be eaten, with the less palatable ones only being used under extreme conditions. During droughts, groups without food would be looked after by adjacent peoples who possessed food, though under very severe conditions everyone would suffer.

Given the isolated nature of their lands the Walbiri persisted with their traditional way of life longer than many other aboriginal tribes in more peripheral locations. The Walbiri first seem to have come into contact with individual Europeans during explorations of the 1860s and with larger groups during the construction of the Overland Telegraph Line between 1870 and 1872. In the last two decades of the nineteenth century, large cattle stations were opened up on the northern boundaries of Walbiri territory. At the same time a small-scale mining boom took place. By the early years of the twentieth century, conflict between the pastoralists and the Walbiri had occurred, which led to the abandonment of some cattle stations. In other cases, though, young aborigines

Table 2.1: Food sources of the Walbiri — fauna

Type	Uses
Anas superciliosa (black duck)	An uncommon but highly prized food
Artamus spp. (wood-swallow)	Rarely eaten
Barnardius zonarius (Port Lincoln parrot)	Often eaten
Calyptorhynchus banksi (black cockatoo)	Eaten; plumes prized
Cracticus sp.? (butcher-bird)	Occasionally eaten
Dendrocygna arcuata (whistling tree-duck)	Highly prized food
Dromaius novae-hollandiae (emu)	Highly prized food; plumes highly valued
Egretta alba (white egret)	Occasionally eaten: plumes prized
Epthianura tricolor (crimson chat)	Rarely eaten; ritually important
Eupodotis australia (bustard)	Highly prized food
Eurostopodus guttatus? (nightjar)	Rarely eaten, ritually important
Falco berigora (brown hawk)	Occasionally eaten; plumes prized
Falco hypoleucus? (grey falcon)	Rarely eaten
Fulica atra (black coot)	An uncommon but prized food
Geopelia cuneata (diamond-dove)	Occasionally eaten
Grallina cyanoleuca (magpie-lark)	Rarely eaten
Gymnorhina sp.? (magpie)	Rarely eaten
Halcyon pyrropygius? (kingfisher)	Rarely eaten
Haliastur sphenurus (whistling eagle)	Occasionally eaten; plumes prized
Histriophaps histrionica? (flock-pigeon)	Prized food
Hylochelidon ariel (fairy-martin)	Rarely eaten
Kakatoë leadbeateri (Major Mitchell cockatoo)	Occasionally eaten; plumes prized
Kakatoë roseicapilla (galah)	Occasionally eaten; plumes prized
Kakatoë sanguinea (corella)	Occasionally eaten; plumes prized
Leptolophus hollandicus (cockatiel)	Occasionally eaten
Lophophaps plumifera (spinifex pigeon)	Prized food
Megalornis rubicundus (brolga)	Highly prized food
Melopsittacus undulatus (budgerigar)	Frequently eaten
Merops ornatus (bee-eater)	Occasionally eaten
Ninox sp. (owl)	Occasionally eaten; plumes prized
Notophys pacifica? (heron)	Occasionally eaten
Notophyx novae-hollandiae (blue crane)	Occasionally eaten
Ocyphaps lophotes (crested pigeon)	Prized food
Oreoica gutturalis (crested bellbird)	Rarely eaten
Pelecanus conspicillatus (pelican)	Eaten, but rarely encountered
Podargus sp.? (frogmouth)	Rarely eaten; ritually important

Table 2.1: Food sources of the Walbiri — fauna (*continued*)

Type	Uses
Podiceps ruficollis (grebe)	Prized food
Pomatostomus rubeculus (babbler)	Rarely eaten
Querquedula gibberifrons (grey teal)	Prized food
Smicrornis sp.? (weebill)	Occasionally eaten
Taeniopygia cantanotis (zebra finch)	Occasionally eaten
Threskiornis molucca? (ibis)	Rarely eaten
Threskiornis sp.? (ibis)	Rarely or never eaten
Turnix velox? (quail)	Frequently eaten
Tyto sp.? (owl)	Rarely eaten
Uroaetus audax (wedge-tailed eagle)	Rarely eaten; plumes prized
Bettongia sp.? (wallaby)	Prized food
Canis dingo (dingo)	Rarely or never eaten
Canis familiaris (dog)	Rarely or never eaten
Dasycercus cristicauda (crest-tailed mouse)	Often eaten
Dasyurinus geoffroii (wild-cat)	Rarely eaten
Felis catus (feral cat)	Occasionally eaten
Largochestes sp. (wallaby)	Prized food
Macropus robustus (euro)	Highly prized food
Macropus rufus (kangaroo)	Highly prized food
Macrotis sp. (bandicoot)	Prized food; tails used for decorations
Mus musculus? (mouse)	Occasionally eaten
Notomys sp.? (hopping-mouse)	Occasionally eaten
Notoryctes typhlops (mole)	Prized food
Nyctophilus sp.? (bat)	Rarely eaten
Oryctolagus sp. (rabbit)	Prized food
Perameles sp. (bandicoot)	Prized food
Petrogale sp.? (wallaby)	Prized food
Phascogale macdonnellensis (fat-tailed mouse)	Frequently eaten
Pseudomys sp.? (bush-mouse)	Frequently eaten
Tachyglossus aculeata (spiny ant-eater)	Highly prized food
Trichosurus vulpecula (possum)	Highly prized food
Amphibolurus barbatus (bearded dragon)	Frequently eaten
Amphibolurus maculatus (lizard)	Frequently eaten
Amphibolurus reticulatus (reticulated dragon)	Frequently eaten
Amphibolurus sp.? (lizard)	Frequently eaten
Aspidites melanocephalus (black-headed snake)	Frequently eaten
Ceramodactylus sp.? (gecko)	Frequently eaten

Table 2.1: Food sources of the Walbiri — fauna *(continued)*

Type	Uses
Chiroleptes sp.? (burrowing frog)	Squeezed to produce potable liquid
Diplodactylus sp.? (gecko)	Rarely eaten; ritually important
Diplodactylus strophrurus (gecko)	Frequently eaten
Dtella sp. (gecko)	Frequently eaten
Egernia kintorei?	Frequently eaten
Gymnodactylus sp.? (gecko)	Frequently eaten
Liasis childreni (snake)	Frequently eaten
Limnodynastei spencerii (burrowing frog)	Squeezed to produce potable liquid
Lygosoma sp.? (lizard)	Frequently eaten
Nephrurus aspa (gecko)	Frequently eaten
Nephrurus levis (gecko)	Frequently eaten
Physignathus sp.? (lizard)	Frequently eaten
Python sp.?	Prized food
Sphenomorphus sp. (skink)	Frequently eaten
Tiliqua occipitalis? (skink)	Frequently eaten; fat a purgative
Typhlops endoterus (snake)	Prized food
Varanus acanthurus (goanna)	Prized food
Varanus giganteus (pirinti)	Prized food
Varanus sp. (goanna)	Prized food
Cerambycidae larvae (witchetty-grub)	Prized food
Cossidae larvae? (witchetty-grub)	Prized food
Eutermes sp.? (flying ant)	Occasionally eaten
Eutermes sp.? (white ant)	Occasionally eaten
Melophorus inflatus? (honey-ant)	Prized food
Melophorus sp.? (honey-ant)	Prized food
Psyllid lerp (manna)	Frequently eaten
Trigona sp. (wild bee)	Honey prized food; wax for adhesive
Grasshoppers	Occasionally eaten
Caterpillars, grubs	Some eaten
Cicadas, crickets	Occasionally eaten
Weevils, lice	Occasionally eaten

Source: M.J. Meggitt, *Desert people — a study of the Walbiri aborigines of central Australia*, University of Chicago Press, Chicago, 1962, Table 3, pp. 10–15.

Table 2.2: Food sources of the Walbiri — flora

Type	Uses
Acacia aneura (mulga)	Edible seeds; wood for implements
Acacia coriacca (wattle)	Edible seeds; wood for implements
Acacia dictyophloba (wattle)	Leaves used medicinally
Acacia kempeana (witchetty-bush)	Edible seeds; trunk harbours witchetty grubs
Acacia notabilis (wattle)	Edible seeds, gum; wood for implements
Acacia sp.? (wattle)	Edible seeds
Acacia sp.? (wattle)	Trunk harbours witchetty grubs
Acacia sp.? (wattle)	Edible seeds; wood for implements
Acacia spondylophylla? (wattle)	Trunk harbours witchetty grubs
Acacia tetragonophylla? (wattle)	Edible seeds
Aristida sp.? (mulga-grass)	Edible seeds
Atalaya hemiglauca (whitewood)	Trunk harbours witchetty grubs
Bauhinia hookerii	Nectar from flowers eaten
Boerhavia diffusa (creeper)	Edible roots
Canthium latifolium (plum-bush)	Edible fruit
Capparis mitchelli (wild-orange)	Edible fruit
Capparis sp.? (wild-orange)	Edible fruit
Carissa lanceolata (conkerberry)	Edible fruit
Clerodendrum floribundum	Edible roots
Cyperus rotundus (onion grass)	Edible roots
Duboisia sp.? (pituri)	Chewed as narcotic
Eremophila freelingii	Edible flowers, leaves
Eremophila latrobei	Edible flowers, leaves
Eremophila sp.?	Edible flowers
Eremophila sp.?	Edible flowers
Eucalyptus camaldulensis (river red-gum)	Edible flowers; wood for implements
Eucalyptus sp.?	Medicinal use of leaves
Eucalyptus sp.?	Trunk harbours wild bees
Eucalyptus terminalis (blockwood)	Trunk harbours wild bees
Ficus platypoda (wild-fig)	Edible fruit
Grevillea sp.?	Leaf-ash chewed with tobacco; wood for implements
Hakea intermedia	Edible flowers
Hakea lorea	Edible flowers
Ipomea costala ('yam')	Edible tubers
Ipomea muelleri	Edible seeds
Loranthus sp. (mistletoe)	Edible flowers
Marsdenia australis (bush-banana)	Edible leaves, fruit
Melothria micrantha	Juice used medicinally
Nicotiana ingulba? (tobacco)	Leaves and stems chewed as narcotic
Nicotiana sp.? (tobacco)	Leaves and stems chewed as narcotic
Portulacca sp.? (pig-weed)	Edible seeds, stems
Santalum lanceolatum (plum-bush)	Edible fruit
Santalum sp.? (plum-bush)	Edible fruit

Table 2.2: Food sources of the Walbiri — flora (*continued*)

Type	Uses
Scleroderma sp.? (truffle)	Edible
Solanum ellipticum (desert raisin)	Edible fruit
Solanum sp.?	Edible fruit
Vigna lanceolata ('yam')	Edible roots

Source: M.J. Meggitt, *Desert people — a study of the Walbiri aborigines of central Australia,* University of Chigaco Press, Chicago, 1962, Table 2, pp. 5–9.

went to work on stock camps and so gained access to metal axes, knives and blankets which were highly prized.

The great turning point in the life of the Walbiri was the very severe drought which occurred between 1924 and 1929. At this time conditions became so severe that the Walbiri could no longer subsist in the desert. As a result the tribes were dispersed and ended up seeking food at cattle stations and in towns such as Alice Springs. Having gained this widespread contact with Europeans and being introduced to new foods and a level of technology which was far superior to anything they had known, the Walbiri never returned to the bush in other than very small numbers once the drought was over. From then onwards they became dependent on white society for their survival. By the mid-1950s, two-thirds of the tribe lived in settlements supervised by the Natives Affairs Branch and the rest lived on cattle stations.

2.2 MODERN SOCIETIES

2.2.1 Introduction

Modern states can be defined as beginning with the advent of the industrial revolution in the United Kingdom from about 1750 onwards. This industrialisation process, like that of agriculture before it, then diffused throughout Western Europe, North America and Japan by the end of the nineteenth century. Associated with it was the concentration of people into urban areas, as agricultural workers left the countryside for the real or perceived benefits of urban life. However, it is only during the twentieth century, and in many cases only after the Second World War, that industrialisation began to have an impact on the world's drylands.

The late nineteenth and twentieth centuries have seen the creation of many new nation states growing from the former empires of the European colonial powers. Many of these states were located in the arid zones of Africa and Asia. In many countries of the world, drylands exist, but often they do not form the majority of a nation's land area. Not surprisingly, the attitude of a country's government to its drylands will depend on the extent of these lands and the degree to which other areas of the country are capable of sustaining the national economy. Given this situation, it is obviously useful to employ a system which classifies the nations of the world in terms of the importance of their arid regions. Such a classification has been devised which divides the countries with significant arid areas into five main groups (Heathcote, 1983; Paylore and Greenwell, 1979) (Table 2.3).

In the top three categories, at least half of a country's area is defined as arid and semi-arid. Within this category there are 39 nations, of which 25 were created in the post-Second World War period. Nine countries are classed as semi-arid, with between a quarter and one-half of their lands suffering from water deficiency. The USA is located in this category. Finally, there is a group of 18 countries which possess major regions of drylands, but these account for less than one-quarter of the total area.

It is easy to quote population figures for individual countries, but much more difficult and in some cases almost impossible to estimate the number of people actually living in dryland environments within individual countries. Thus, for example, the total population of the arid states listed in Table 2.3 is 3,415·2 million in 1985. Heathcote (1983) calculated that in 1979, approximately 650 million people out of a world population of 4,327 million actually lived within dryland environments. This represents only about 15 per cent of the total. Dryland populations are, therefore, very much a minority group in a world context.

Of what are described as the 'core' arid nations, Egypt has by far the largest population with approximately 48 million people in 1985. Egypt is unique in the fact that it possesses the River Nile. Many of the other countries in this category are also unusual in that they possess very large oil reserves, the revenues from which have permitted a form and pace of development which has not been possible elsewhere. In contrast, other countries such as Mauritania, Djibouti and South Yemen remain amongst the poorest in the world with access to few natural resources.

Table 2.3: Arid nations of the world

Description	Percentage of nation arid or semi-arid	Country	Population, 1985 (millions)
Core	100	Bahrain	0·4
		Djibouti	0·3
		Egypt	48·3
		Kuwait	1·9
		Mauritania	1·9
		Oman	1·2
		Qatar	0·3
		Saudi Arabia	11·2
		Somalia	6·5
		South Yemen	2·1
		UAE	1·3
		Sub-total	75·4
Predominantly arid	75–99	Afghanistan	14·7
		Algeria	22·2
		Australia	15·8
		Botswana	1·1
		Cape Verde	0·3
		Chad	5·2
		Iran	45·1
		Iraq	15·5
		Israel	4·2
		Jordan	3·6
		Kenya	20·2
		Libya	4·0
		Mali	7·7
		Morocco	24·3
		Namibia	1·1
		Niger	6·5
		North Yemen	6·1
		Pakistan	99·2
		Senegal	6·7
		Sudan	21·8
		Syria	10·6
		Tunisia	7·2
		Upper Volta	6·9
		Sub-total	350·0
Substantially arid	50–74	Argentina	30·6
		Ethiopia	36·0
		Mongolia	1·9
		South Africa	32·5
		Turkey	52·1

Table 2.3: Arid nations of the world (*continued*)

Description	Percentage of nation arid or semi-arid	Country	Population, 1985 (millions)
		Sub-total	153·1
Semi-arid	25–49	Angola	7·9
		Bolivia	6·2
		Chile	12·0
		China	1,042·0
		India	762·2
		Mexico	79·7
		Tanzania	21·7
		Togo	3·0
		USA	238·9
		Sub-total	2,173·6
Peripherally arid	<25	Benin	4·0
		Brazil	138·4
		Canada	26·4
		Cent. Af. Rep.	2·7
		Ecuador	8·9
		Ghana	14·3
		Lebanon	2·6
		Lesotho	1·5
		Madagascar	10·0
		Mozambique	13·9
		Nigeria	91·2
		Paraguay	3·6
		Peru	19·5
		Sri Lanka	16·4
		USSR	278·0
		Venezuela	17·3
		Zambia	6·8
		Zimbabwe	8·6
		Sub-total	663·1
		Total	3,415·2

Sources: Paylore and Greenwell, 1979; Heathcote, 1983; population data from *World Population Data Sheet 1985*.

2.2.2 Development planning

A feature of many arid nations in the post-Second World War period has been the growing importance of development planning.

This has effectively been a process of Westernisation by which urban/industrial growth is seen as the key to rapid economic growth. As a result of this policy most investment has been concentrated in the industrial and transport sector, and very little has gone into agriculture. Such programmes have brought quick returns and obvious benefits for middle- and working-class dwellers of urban areas, whose political influence is growing. In contrast, agricultural development programmes take a long time to mature and the benefits are not always clearly seen. Moreover, the rural populations, despite their large numbers in many countries, are poor, and perhaps more importantly, lack political influence. What has happened, therefore, is that urban areas have tended to be subsidised, particularly in such areas as food prices, while rural wealth is diverted into towns and cities by government-controlled pricing policies.

Under the Shah, Iran embarked upon a series of development plans from the late 1940s. The first, a Seven-Year Plan (1949–56), was to devote 25 per cent of its income to agriculture, 24 per cent to transport and 19 per cent to industry. However, with the nationalisation of the Anglo-Iranian Oil Company in 1953 and the subsequent boycott of Iranian oil by Western nations, income to the country dropped sharply and the objectives of the plan had to be revised.

The Second Plan (1956–62) was the one that laid the basis for the country's rapid economic growth, following the resumption of oil revenues in 1954. Oil exports had ceased in 1953 with the nationalisation of the Anglo-Iranian Oil Company and the boycott of Iranian oil by Western countries. Exports began again in 1954 when agreement was reached between Iran and a consortium of British, Dutch, American and French companies. The Plan concentrated on infrastructure provision, including the construction of large multi-purpose dam schemes to provide water for urban, industrial and irrigation use, as well as generating hydro-electric power. In turn there followed a Third Plan (1962–8), a Fourth Plan (1968–73), and a Fifth Plan (1973–8), all of which attempted to encourage the economic growth generated earlier.

The basic aim of the Shah's policy was rapid industrialisation while oil revenues were available, and as a result, the agricultural sector received little investment in the 1960s and 1970s. Indeed, if major water resource projects are excluded, agriculture only received 11·7 per cent of the total investment in the Third Plan and only 8 per cent in the Fourth Plan (Lawless, 1985). For the Fifth

Plan allocations to agriculture were increased, though given the growing instability in the country, they were never delivered and by the time the plan ended only amounted to 12 per cent.

The Fifth Plan was an interesting one in so far as during its life, oil revenues rose sharply, as a result of the OPEC countries' quadrupling of the price of oil following the 1973 Arab–Israeli war. The enormous quantities of money which were flowing into the country led to the Plan being significantly revised to a new budget of US$69 billion to make use of the extra funds (Halliday, 1979). However, the government planning mechanisms were not able to cope successfully with the flow of funds which were available. A massive construction boom occurred in all the major cities, but particularly in Tehran, and this drew in thousands of rural workers from throughout the country. When the effects of world recession began to bite in 1978, these unemployed workers provided the manpower for the mass demonstrations in Tehran which led to the overthrow of the Shah in 1979.

The revolution and the coming to power of Ayatollah Khomeini in Iran meant that the objectives of the Sixth Plan were abandoned. The new government reassessed all its policies and in its first plan called for low growth rates and an emphasis on small-scale industrial projects. In agriculture the aim was to become self-sufficient in basic foodstuffs. Although extra resources were allocated to this sector, no immediate results occurred and during the 1980s the bill for food imports has continued to grow.

In contrast to Iran, Oman was one of the last countries in the Middle East to embark upon a development plan, with its First Plan covering the period from 1976–80. The present ruler, Sultan Qaboos, came to power in 1970 and since that time economic development has occurred at a rapid pace. Before 1970 the country was effectively isolated from the world economy as a result of the policy of the ruler, and traditional ways prevailed everywhere. The factor which precipitated change was the discovery of oil in 1962 and the first export shipment in 1967. As a result of this the state revenues increased from about 1 million Omani rials in 1967 to over 450 million rials ten years later (Bowen-Jones, 1978). By 1977 oil revenues were accounting for 95 per cent of the income of the government.

The chief objective of the First Plan was economic diversification aimed at encouraging new sources of income which would supplement and perhaps eventually' replace oil revenues, and another aim was that of reducing regional differences in living standards. Like many other countries modern economic development in Oman began around the capital (Muscat) and over the years has tended to remain there, producing what is in effect a self-generating boom. Even by the mid-1970s it had already led to immigration from other rural areas and the influx of workers from other parts of the Middle East region.

Table 2.4: Sectoral distribution of investment in the First Oman Development Plan, 1976–80 (million Omani rials)

Sector	%	Total
Petroleum and mining	16·5	154·3
Agriculture and fisheries	4·4	41·0
Manufacturing	4·2	39·7
Trade and tourism	1·3	11·9
Financial institutions	1·8	17·0
Economic infrastructure	57·7	538·7
Social infrastructure	14·1	132·2
Total	100·0	934·8

Source: Bowen-Jones, 1978.

The sectoral distribution of investment for the period of the First Plan is given in Table 2.4. On the government side the figures clearly reveal the massive investment in infrastructure provision, accounting for almost 58 per cent of all expenditure. The most important single item was road construction, followed by electricity generation and water provision.

What might seem rather surprising at first are the very low levels of investment in manufacturing, and agriculture and fisheries, both of which received less than 4·5 per cent of total funds from government sources. However, it should be remembered that without a sound infrastructure economic growth in other parts of the economy is unlikely. Manufacturing fared better from private investment, receiving 17·5 per cent of the total. It is also worth noting that in 1976 the contribution of agriculture and fisheries to GDP was only 2·4 per cent and that of manufacturing only 0·4 per cent. With such a small original base large investments in sectors like these could well be wasted.

Prior to the discovery of oil, agriculture provided work for the majority of people in Oman. However, the available resources of soil and water have always been meagre and so there was little opportunity for new projects to open up large areas of land which have not been previously cultivated. Agricultural investment in the First Plan therefore concentrated on upgrading and extending existing schemes rather than developing new ones. Part of the problem about development in Oman was that the nature of the resources available were not known with any certainty. Even population numbers were not known, though it was believed there were about three-quarters of a million people. This necessitated surveys being undertaken and coherent development strategies being drawn up.

The case studies of Iran and Oman permit a pattern of development to be outlined which is applicable to all the developing countries of the arid zone. In all of them economic development planning began in a coherent manner between the late 1940s and the late 1970s. Everywhere the aim was basically the same — namely, that of transforming what were traditional agricultural societies into modern economies with significant industrial production. The countries have met with varying degrees of success, dependent on such factors as when development began, the size of the financial resources available, the initial state of the economy, and the efficiency with which the resources were utilised.

Similar patterns of economic development can be identified in many countries of the arid zone. The first stage normally involves heavy investment in infrastructural provision, which is really a prerequisite for further growth. The major call on investment in the early stages is for road network construction to replace mud and gravel roads, which is highly expensive. The great benefit of such a network is that it provides rapid and year-round access to regions which were formerly extremely isolated. Any resources such regions possess can now become part of the national economy and perhaps even of the world economy.

Besides road building, the construction of utility networks for electricity, water and oil/natural gas also soak up huge funds in the early stages of the development process. In themselves they are not wealth-creating, but without them industry is not able to prosper. The environmental impact of such schemes can be considerable, with major dam projects and their associated irrigation networks having the greatest effect.

2.2.3 Population growth

Superimposed on the development process are other changes

which are taking place. Perhaps the most important here has been the spread of health care and simple hygiene precautions. The net effect of such measures has been a rapid population growth as infant mortality has fallen and life expectancy has begun to rise. Such trends were observed in some countries from the early years of the twentieth century, but the impact has been especially marked in the post-Second World War period, at a time when development planning was just beginning.

Figure 2.5 The demographic transition model. This illustrates the way in which population grows as traditional societies become westernised

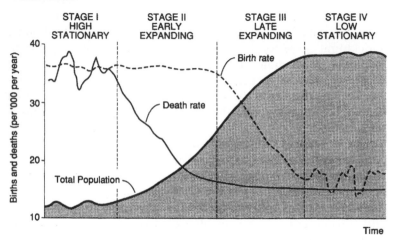

The management strategy of a government towards its environment and to economic planning depends on its perception of the resources it has available and the changes which are likely to occur in the future. As resource use is linked to people, an understanding of population change is a crucial aspect of any planning. It is now widely accepted that the processes of industrialisation and urbanisation have greatly affected the ways in which population increases. This has led to the establishment of a demographic transition model which seeks to systematise the ways in which populations develop (Figure 2.5). As such it provides a useful tool for comparing the stages of development of developing countries. An analysis of the population characteristics of the three most arid categories of Table 2.3 reveals that almost all are in the early expanding or late expanding stages. In effect this means that their populations are likely to continue to rise rapidly in the near future, putting an ever-increasing strain on available resources. At the present time the total population of these countries is 580 million, but by the year 2020 numbers are expected to increase to 950 million.

The nature of population changes which have occurred to date are often difficult to follow because of the lack of accurate population data going back to the nineteenth century. However, from the evidence available it can be seen that severe pressure on resources in arid lands only dates from the 1920s, when populations first began to rise (Figure 2.6). What is especially

Figure 2.6 Population growth in selected arid countries and states in the USA from the mid-nineteenth century to the present day

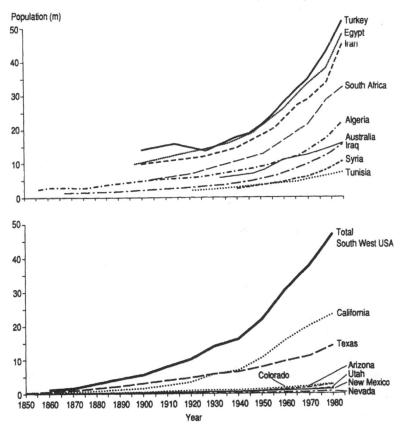

noticeable is the much more rapid increase in population growth in the post-Second World War period. It is this recent rise in numbers which is currently hindering all attempts to develop the arid lands in a rational and organised manner.

This growth of population has contributed to another phenomenon, which is the movement of rural dwellers into urban areas. This is a complex process which reveals different key factors from one country to another. However, it is possible to isolate certain features which are found in many places. The first is growing pressure on available resources in rural areas as population numbers increase. This can be thought of as a 'push' factor. At the same time economic development usually provides maximum economic growth in the capital city, and to a lesser extent, in the large urban centres. New building projects for industry, commerce and infrastructure provision demand large manual labour forces which often cannot be satisfied locally. Such jobs, which are well paid when compared with rural occupations, attract people from the countryside. This is a 'pull' factor. Initially, it is the young males who come to the urban areas, but families follow soon afterwards. Inevitably this leads to housing shortages, which are met by the creation of shanty towns on the margins of the larger urban centres. Such settlements have few amenities and usually lack adequate water and sewage facilities. Health conditions are poor and poverty widespread. Not surprisingly, they are often the breeding grounds of political unrest against the established government.

In rural areas, despite the drift to the towns, population numbers have often increased markedly in absolute terms over the last 40 years. This fact is often disguised by the way in which statistics are presented. In just about every country rural populations expressed in percentage terms are falling when compared with their urban equivalents. What is happening, though, is that urban growth rates, brought about by both natural increase and in-migration, are far outstripping the actual growth rates of the rural areas.

Almost without exception, all the developing arid countries followed a policy of almost benign neglect as far as their agricultural sectors were concerned, until the late 1970s. This was partly because investment could be spent more effectively elsewhere on infrastructure provision, and partly because the planners felt that it would be the urban areas which would provide the self-sustaining growth necessary for the economy to function as a modern state. The result has been that, given the rapid population growth which they have experienced, many countries have become more and more dependent on imports for their food needs. This has soaked up urgently needed foreign exchange and in some cases led to substantial foreign debts (Beaumont and McLachlan, 1985). The dangers associated with neglecting the agricultural sector were widely recognised by the late 1970s, though no solution to the problem of large food imports has been reached.

In the Middle East 20 of the 21 countries which are usually defined as belonging to the region are classified in the top three groups in terms of their aridity rating (Table 2.3). The only exception is the Lebanon which is placed in category 5, 'peripherally arid'. Indeed, of the 20 countries all are in categories 1 and 2 with the exception of Turkey, which is placed in the 'substantially arid' group. The Middle East, therefore, provides an ideal region for allowing generalisations to be made about what is going on in a substantial part of the arid zone of the world.

Despite the fact that almost all of the countries of the Middle East have gone in for some form of planning of their agricultural sector, the fact remains that in the 1980s the region has one of the most rapidly growing regional food deficits in the world. Yet even as late as the 1940s the region was a net food exporter. What is quite remarkable is that the region which provided the domestication of cereals some 10,000 years ago was, by 1980, importing about 30 per cent of its cereal needs (Beaumont, 1985b, p. 315).

When it is remembered that in 1980 the population of the Middle East was about 270 million and that by 2000 it is estimated that it will be approximately 420 million, the true magnitude of the food problem facing these countries can be seen. Already, by 1980, Iran was importing between 30 and 40 per cent of its food requirements, Egypt 40–5 per cent, Jordan more than 50 per cent, and Saudi Arabia as much as 75 per cent (Weinbaum, 1982).

From these figures it should not be thought that no improvements in agriculture have occurred in the post-war period, for that is not the case. In the 21 countries of the Middle East, cereal production in the late 1940s and early 1950s was about 25 million tonnes, and by 1982 this had risen to 59 million tonnes (Beaumont, 1985b). This was largely the result of an expansion in the cultivated area: for example, over the same time period, the area of wheat increased from 13·5 million to 23·9 million ha and that of barley from 7·5 million to 10·2 million ha. While in some countries sharp increases in yields did occur, others saw relatively little change.

What is particularly worrying, though, is that despite the increases in cultivated area and increases in yields, *per capita* cereal production has only increased from 241 kg in 1950 to 245 kg in 1980 (ibid.). In certain countries, such as Algeria, Iraq, Jordan and Morocco, cereal production in *per capita* terms has registered a marked decline.

2.2.4 Agriculture and land reform

In most countries agriculture has the greatest environmental impact, though locally, other land uses may be more important. With the modernisation of agriculture, the most important change has been in the scale of operation, for machinery is more effectively used on larger holdings. However, in many cases it is only the newly developed land areas which possess the large field sizes, whereas in the smaller and more traditional areas, field sizes remain small and irregular in shape. Indeed, in such cases the machinery employed is also often small-scale, with two-wheel tiller cultivators being preferred to the Western four-wheel tractor. This is partly a reflection of the lower capital investment needed, but also of the fact that such machinery is highly suited to such conditions.

An interesting point is that environmental impact is often not greatly altered as a result of management strategies. Thus, for example, land reform has been a major policy initiative of many governments in the arid lands. Yet despite changes which have taken place, the pattern of crops which are grown has not changed greatly, with one or two important exceptions. The traditional crops, cereals of one type or another, continue to dominate national economies. In some countries, cash crops, like cotton or tobacco, have gained greater significance.

Land reform has been undertaken for a variety of reasons, but it has often been linked with a government's attempt to maintain political control. In both Egypt and Iraq in the 1950s military coups overthrew monarchies and one of the first actions of the new governments was the introduction of a land reform programme aimed at destroying the power of the major land owners who had been strong supporters of the monarchy. Prior to the revolution in Egypt less than 6 per cent of the landowners held 65 per cent of the land (Saab, 1967). In Iran a land reform programme was carried out by the Shah with two major political aims. The first was to limit the power of the large landowners who might conceivably challenge the power of the Shah, and the second was to create a new class of landowners who would have a vested interest in supporting the government. Ideas about social justice and equality played a very small part in this land distribution process (Hooglund, 1982).

To rid agricultural development from what he considered to be the constraints of peasant mentality, the Shah of Iran established a number of agribusinesses in Khuzestan in the early 1960s. Six private companies covering a 100,000-ha (250,000-acre) area were established on land fed with irrigation water from the then newly constructed Dez Dam. After initial success all these businesses collapsed owing to a host of environmental and economic miscalculations. A classic problem with one of the farms was land-levelling operations which led to the removal and burial of most of the top soil, leaving behind infertile and almost unworkable subsoils. Equally, the large size of the operations proved an impediment and the hierarchical organisation was not able to make decisions quickly enough for local conditions to be exploited efficiently. Halliday (1979) argues that in California, where similar conditions prevail, the optimum size for an irrigated farm is slightly less than 300 ha.

Many land reform programmes seem to have encouraged the creation of a group of rich peasants. In many cases land was only awarded to people who were already cultivating the land as tenant farmers. This was because it was felt that such people possessed the necessary skills needed to increase agricultural production and also because in most situations there was just not enough land for it to be distributed to labourers and other groups. It should be remembered that at least 40 per cent of rural families in many parts of the arid zone did not farm any land, either as tenants or owners, prior to many land reform programmes. Despite this limitation in terms of those who benefited, a common complaint of many land reform programmes has been that the land parcels actually allocated to individuals have been too small to be economically viable.

Some land reform programmes have given rise to the establishment of new rural institutions, of which the most important has been the government managed co-operative. These were intended to provide credit, the means for production and help with marketing. In many countries such co-operatives have not lived up to their expectations. Credit has been insufficient and machinery supply and maintenance a problem. Even more important, though, is the fact that farmers do not manage their lands individually. As a result, a feeling of alienation has occurred, with the co-operative officials viewed in much the same light as the landowners they replaced. In other cases state farms have been established in which members are paid either a fixed wage or in relation to the number of hours worked. With such organisations the workers do not feel a close tie with the land and so innovation and commitment have been lacking.

2.2.5 The future

Given the rapid urbanisation and population growth which have occurred in dryland countries, it seems increasingly likely that government policies will continue to be geared towards maintaining low urban food prices in order to minimise unrest. This will be achieved by giving farmers low prices for their crops and by subsidising staple food products such as cereals. Inevitably this will mean that wealth is being extracted from rural areas and dispersed in cities and towns. Unfortunately it is not being used for productive investment, but rather for political aims to maintain urban tranquility. If such a policy continues it can only mean that agriculture will be drained of the funds required for modernisation.

Almost everywhere the urban areas have a privileged position. They are the centres of political influence and life which is often subsidised from central government funds. These cities, though, have often proved unsuccessful as generators of industrial development. Even in capital cities very few industrial firms have been established and the range of products produced is small. Part of the problem lies in the fact that the rural areas do not possess sufficient funds to generate market activity, and even in the urban centres the vast majority of people remain dismally poor.

3

Intensive Use of the Dryland Environment

3.1 URBANISATION

3.1.1 Introduction

Rapid urbanisation is a feature of the nineteenth and twentieth centuries. It first occurred during the mid- and late nineteenth century in the industrialised nations of Western Europe and North America. In Western Europe this process has slowed down during the twentieth century, although in the USA the process continues apace with the growth of the cities of the Sun Belt (the southern states of the USA from Florida to California). In the developing world, including many of the arid nations, rapid urban growth has been a phenomenon of the post-1950 period, though in some countries the process began in the inter-war period.

Table 3.1: Number of dryland cities of more than 100,000 population in different states

Country	Number of cities of more than 100,000
India	62
USSR	55
USA	42
Mexico	22
China	19
Pakistan	19
Egypt	18
Iran	12
Morocco	10
Turkey	10

Source: Cooke *et al.*, 1982.

The result is that by the 1970s there were more than 350 cities in drylands with more than 100,000 people (Cooke *et al.*, 1982, p. 3). Of this total, 208 were located in Asia, 94 in the Americas, 53 in Africa, and none in Australia. On a country basis, India, the USSR and the USA possess the largest numbers (Table 3.1). Of these dryland cities 69 have more than 500,000 people and 30 have more than one million. Cairo, Peking and Los Angeles stand out as the biggest of the really large urban agglomerations (Table 3.2).

Table 3.2: Major urban dryland centres with more than 2 million people

Continent	Estimate	City proper	Urban agglomeration
Africa			
Cairo (Egypt)	1974	5,715,000	–
Alexandria (Egypt)	1974	2,259,000	–
Asia			
Peking (China)	1970	7,570,000	–
Tientsin (China)	1970	4,280,000	–
Teheran (Iran)	1973	4,002,000	–
Delhi (India)	1971	3,287,900	3,647,000
Karachi (Pakistan)	1972	3,498,634	–
Lahore (Pakistan)	1972	2,165,372	–
America			
Los Angeles (USA)	1975	2,727,399	7,032,075
Santiago (Chile)	1970	3,273,600	3,350,680
Lima (Peru)	1972	2,833,609	3,303,523

Source: Cooke *et al.*, 1982.

3.1.2 Growth rates of cities

A great problem in the study of cities is that data sources are extremely variable as to both the volume and the accuracy of the information available. In some countries like the USA, detailed and accurate statistics are available, while in many Asian and African countries census figures are of variable quality.

Between 1950 and 1970 the growth rates of dryland cities were well in excess of the average rate of population growth of the

Table 3.3: Urban populations and growth rates in the Middle East, 1960–80

Country	Urban population as percentage of total population		Average annual urban growth rate (%)	
	1960	1980	1960	1980
Sudan	10	25	6·9	6·8
Yemen A.R.	3	10	7·5	7·2
Morocco	29	41	4·2	4·6
Syria	37	50	4·8	5·0
Turkey	30	47	5·1	4·6
Algeria	30	44	3·9	5·8
Iran	34	50	4·7	4·9
Iraq	43	72	6·2	5·4
Saudi Arabia	30	67	8·4	7·6
Libya	23	52	8·0	8·3
Kuwait	72	88	10·4	7·4

Source: World Bank (1981)

countries in which the cities were located. In the decade from 1960 to 1970, two-thirds of dryland cities for which data were available revealed growth rates of 3 per cent or more, while about a quarter of the total increased at rates in excess of 5 per cent (Cooke *et al.*, 1982, p. 19). One or two cities, such as Tijuana (Mexico) and Tucson (Arizona), registered growth rates greater than 10 per cent over the decade. In nearly all cases in the developing countries the capital city has recorded the most rapid increase in numbers. Particularly rapid urban growth has occurred in the arid nations of the Middle East and North Africa, where many countries have witnessed annual growth rates of more than 4 per cent between 1960 and 1980 (Table 3.3) (Clarke, 1981, 1985).

3.1.3 Urban management problems

From the environmental management viewpoint the growth of urban systems imposes increasing strains on available infra-structures and makes even greater demands on available resources. The most obvious demand is for land on which to construct houses, offices, shops, factories and roads. In many cases this land is of high agricultural quality and so the loss to the agricultural

Plate 3.1 Modern road development on an alluvial fan in Las Vegas, Nevada.

Plate 3.2 Traditional irrigation canal on the Varamin Plain, Iran.

Plate 3.3 Traditional well, Battina coast, Oman.

Plate 3.4 Traditional means of diverting water into irrigation canals along major rivers in Iran.
Timbers were used to form a framework and then these were covered with brushwood to back up the water so that it could enter the irrigation canal.

Plate 3.5 Groundwater extraction by deep well in the Isfahan Oasis.

Plate 3.6 Karaj Dam, Iran.
The snow covered Elburz Mountains are seen in the background.

Plate 3.7 Pre-formed concrete canalets carrying irrigation water to an agricultural project in western Turkey.

sector can be significant and permanent. Perhaps even more important are the flows into and out of the urban systems. McHale (1972) estimated that for a city of one million people in the USA, daily inputs in excess of 600,000 tons would be required (Table 3.4). Of these the largest is the supply of water for domestic and industrial use. This also gives rise to the largest output — sewage. To handle these enormous quantities a complex infrastructure of water treatment works, pumping stations, sewage plants and pipeline systems have to be constructed at great expense.

Table 3.4: Magnitude of city metabolism for a city of one million population in the USA

Input		Daily flow (tons)	Output	
Water		625,00	Sewage	500,000
Fuels			Refuse	2,000
Coal	3,000			
Oil	2,800			
Nat. gas	2,700		Air Pollutants	
Motor	1,000		CO	450
Sub total		9,500	Sulphur dioxide	150
			Particulates	150
Food		2,000	Nitrous oxides	100
			Hydrocarbons	100

Source: McHale, 1972.

The amount of fuel used will depend to some extent on the climatic region of the city and its level of economic development. Fuel for transport systems, particularly in the developing world, seems destined to increase. Food is obviously a necessity for life and has to be brought into urban areas in large quantities. It is interesting that as standards of living rise the amount of packaging material increases substantially, so necessitating the movement of larger amounts of material. Opposing this tendency, though, is the fact that more foods are being processed prior to sale and the non-edible parts removed. However, most of the processing takes place in urban areas.

The most intractable environmental problems are caused by the generation of waste products. In a city of one million people approximately half a million tons of sewage are generated each day. In former times, before significant industrialisation, the major problem was the impact of large quantities of biodegradable organic matter on the oxygen content of receiving water bodies. At certain times of year in drylands, naturally occurring water flows are often very low indeed and under such conditions it was not unusual for the oxygen level to be so depleted that anaerobic conditions developed. The other difficulty with sewage disposal is that of disease potential associated with the pathogenic bacteria and viruses found in the waste waters. However, with industrialisation a host of new chemicals has been introduced, many of which cause problems if the water is to be used later for either domestic supply or for irrigation.

Each day approximately 2,000 tons of refuse are produced in a large city, consisting of household, commercial and industrial wastes. It is varied in content but is usually dry. The normal method of disposal is to use it for landfill, though in some countries it is incinerated to reduce bulk. This can be expensive and can also produce serious air pollution.

About 1,000 tons of pollutants are issued daily into the atmosphere. The types of pollutant depend on the climatic regime of the city and the standard of living of the people. In countries like the USA the main pollutants are those generated by automobiles, though in the inland desert cities an additional burden is produced in winter from central heating boilers and furnaces. In the developing countries, although there are fewer motor vehicles per head of the population, these vehicles often produce much more pollution owing to bad maintenance and the lack of emission control devices. In addition, industrial pollution can be severe from the use of primitive techniques. In Tehran, for example, severe particulate and smoke pollution used to be produced from the many brickworks which existed in the south of the city.

During a period of urbanisation the spurt of construction activity places high demands on local building materials, particularly sand and gravel, building stone and cement (Fookes and Higginbottom, 1980). Such demands may not be capable of being met locally, so necessitating the importation of this material from elsewhere.

All too often city growth in conditions of rapid urbanisation in the Third World takes place in the context of lax or even non-existent planning regulations (Drakakis-Smith, 1987). This can mean that one is dealing with an uncontrolled and often uncontrollable situation as poor peasants flood into the largest cities. Certainly in many parts of the developing world 'shanty' towns grow up on the margins of larger urban centres in an unstructured and often illegal manner. These communities generally lack basic amenities like running water and sewage removal, and as a consequence the dangers from disease are very great.

Cairo has grown by eight million people between 1970 and 1985, though the lack of statistics makes it difficult to be precise about numbers and the extent to which migration has supplied the extra people. In Cairo large slum communities have developed to the west of the Nile and also to the north of the old city. All available land has been taken over and it is claimed that more than half a million people are living in cemeteries and at least a similar number on the roofs of buildings. In such areas planning and building regulations do not exist. Water provision and sewage facilities are of the most basic kind and disease is rampant. Buildings are thrown up without an awareness of the need to comply with any standards and with increasing population pressures new buildings are constructed on top of existing ones. The structures are often inadequate and therefore collapse, with the inevitable deaths and injuries. Under- and unemployment are endemic and many people only survive in such conditions because of subsidised foodstuffs. Yet organisations like the World Bank and the International Monetary Fund want these subsidies eliminated as they believe that they will eventually lead to the collapse of the agricultural system. How and even if one can cope with such an urban system is problematical. Conditions are getting worse and it would seem inevitable that some form of unrest will occur against the appalling conditions under which the people are living. However, in a country like Egypt, which is suffering from severe financial constraints, it is difficult to see exactly what can be done to alleviate the problems.

3.1.4 Geomorphological hazards

Even in more advanced societies planning procedures rarely pay enough attention to the potential geomorphological hazards which are likely to affect settlement in a given area (Table 3.5). Perhaps somewhat surprisingly, many of the more intractable problems are related to water in excess. Dryland cities are frequently built on or near stream courses and, given the nature of arid rainfall, these

Table 3.5: Geomorphological hazards in drylands

Flooding of valleys, fans and playas
Gullying
Hydrocompaction
Surface subsidence due to water abstraction
Sedimentation (fluvial)
Salt weathering
Piping
Landslides and related slope-failure phenomena
Sabkha inundation
Aeolian deposition, dune encroachment, dune reactivation, dust
Calcretisation
Desiccation phenomena

Source: Cooke *et al.*, 1978.

fluvial systems are subject to infrequent but often devastating floods.

Flood damage is common in dryland urban environments. Part of the problem is one of perception, since the dominant dryness of the climate produces a sense of security that flooding is only a remote possibility. This, coupled with the fact that the return period of major flood events may be in excess of 20 years, often means that the effects of the last major flood are dim in the memories of the inhabitants of a city. Besides this, the rapid pace of urban growth, added to the fact that many of the cities' new residents have often come from elsewhere in the country, tends to further reduce the perceived flood hazard. This can encourage the construction of buildings in flood-prone areas due to inadequate planning legislation.

The nature of a flood event is easy to describe. Rainfall usually occurs on highlands adjacent to the urban area. The water is collected by small streams into the major channel and then transmitted downstream to an alluvial plain or fan on which the city is located. A feature of such flood events in drylands is the rapidity of their occurrence which makes it difficult to introduce an adequate flood warning system. The simplest way to reduce flood damage is to zone the areas at risk and to prevent constructional activity within them. Washes across alluvial fans are obviously high-risk zones, yet it is surprising how many cities in drylands have permitted the development of such areas for housing. Planning regulations often do not exist in the cities of many developing countries and so the potential for damage and loss of life is high. Besides the behavioural approach described above, there is the structural approach to flood control in urban areas. This is a very high-cost approach and involves the construction of flood storage reservoirs, embankments and the lining and widening of channels. It can usually only be practised by rich countries in the West. One of the best-known flood control systems is found in Los Angeles and is described in Chapter 12.

These same rains which produce floods along low-lying valleys can cause slope instability in uplands. Such regions often have deep soil profiles and little vegetation. Following heavy rains the soils become saturated, causing slope failure, mudflows and landslides. This can destroy property and other structures, as well as delivering large quantities of sediment into the rivers. The actual process of urbanisation itself can produce severe environmental problems, two of the most important of which are increases in the speed and volume of runoff and the production of heavy sediment loads. Some of the best documented examples of this are found in California (Knott, 1973). On the peninsula to the south of San Francisco rapid urbanisation took place between 1950 and 1970. Part of this area, located to the south of Daly City, was in the Colma Creek watershed. The Colma Creek basin covers 42 square km and rises to 400 m in the San Bruno Mountains (Figure 3.1). In the north-eastern part are relatively resistant sandstones and shales, while along the western margin less resistant marine sediments of Tertiary and Quaternary age outcrop. The climate is typically Mediterranean, with warm and dry summers and mild and humid winters. Average rainfall over the basin varies from 250 mm to 460 mm per year and approximately 85 per cent of the total falls in the five months from November through March.

The key factor in the basin is the urbanisation of the lower parts which took place between 1946 and 1970, and which led to a remarkable change in land use, with the urbanised area increasing from 15 to almost 55 per cent, and the agricultural area falling from 70 to 3 per cent (Table 3.6). Field analysis in the late 1960s revealed that the frequency of flooding had increased and that massive erosion of sediment was occurring from the areas undergoing urbanisation. Four land use types with vastly different sediment yield characteristics were recognised. These were open space areas, which possessed native or human-established vegetation; urban areas, classified as residential, commercial and industrial; agricultural areas, which had been under cultivation for

Figure 3.1 Urbanisation of the Colma Creek watershed, California

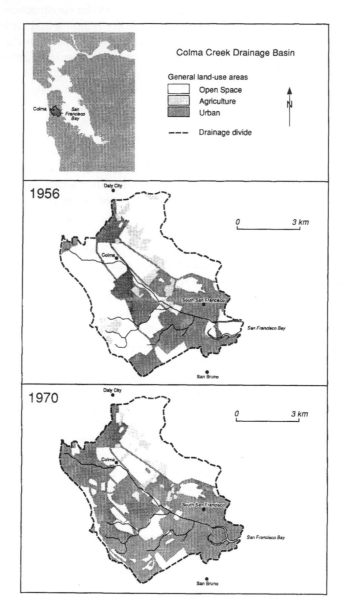

Source: Redrawn from Knott J.M., 1973, *Effects of urbanisation on sedimentation and floodflows in Colma Creek Basin, California*, United States Geological Survey, Open File Report.

Table 3.6: Land use changes in the Colma Creek basin, California

Land use	1946	1956	1970
Agriculture	70	12	3
Open space	15	51	37
Urban	15	34	54
Under construction	0	30	60

Source: Knott, 1973.

many years; and finally construction areas where urbanisation was actively occurring.

Annual sediment yields for the different types of land use revealed that the agricultural and construction zones had erosion

Table 3.7: Erosion rates from areas of different land use in Colma Creek, California, 1969–70

Storm size	Land use	Tons/square mile	Index
Moderate	Open space	4·5	1·0
	Urban	10	2·5
	Agriculture	270	63
	Construction	350	81
Large	Open space	9·0	1·0
	Urban	23	2·6
	Agriculture	620	69
	Construction	790	88

Source: Knott, 1973.

rates which were up to 80 times greater than those for open spaces (Table 3.7) (ibid.). What is quite remarkable is that in 1969 the construction areas only covered about 14 per cent of the total area of the basin, yet they contributed 72 per cent of the sediment load. The message for the future is clear — namely, that the erosion rates are dependent on land use conditions with relatively small percentages of critical land areas being able to supply large sediment volumes. By the end of the century the urbanisation process in this part of California may well be completed; under these conditions sediment yields are likely to fall dramatically.

In many dryland urban environments, groundwater is extracted for human use, but with certain types of geological strata this activity can cause serious problems. When water is withdrawn, the sediment particles pack closer together causing the ground surface to subside. If this is continued over a long period, it can damage the foundations of buildings as well as rupturing pipeline and other infrastructure systems. The centre of Mexico City, situated on an old lake bed, provides an excellent example of such conditions.

3.1.5 Constructional limitations

Disintegration of building materials through salt action is a common phenomenon in drylands. Salt is introduced into buildings in three main ways. First, it might be present in materials, such as sand and gravel, used in the construction process. Secondly, it can be formed in structures as a result of chemical changes in the original materials, and finally it can reach buildings and structures directly from groundwater or associated capillary rise. The impact that salt action will have depends on the range of environmental conditions experienced by the site, the type of materials used, and the level of workmanship (Fookes and Collis, 1975).

Concrete seems to be subject to more rapid deterioration and cracking phenomena under arid conditions than in other environments. It is subject to internal attack as a result of salts (such as chlorides and sulphates) being incorporated into the constituent materials. Salt attack on steel reinforcing bars results in a volume increase of between three and seven times which puts pressure on the surrounding concrete and leads to its disintegration. When construction is occurring in coastal locations, sands and gravels are easily contaminated by airborne salts or from groundwater (Fookes and Higginbottom, 1980a and b).

Concrete is also subject to external attack by both physical salt weathering and sulphate attack. Physical salt weathering results from crystal growth in small surface cracks, with the salts usually reaching the structure either directly from saline groundwater or by capillary rise. It is normally restricted to a zone from just below the ground surface to about 2 m above it. Sulphate attack is due to capillary rise of water and salts from underlying saline groundwater. A number of chemical changes in the concrete take place, the most important of which increase the volume of the material. The result is the heaving and break-up of floors and walls.

Environments in which saline groundwater is close to the surface can give rise to what engineers call 'aggressive ground conditions'. These are essentially salt weathering phenomena,

Figure 3.2 Groundwater quality variations at Suez, Egypt (after Sir W. Halcrow & Partners for Ministry of Housing and Reconstruction, Egypt, 1978)

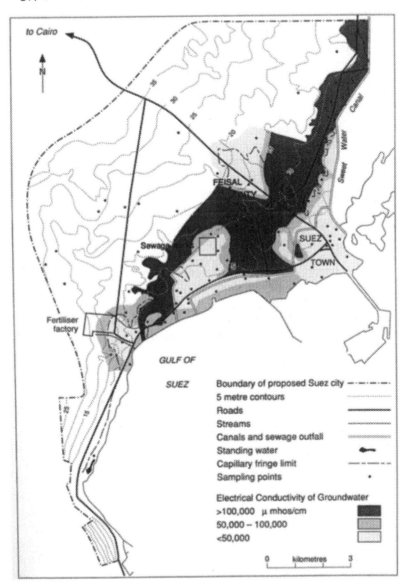

which can attack buildings and structures. Such conditions predominate in coastal locations and in other areas where local geological conditions force groundwater to the surface. The problem can best be overcome by not building in areas where aggressive ground conditions are found, or by the use of protective measures, such as membranes, which keep the salts away from the materials likely to be damaged. Protection like this can add substantially to construction costs.

Saline water can rise above the water table through a process known as capillary rise. The capillary fringe which is produced is usually less than a metre in height in gravels, but can attain heights of up to 4 m in clay soils (Cooke et al., 1982, p. 169). This means that foundations can be affected in heavy soils even when the water table is a number of metres below the ground surface. On Bahrain Island, in the Persian Gulf, rapid urban growth has occurred in the northern part of the island. Surveys of aggressive ground conditions have shown that they can occur almost anywhere below the 10 m contour line (Brunsden et al., 1979). In this hazard area the capillary fringe in which salts are concentrated as a result of evaporation can be up to 3 m above the water table.

At the new town site of Suez in Egypt investigations have shown that 30–40 per cent of the site is subject to aggressive ground conditions (Cooke et al., 1982, p. 189). Three zones with varying ground conditions have been recognised (Figure 3.2). The main problem area is where the groundwater is within 2·5 m of the surface. Throughout this zone the foundations of buildings are likely to be below the groundwater table and so the potential danger from salt damage is very high. If at all possible this land should not be used for building purposes. Where the groundwater was between 2·5 and 5 m below the surface it was likely that the foundations of the larger buildings would be below the water table and even the foundations of smaller buildings would be likely to enter the capillary fringe. All foundations would therefore be at risk and so some form of protection would be needed. The final zone was where the water table was at least 5 m below the ground: here, aggressive ground conditions were considered to be of minimal impact.

3.1.6 Tucson, Arizona, USA — water for a desert oasis

The city of Tucson provides an excellent example of the growing pressure on water resources associated with rapid urbanisation in an arid environment (Martin *et al.*, 1984). It also reveals how attitudes to water as a resource have changed dramatically over the last two decades. Tucson is a desert oasis situated in the broad alluvial valley of the Santa Cruz River and receives an annual precipitation of 280 mm. Rapid city growth has taken place in the post-Second World War period. In 1940 the population was only 60,000; by 1965 it had reached 230,000, and by 1985 was close to 500,000. If present growth rates are maintained the population will attain 800,000 by the year 2000, and perhaps as high as 1·8 million in 2035.

Given the desert nature of the climate, surface water supplies are virtually non-existent, and so recourse has had to be made to groundwater. In the nineteenth century, perennial flow did occur along sections of the Santa Cruz River, but with the increasing development of groundwater these flows soon ceased. It is estimated that up to 1945 groundwater abstraction was less than the amount of water transpired from the original riverine vegetation. Subsequently increasing demands have meant that withdrawal of water is now vastly in excess of recharge; and as a consequence, groundwater levels are falling in some areas by as much as 1·5 m per year.

The vast amounts of alluvial material in the Tucson basin along the Santa Cruz River contain enormous volumes of groundwater. Nevertheless, although everyone accepts that the overdraft situation is getting worse, there is as yet no physical shortage of water. Currently about three-quarters of all the water used is fossil water which is being withdrawn from storage.

The Tucson basin and the adjacent Avra valley are fertile agricultural areas and in the early 1980s about 65 per cent of groundwater abstracted was used for irrigation. This reveals that the city of Tucson can relatively easily obtain extra supplies of water by buying up farms with their associated water rights. Such activity has been going on for some time in the Avra valley, with the result that the city in 1980 was obtaining about one-quarter of its supply from this source.

Water supply in the urban region is the responsibility of the City of Tucson Water Department. Until the 1970s the organisation followed the traditional water engineering view of supplying as much water as was demanded by the community. Indeed, water consumption was almost encouraged as water rates were low and annual increases had often been less than the rate of inflation.

During the 1960s increasing concern over the future water supply to Phoenix and Tucson led to the development of the Central Arizona Project. This controversial scheme is designed to bring Colorado water along a large aqueduct, first to Phoenix and then to Tucson. By the time the water gets to Tucson it will be extremely expensive and of a lower quality than that of the existing groundwater. By the early 1990s the Central Arizona Project is expected to be delivering about 228 million cubic m (185,000 acre feet), but it will increase to 295 million cubic m (239,000 acre feet) by 2025 (see Chapter 6).

Worry about groundwater use led to the state legislature enacting the 1980 Arizona Groundwater Management Act, which called for a balance between groundwater recharge and abstraction rates by 2025. However, to date there has been felt little need to reach this position quickly, as it is perceived that adequate resources still remain in the ground. Part of the difficulty stems from a policy uncertainty as to what exactly the overall aims are. On the one side there is a feeling of a need to conserve water for the future, but at the same time rapid development of urban areas is being encouraged.

During the 1970s significant political changes occurred in Tucson which led to a complete re-evaluation of water management policy (Martin et al., 1984). In the 1960s and early 1970s, as the city of Tucson grew rapidly, the water service network greatly increased in size, but the extension of the network capacity and the opening up of new water sources could not keep up with the needs of rapid economic development. Moreover, the new suburbs were often located in foothill zones where pumping costs were high. The great difficulty with the Tucson system was that it had to possess the capacity to meet the peak summer demands. In the early 1970s the peak-day demand was double the average annual daily demand, whilst the peak hour of the peak day was three-and-a-half times the average annual hourly demand. Naturally this meant that a high proportion of the capital investment was devoted to providing capacity which was used on relatively few occasions.

Before the mid-1970s the policy of the City of Tucson Water Department had been to meet all demand needs. Between 1950 and 1970 the number of consumers had increased fourfold, yet water rates were kept low and even decreased as larger quantities of water were used. By the early 1970s the Water Department was becoming so worried about the potential escalation of investment costs that it commissioned a firm of consulting engineers to investigate the problem. The report recommended an average water rate increase of 30 per cent, together with an increasing rate for when large quantities of water were used. This policy was implemented in February 1974 with surprisingly little in the way of complaints or changes in the pattern of water use. In the summer of the same year Tucson registered its highest *per capita* water use ever, with annual figures for 1973–4 reaching 777 litres/*capita*/day (205 US gallons *per capita* per day). In June, during a particularly hot and dry spell, the pumping capacity was unable to cope with the peak demand and some of the higher parts of the city were without water, posing an extremely serious fire hazard.

During the early 1970s environmental groups in the Democratic Party were of growing importance in Tucson and in 1976 they gained control of the city council. Their political platform was made up of no-growth policies and emphasis on the conservation of resources. Interestingly their coming to power coincided with the publication of a new report on what should be done to improve the Tucson water supply situation. By this time the staff of the Water Department also firmly believed that a radical solution was needed because the continual expansion of the system was proving difficult to operate economically with the current rate structure and the need to supply peak demands.

This report, known as the Carollo Report, proposed a six-year investment programme to remedy the inadequacies of the system and an average rate increase of over 40 per cent. It also advocated an increasing block structure rate to discourage wasteful usage. This was, for the first time, an attempt to manage water demand. Even more important, though, it proposed that the operating and capital costs should be recovered from the consumers responsible for the the costs. In effect this meant higher rates for people living in elevated areas where pumping costs were considerable and also for new consumers where pipeline construction was needed. The attitude of the council is summed up in a quotation from their minutes: 'Those people who want to have a green jungle around their house even though they live in the desert, will have to bear that economic cost' (Tucson City Council Minutes, 7 June 1976, p. 8).

There was still the problem of how to reduce peak demand, which imposed such a financial burden on the system. In 1976 it was estimated that only 30 per cent of the capacity of the system was needed to supply the average annual water demand; the rest was needed for peak demand only.

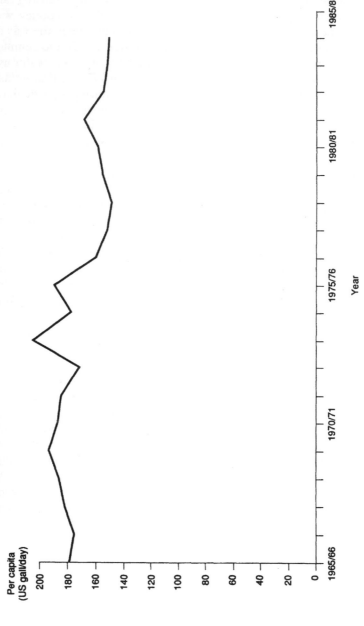

Figure 3.3 Per capita water use (U.S. gallons/day) in Tucson, Arizona

Source: Data from Tucson Water.

The newly elected environmentally conscious council enthusiastically brought in the proposed water rates in July 1976. The effect was immediate: in some of the high water lift areas, water bills quadrupled and in many parts they doubled. Public protest was so great that a procedure was set up to get the councillors dismissed from office and replaced by members who would abolish the new water rating system. The campaign proved successful and in January 1977 the new councillors took office.

What is remarkable is that the newly elected councillors were quickly converted to the merits of the new water system and actually increased the rates by a further 10 per cent. At a public hearing even the Chamber of Commerce was converted to the idea that people should pay for the cost of the service to provide them with water. In March 1977 the new council confirmed the 'cost of service' principle and even committed itself to three more water rate increases over the next three years.

The council quickly realised that the new rate structure, which attempted to manage demand, would only succeed if the peak demand could be reduced. It therefore established a public relations campaign called 'Beat the Peak', which tried to get the public to cut down their peak demands by not watering lawns more than once every other day, not watering at all on Wednesdays, and never watering between 4 and 8 p.m. This was not put over as a conservation measure, but rather one which would save on capital expenditure by not requiring such a large pumping capacity to supply much lower peaks. As such it would be saving money on bills in the future. The programme does seem to have been successful, and peak demand has been reduced considerably. Thus, for example, the peak daily demand in 1976–7 was 496,214 cubic m (131·1 million gallons), while in 1978–9 it had fallen to 428,462 cubic m (113·2 million gallons). The *per capita* demand of the early 1980s is still lower than at any time since the early 1960s (Figure 3.3).

The effects of the new water rating policy can be seen by looking at the water consumption of single family dwellings between 1974 and 1979 (Figure 3.4). In terms of annual usage there is an approximately 20 per cent fall between 1974 and the late 1970s. When examined in more detail it can be seen that the greatest absolute decline in water use has occurred in the summer months, though reductions have also taken place in winter as well. This fall in summer use is to a large extent explained by changes in garden layout and management. What has happened is that high water

121

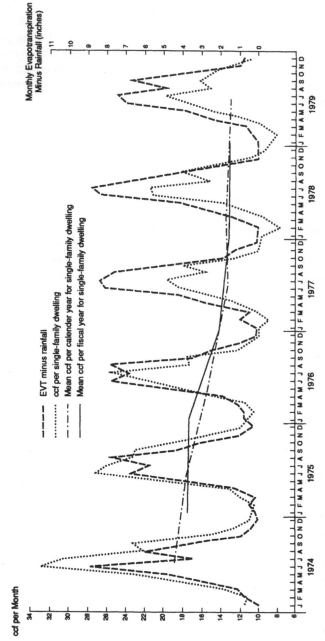

Figure 3.4 Monthly water use by single family dwelling in Tucson, Arizona in relation to evapotranspiration minus precipitation (ccf = 100 ft³)

ccf per Month

- - - - - EVT minus rainfall
············ ccf per single-family dwelling
– · – · – Mean ccf per calender year for single-family dwelling
———— Mean ccf per fiscal year for single-family dwelling

Monthly Evapotranspiration
Minus Rainfall (inches)

Source: Redrawn, by permission, from William E. Martin, Helen M. Ingram, Nancy K. Laney, and Adrian H. Griffin, *Saving Water in a Desert City.* © 1984 by Resources for the Future, Washington, D.C.

demanding plants, such as grasses and certain trees and shrubs, have been replaced by species more suited to the desert environment. As a result the greenery which used to characterise the suburbs is disappearing and being replaced by a much browner environment. This might not be as aesthetically pleasing, but it is cheaper in water terms and also requires much less expenditure of effort to maintain.

Perhaps the most important point emerging from this study is the change in attitude to water as a resource. Up to the early 1970s it was viewed as a resource which was too important to be allocated in strict monetary terms, and was therefore allocated by regulations and rates drawn up by the Water Department. These rates greatly undervalued water and as a consequence it was wastefully used. Since the mid-1970s the price mechanism has been accepted as a valuable tool in the allocation of water resources and a means of curtailing rising demand. The public has also been educated as to the effects which peak demands have on the cost of the distribution system and has learnt to reduce both demand peaks and overall use levels.

Table 3.8: Consumptive water use patterns, Tucson Active Management Area, 1980

Sector	Acre feet	% of total consumption
Municipal	60,000	19·2
Industrial	30,000	9·6
Mining	53,000	16·9
Agriculture	170,000	54·3
Total	313,000	100·0

Source: Martin *et al.*, 1984, p. 12.

By 1980 municipal and industrial uses of water accounted for about 30 per cent of total consumptive use in Tucson (Table 3.8). Agriculture still remains the pre-eminent water consumer accounting for over one-half of all use. In the future it seems inevitable that urban water demand will only be met by continued cessation of irrigation in the Avra valley and the transfer of this water to the Tucson region. By the year 2025 it is estimated that the irrigation acreage in the region could fall by about one-third compared with the 1980 value.

The key to the future use of water does seem to be extra water supplies being provided by the Central Arizona Project, which will help to reduce but not eliminate the mining of the groundwater reserves. In 1980 mined groundwater supplied about 61 per cent of the total demand of 1·452 million cubic m per day (430,000 acre feet per year). By the year 2000 it is hoped that this will be reduced to about 19 per cent in a total demand of 1·726 million cubic m per day (511,000 acre feet per year), with the Central Arizona Project providing 36 per cent (SAWARA, undated).

Experimental work to reduce domestic water use has also been carried out in Tucson by the University of Arizona, working in conjunction with a number of other agencies. The main part of this work has been the design of a house, called the Casa del Agua, to illustrate how substantial water savings can be achieved without any major changes in lifestyles. It is claimed that the house will be able to reduce the daily domestic demand *per capita* by two-thirds

Figure 3.5:   Water use in the Casa del Agua, Tucson, Arizona
This is an experimental and demonstration house designed by the University of Arizona to permit a high standard of living yet use the minimum amount of water possible

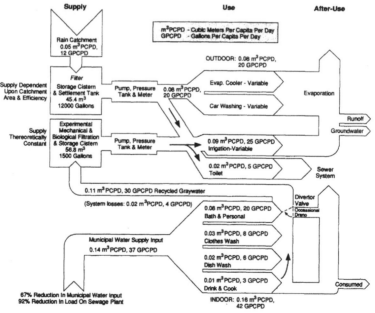

Source: Redrawn from Brittain R., 1986, 'Casa del Agua', *Arid Lands Newsletter*, No. 24, pp. 2–14. Reproduced by permission of the Office of Arid Land Studies, University of Arizona, Tucson.

from 397 litres/day (105 gallons/day) to only 140 litres/day (37 gallons/day) (Brittain, 1986). The volume of sewage generated is reduced by 90 per cent.

The basic objective of the project is to make maximum use of all available water. All rain-water from the house roof is collected in a large underground tank from where it can be used for activities such as evaporative cooling and toilet flushing (Figure 3.5). The system also makes extensive use of greywater — that is, water generated from bathing, washing of clothes, and dish washing. Following simple mechanical and biological treatment this water will be used for garden irrigation and toilet flushing. Only toilet residues are returned directly to the sewer system. Water saving appliances have been used throughout the house: for example, low-flush toilets with 4 litres (one US gallon) capacity, compared with the standard 19 to 27 litre (5 to 7 gallons) capacity, will alone save at least 76 litres (20 gallons) a day *per capita*. Similarly, low-flow showerheads can probably reduce demand by a further 45 litres (12 gallons) *per capita* per day. The house was only commissioned in late 1985 but has already been able to show that significant water savings can be made, though the investment costs are high.

3.2 IRRIGATION

3.2.1 Introduction

Throughout the world agricultural activities utilise far more water than industry. Although water is needed for a variety of purposes in agriculture, irrigation is by far the most important single use. Over the years large areas of the natural environment have been modified and manipulated to form relatively stable, artificial eco-systems which are capable of producing food crops (White, 1978). In semi-arid regions people discovered that crop productivity could be increased by supplying extra water during times of moisture stress. It was also found that with the aid of irrigation food crops could be introduced into areas where intense aridity prevented their indigenous development. Early examples of this are found in the irrigation civilisations of the Nile valley and the Tigris–Euphrates lowlands. Today, irrigation is most highly developed in the arid and semi-arid parts of the world, though in recent years it has been expanded markedly in temperate zones. In temperate areas irrigation is supplemental in nature, in so far as some crops could be successfully grown without additional water supplies. The effect of adding this extra water can, however, greatly increase yields by ensuring that crop growth continues at the maximum rate even during periods when precipitation amounts fall below normally expected values.

The date when irrigation was first used is unknown. However, the town of Jericho in the Dead Sea lowlands is probably one of the earliest sites in the world where irrigation was practised (Kenyon, 1969–70). It is thought that this settlement had a population in excess of 2,000 people and extended over an area of about 4 ha. The scale of irrigated agriculture in Jericho is, however, dwarfed by that associated with the civilisations which grew up in the Tigris–Euphrates lowlands and the Nile Valley about 6,000 years ago. These systems, together with later developments along the Indus, Yellow and Yangtse Rivers, have been described in detail by Wittfogel in his book *Oriental despotism*. In all these cases, people were able to prosper by the evolution of a highly centralised, bureaucratic society devoted to a policy of water control and the development of irrigated agriculture. The hydraulic engineering capability of these peoples was of a standard which has only been improved upon since the beginning of the nineteenth century. The fact that these societies were able to accomplish such feats of engineering depended largely upon their possession of a cheap and abundant energy source — slave labour. It was only by maintaining this energy supply, often by conquest, that the complex system of canal and headworks could be kept functioning.

Irrigation in the Tigris–Euphrates lowlands was a major challenge to human skills. Here were two rivers to be controlled, both of which experienced huge snowmelt floods in late spring and early summer, when the crops were already growing in the fields. As a result flood damage had a devastating effect on the agricultural economy. Two other problems also dogged the hydraulic engineers of Mesopotamia: siltation meant that the channels and banks of the major rivers often rose above the level of the surrounding lands, making adequate drainage of the fields almost impossible; and silt deposition in the irrigation canals reduced their carrying capacity and required periodic cleaning operations. Associated with the poor drainage of the alluvial soils was the problem of salinity build-up. As a result of high water tables and inadequate drainage, large areas of once fertile soils deteriorated into waste lands (Jacobsen and Adams, 1958).

In the Tigris–Euphrates lowlands one of the most impressive irrigation systems was that devised to irrigate the Diyala Plains during the Sassanian Period (226–637 AD) (Adams, 1985). A major feature of this system was the Nahrawan Canal, which was more than 300 km in length and more than 50 m in width. A key factor in the success of the irrigation network was the fact that the Tigris was utilised as a main drain to carry away the excess waters delivered by the Nahrawan Canal in the east and the Euphrates in the west.

The reasons for the decline in agricultural productivity in the Tigris–Euphrates lowlands are still the subject of much debate. Some claim that it is part of a natural environmental cycle of decline and recovery, while others argue that environmental degradation only occurred when the power of the central government declined and the essential maintenance work to the canal and drainage networks was not carried out. Once a certain level of degradation was reached, reclamation became impossible and the land went out of cultivation. In turn this would place greater strain on the central government, which would have to devote a larger proportion of its available energies to preserving its own power. Eventually the social system collapsed, a victim of a harsh and demanding physical environment.

Along the Nile valley a simple system of basin irrigation evolved about 5,500 years ago and this was maintained until the nineteenth century (Hamdan, 1961). It proved extremely easy to control. Breaches in the natural levees of the river were cut, through which the rising floodwaters of the Nile poured. These waters were led into the basins where the soil moisture was recharged and the suspended Nile silt deposited. After a number of days any remaining water was drained back into the river. With the flood discharge peaking in the late summer only a winter crop could be grown on the soil recently moistened and fertilised by the Nile waters and silts. Once the crop was harvested, the land was left fallow until the succeeding flood.

The fact that irrigated agriculture was practised continuously for thousands of years along the Nile valley suggests that it must have been an ecologically balanced system. This is easy to understand when one considers that humans did very little to alter what was a naturally occurring phenomenon — namely, the seasonal flooding of the riverine lands. Given the timing of the flood, all that was necessary was to plant seeds on the moistened soil and leave nature to do the rest. All that had been done, in fact, was to replace the

127

natural vegetation with a food crop. The amount of modification of the environment to achieve this was minimal. The only problem facing the farmers of the Nile valley was the height of the flood: this controlled the area of land which was inundated and hence the size of the crop which could be produced. When the flood level was low, famine was often a consequence.

3.2.2 Irrigation water sources

Water for irrigation has to be obtained from either surface or groundwater sources, or a combination of the two. In many parts of the world, traditional methods of irrigation are still employed, though every year new techniques are rendering certain methods redundant. This is particularly the case with groundwater. Here the problem has always been that of lifting water from the water table to the ground surface. In most arid lands where irrigation is practised this was achieved largely by the use of animal power. In its simplest form, a walkway was constructed, sloping away from a well, down which the animal (usually a buffalo or a camel) walked, pulling a bucket from the bottom of a well. When the bucket reached the ground surface, the water contained within it was poured into an irrigation canal and so led off to the fields. In a more sophisticated form of groundwater extraction the animal would move continuously around a circular path pulling a long wooden beam. The beam was connected through a series of gears to buckets attached to a continuous rope. This method allowed much larger volumes of water to be abstracted, but did require a more advanced level of technology.

With the advent of cheap energy sources, whether in the form of electricity or oil, the motive power of animals was quickly replaced by electric motors or diesel engines on existing wells. The increased energy which was made available with these mechanical aids meant that water could be extracted from wells at a rate in excess of water recharge. In many parts of the world this has led to a rapidly declining water table and the expensive necessity of deepening of wells.

An unusual method of groundwater extraction, which is still in widespread use in the Middle East, is the *qanat* or *falaj*. This consists of a gently sloping tunnel cut through alluvial material which transmits the water from beneath the water table to the ground surface (Beaumont, 1971) (Figure 3.6). The major

Figure 3.6 A qanat: cross-section and plan

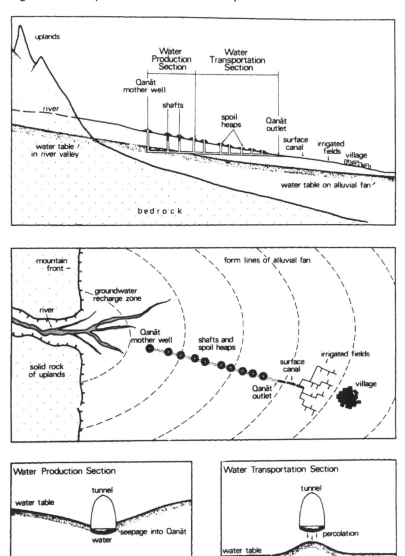

advantage of the *qanat* over the well is that once constructed it will transport groundwater to the surface without any further energy inputs apart from cleaning and maintenance operations. This can be a problem, however, as much water is allowed to run to waste during the winter months. Even during the growing season water flowing from the *qanat* during the night will often be wasted unless a storage pond is constructed.

Qanat systems are found throughout the Middle East, but reach their maximum development in Iran. Here many hundreds have been constructed in alluvial fans bordering the Great Kavir of the central plateau. The magnitude of some of these systems is remarkable. In eastern Iran, mother-well depths of over 290 m have been reported, while around Yazd *qanats* can be up to 50 km in length (Beaumont, 1971; English, 1968). In general most *qanats* are much smaller, with mother-well depths of between 10 and 50 m and with lengths ranging from 1 to 5 km. Perhaps somewhat surprisingly, given the immense effort which has gone into their construction, water discharges in Iran are relatively low with most values falling between 0 and 80 cubic m/hr.

3.2.3 Irrigation systems

Irrigation systems can be divided into two basic types. The first is surface irrigation, either flood (basin) or furrow, in which the water always maintains a free water surface in contact with the air. With this method gravity is the sole driving force. As has already been noted, surface irrigation systems date back at least 10,000 years, with few major changes in the methods of water transport and field application before the twentieth century. Perhaps the most significant advances have been the introduction of reinforced concrete as a constructional material and earth-moving equipment powered by internal combustion engines. These have enabled large dams and canals to be built and so have opened up a new era of environmental manipulation.

The second type is pressure irrigation. This involves the transmission of water along pipeline networks and eventual distribution to the crop via sprinkler or trickle/drip systems. Hydrostatic pressure of between 1 and 3 atmospheres propels the water and this is generated by pumping or gravity sources. Pressure irrigation systems have only become possible with the technological advances which permitted the fabrication of small-diameter pipes at relatively low costs. On a world scale, pressure irrigation systems have only become of importance since the Second World War.

Figure 3.7 Wetting stages of a field during surface irrigation

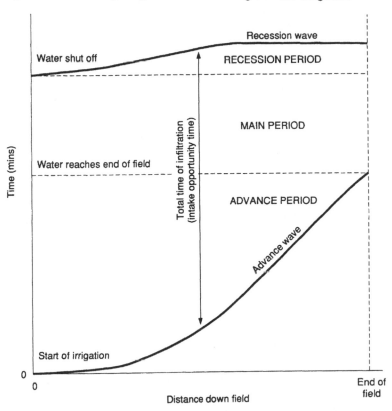

Source: Redrawn from Withers B. & Vipond S., 1974, *Irrigation — design and practice*, Batsford, London. Reproduced by permission of Batsford Ltd.

In terms of the economics of their operation there are marked differences between surface and pressure irrigation systems. With a surface system, at any one time a minimum depth of water has to be applied on to the area which is being irrigated. Of necessity this means that large parts of the irrigated field will receive water in excess of the needs of the growing crop (Figure 3.7). As a result the system is wasteful of water, often to a high degree in the case of large fields or long furrows. Another constraint on surface irrigation is that each water application is associated with a fixed cost. This means that the system is usually operated in such a manner as to minimise the number of water applications and to

131

maximise the moisture storage within the soil profile of the irrigated field. A repeated application of water is only necessary when the soil moisture content has been depleted to almost zero.

Over the last few years innovations have made surface irrigation methods much more efficient. Laser levelling techniques have permitted the introduction of a highly efficient form of basin irrigation. Special earth-moving machinery which picks up a signal from a horizontal laser beam can level fields to an accuracy of a few centimetres over distances of 200 m. Such technology was introduced into the south-west USA in the late 1970s in conjunction with automated field inlet gates, to produce sophisticated control over one of the simplest irrigation methods. It has proved highly successful.

Pressure systems have very high capital costs for installation as long pipe-runs have to be available so that water can be delivered to all parts of the cropped area. In general costs rise markedly with increasing pipe diameter. Relative to investment costs, the operating costs of pressure systems are low. This has caused such systems to be operated with relatively small water dosages being added at each application. Such an approach means that the water needs of the growing crop can be more closely matched and wastage of water is consequently minimised. Given the relative ease with which irrigation can be achieved, the use of the pressure system means that soils with only low moisture storage capacities can easily be managed for crop production. Even more important is the fact that pressure irrigation can be utilised in areas where surface irrigation systems would be quite impracticable owing to excessive ground slopes.

Trickle irrigation systems suffer from high operating costs, largely because the plastic pipes from which they are usually constructed have only a short life-expectancy. They are also subject to clogging of the outlet points by the precipitation of calcium and iron salts or by algal growths. Within the pipes themselves the deposition of silts and clays can greatly reduce water flows. Careful management of trickle systems using filtration and various forms of chemical treatment can reduce the clogging problem, but as a consequence operating costs can rise sharply. If saline waters are used in trickle systems they can lead to serious salt build-up, as insufficient water is normally present to permit adequate leaching to occur. A feature of trickle irrigation has been the rapid way in which it has been adopted by commercial agricultural producers. In 1970 in California only 60 ha were irrigated by this method, whereas three years later this figure had increased to 16,000 ha (National Academy of Sciences, 1974).

Work on irrigation has suggested that crop yields are increased by the frequent addition of measured quantities of water (Rawlins and Raats, 1975). Such schedules produce optimum conditions for growth in the root zone and at the same time can greatly reduce water usage. The implication is that if irrigated agriculture throughout the world is to become more productive and at the same time make more efficient use of water resources, it is imperative that a high frequency application of carefully measured quantities is introduced. This would mean the widespread adoption of pressure irrigation techniques as well as the newer more sophisticated methods of surface irrigation.

Besides considerable savings in terms of the smaller volumes of water required for crop growth, pressure irrigation systems also do not require the same degree of land preparation and drainage facilities, which are needed for surface irrigation systems. Still greater savings can be made by lower rates of fertiliser application, as drainage waters and hence fertiliser movements from the fields are minimised. In some cases fertiliser losses can be reduced by two-thirds, with corresponding energy and monetary savings (ibid.).

Detailed comparisons of drip, sprinkler and furrow irrigation have shown that when equal amounts of low salinity water are used, the drop irrigation method produced yields about 50 per cent higher than for other methods (Bernstein and Francois, 1973). When brackish water was used the yields of plants grown by all three methods of irrigation declined. For drip irrigation the decline in yield was only 14 per cent, compared with 53 per cent for the furrow method and 94 per cent for sprinkler irrigation. In a second experiment the irrigation frequency for the furrow and sprinkler methods was increased to what was regarded as optimum conditions. This resulted in the yield differences between the three methods being reduced to almost zero. Summarising these results it was seen that drip irrigation required about one-third less water than did furrow irrigation to attain maximum yields, with the main water savings occurring when the crop was young. If irrigation using the furrow method was inefficiently applied, the water savings by use of the drop method would be correspondingly greater. It is important to note that sprinkler and furrow methods will produce yields almost identical to the drip method, provided that extra water supplies are utilised.

Figure 3.8 Diagram for the classification and use of irrigation waters

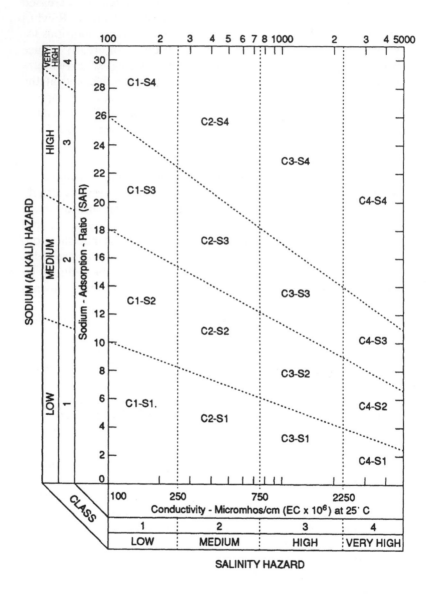

Source: Redrawn from Wilcox L.V., 1955, *Classification and use of irrigation waters*, United States Department of Agriculture Circular No. 969.

3.2.4 Irrigation water quality

The quality of irrigation water can affect the success of a project almost as much as the quantity of water available (Ayers and Westcot, 1985). All waters contain a range of dissolved materials and it is the nature and concentration of these chemicals which are used as a basis for the classification of irrigation waters. The most important measures of irrigation water quality are electrical conductivity and the sodium adsorption ratio (SAR). As the dissolved solids content of water increases, the ability of the water to conduct electricity also increases. Recordings of electrical conductivity therefore provide a measure of the total dissolved solids within a water sample. The SAR reflects the relative concentrations of the monovalent cation sodium, relative to the divalent cations calcium and magnesium. These two measures are used in diagram form to estimate the salinity hazard and the sodium (alkali) hazard of an irrigation water (Figure 3.8). When the salinity of the saturation extract of the soil increases, the growth of the plant is inhibited. Experiments have revealed that there is a close relationship between the conductivity of the saturation extract of the soil and that of the irrigation water. The quality of irrigation water can therefore be monitored to provide an indication of the salinity hazard. Four classes of water are usually recognised in terms of electrical conductivity values, with the highest quality water possessing readings of less than 250 micromhos/cm (Wilcox, 1955).

The sodium (alkali) hazard is more complicated than the salinity hazard as its effects are also related to the salt content of the irrigation water. When the irrigation water has a high dissolved salt content the sodium (alkali) hazard is much greater for any particular value of the SAR (Figure 3.8). Using both the salinity hazard and the sodium (alkali) hazard it is possible to distinguish 16 classes of irrigation water. Of these only C1 – S1 water can be used for most crops without any problems being caused; all other water types may produce harmful results under certain conditions.

3.2.5 Distribution of irrigated agriculture

About 15 per cent of the world's cultivated lands are irrigated, though proportions vary from region to region. A striking feature of irrigated agriculture at the present day is the way in which more than 60 per cent of the total is concentrated in Asia (Table 3.9). Equally impressive is the relative importance of irrigation compared with dry farming in Asia, where more than one-third of the total cultivated land is irrigated.

Table 3.9: Irrigated area by continent, 1984

Continent	Irrigated area (000 ha)	Total cultivated area (000 ha)	Irrigated area as percentage of cultivated area	Irrigated area as percentage of world total
Africa	10,390	184,184	5·64	4·73
N. and C. America	27,414	274,417	9·99	12·48
S. America	7,979	139,188	5·74	3·63
Europe	15,616	140,409	11·12	7·11
Oceania	1,869	49,752	3·76	0·09
USSR	19,485	232,215	8·39	8·87
Asia (excl. USSR)	136,962	456,596	30·00	62·34
World	219,715	1,476,761	14·88	100·00

Source: Data from *FAO production yearbook 1985*.

On a country-by-country basis China is easily the main irrigation nation, accounting for 21 per cent of all the world's irrigated lands (Table 3.10). In China 45 per cent of the cultivated area is irrigated — an exceptionally high figure for such a large country. It should be recognised, however, that reliable data on China are difficult to obtain and that some workers have suggested that the irrigated area is not as large as the UN figures suggest (Sigurdson, 1977). Perhaps surprising to many is the absolute size of the irrigated areas in the two super-powers (the USA and the USSR), placing them well towards the top of the world irrigation league. The relative importance of irrigation in the agricultural systems of these two countries is remarkably similar.

A calculation of how much water is withdrawn for irrigation throughout the world each year is a near impossible task owing to the very large number of variables involved. The only satisfactory way to gain an overall estimate of water withdrawal is to extrapolate from irrigation data known in detail for a particular country. Information from the USA reveals that the amount of water withdrawn for irrigation varied from a minimum value of less than 1,705 cubic m per ha per year (500 gallons/day/acre) in

Table 3.10: Irrigated area by country, 1984

Country	Irrigated area (000 ha)	% of cultivated land
China	45,420	45
India	39,700	26
USA	19,831	10
USSR	19,485	8
Pakistan	15,320	76
Iran	5,730	39
Indonesia	5,450	26
Mexico	5,100	21
Thailand	3,550	18
Japan	3,250	68
Spain	3,145	15
Italy	2,970	24
Afghanistan	2,660	33
Romania	2,612	25
Egypt	2,474	100
Total	219,715	

Source: *FAO production yearbook 1985.*

the wetter regions of the east, to a maximum of more than 19,698 cubic m per ha per year (5,769 gallons/day/acre) in the upper Colorado basin (Solley *et al.*, 1983). These variations are mainly due to climatic differences between the regions concerned, though different irrigation management policies can also be important. An average irrigation withdrawal for the Water Resources Council regions of the USA comes out at 8,830 cubic m per ha per year (2,586 gallons/day/acre).

Using this average value for the USA it is possible to obtain a value for world irrigation water withdrawal of the correct order of magnitude for a total irrigated area of about 220 million ha. The figure arrived at is 1,940 cubic km of water each year. Of this total, 1,210 cubic km are utilised in Asia. This indicates the very severe pressure which is already being exerted on the water resource base of the continent. Although the total value of runoff in Asia is in excess of 13,000 cubic km each year, only 3,400 cubic km is classed as stable runoff, which can be relied upon for beneficial purposes (Lvovitch, 1973). It should be noted that a large proportion of this stable runoff is generated within the USSR in areas where irrigation is not common. Bearing the above in mind, it seems reasonable to assume that the current volume of water used for irrigation in south and east Asia must be close to the annual average runoff of the area. This implies that any further expansion of the irrigation network will require expensive river management structures aimed at increasing the volume of reliable water resources.

3.2.6 Irrigation development in the USA

3.2.6.1 The early years

From the 1860s onwards the American West was opened up to settlement. The optimism of the early settlers was high, but their knowledge of dryland conditions almost non-existent as they were used to farming under the humid conditions of the eastern United States. A period of increasing rainfall seems to have accompanied the settlers into the West and this led to the widespread and dangerous belief that 'rain follows the plough'. In the USA the person most closely associated with the early development of drylands is John Wesley Powell (Darrah, 1969). Best known for his exploration of the Colorado River and his journey through the Grand Canyon, he later went on to become the Head of the United States Geological Survey. Powell, with a wide knowledge of the American South-West, was one of the first people to realise that new attitudes to land in arid regions would have to be developed. In particular he realised that west of the hundredth meridian annual precipitation totals were highly variable and that rain-fed agriculture would always be a risky business. He also understood that in the more arid areas, water was the key to development and that only a limited amount existed:

> About two-fifths of the entire area of the United States has a climate so arid that agriculture cannot be pursued without irrigation. When all of the waters running in the streams of this region are conducted on the land, there will be but a small portion of the country redeemed, varying in the different territories perhaps from one to three per cent. (J.W. Powell, 43rd Congress, 1st Session, H.R., Rep. 612, 10 April 1874, p. 10.)

One of the key Acts opening up the West was the Homestead Act of 1862, which permitted a settler to claim a quarter section of 160 acres (64·75 ha) for his own use. Provided that a building on the land was constructed and lived in, and other improvements made, the settler was allowed to keep the land without cost. While this system worked well in the East it was not suited to the drylands of the West. In 1877 Powell wrote:

> The present land system of the country, whether as to the purchase, pre-emption, or homestead plans, is not at all suitable for the area of the arid region . . . In the whole region, land as mere land is of no value; what is really valuable is the water privilege. (*New York Tribune*, 28 April 1877.)

In this statement Powell clearly showed that he understood the immense significance of water rights in the opening up of the western lands.

Powell's most comprehensive work on the American drylands was his *Report on the lands of the Arid Region of the United States with a more detailed account of the lands of Utah* (1878). The report proposed a classification of the western public lands into five categories: mineral, coal, irrigable, pasturage and timber lands. Most of the text dealt with the last three categories. In particular he suggested ways in which these lands could be used and protected. In effect this meant that Powell was calling for a total rethink of the federal government system of dealing with the arid lands.

He was especially critical of the Homestead Act, 1862, and its 160 acre (64·75 ha) land grant, which he believed was totally insufficient for the drylands. Instead he proposed that four square miles or four sections (2,560 acres or 1036 ha) constituted the minimum necessary for a grazing unit and in many parts the minimum would have to be even larger. He also stated that each farm unit should include potentially irrigable land, thus suggesting that land apportionment should be done on the basis of different land types, rather than on the straight lines of the surveyor. Such views were before their time, however, and they were ignored by the legislators in Washington who had little knowledge and even less understanding of the western lands.

The Desert Land Act of 1877 had been intended to promote the settlement of the arid lands, but in fact proved to be a hindrance. The unit of land was increased to 640 acres (259 ha), but the settler was not allowed to homestead it. Instead he had to pay 25 cents deposit for each acre (0·4 ha) and a further one dollar an acre within three years. To keep the land, part of it had to be irrigated

139

within three years. However, it was soon realised that most of the western lands were not capable of being irrigated, at least economically. Powell also proposed what were in effect irrigation districts, with respect to the irrigable lands of the farming units: 'All the pasturage lands should have water fronts and irrigable tracts and the residents should be grouped and as the lands cannot be economically fenced and must be kept in common, local communal regulation or co-operation is necessary.'

In summary of his ideas, four proposals were submitted to Congress. These were:

(1) all lands for which water is accessible for as much as 320 acres (129·5 ha) shall be classed as irrigable land;

(2) the amount of land allocated to each person shall not exceed 80 acres (32·4 ha);

(3) nine or more persons qualified for homestead entry shall form an irrigation district and make such bylaws for their own government as do not interfere with the general laws of the country; and

(4) the right to the water shall inhere in the land and in the conveyance shall pass with the title of the land. Failure to use water shall after a period of five years cause the right of water to lapse.

However, none of these proposals were taken up by the Congress and the haphazard and often wasteful use of the arid lands continued.

During the 1880s the irrigation movement gained growing support in the United States and in 1888 the Congress authorised funds for an irrigation survey to be undertaken by the US Geological Survey, the then head of which was John Wesley Powell. The survey went ahead rapidly and by 1889 150 reservoir sites had been identified and 30 million acres (12·14 million ha) classified as suitable for irrigation. However, growing conflict between Powell and the legislature occurred and in 1890 the funds available were so curtailed that the irrigation survey was effectively discontinued with its work incomplete.

The irrigation movement continued and even increased in importance following the major drought of 1890, but its supporters, like the early settlers, were over-optimistic in their aspirations. They called on the government to make lands in the public domain available for irrigation and irrigation congresses were held to stir up public opinion and to put pressure on the federal government.

Powell was once again appalled by the lack of knowledge and understanding of the arid lands being shown by the supporters of the irrigation movement. At Los Angeles in 1893 he attended a gathering of the irrigation movement and told the delegates in no uncertain terms that they were being unrealistic in their plans:

> I wish to make clear to you . . . there is not enough water to irrigate all the lands . . . there is not sufficient water to irrigate all the lands which could be irrigated . . . only a small portion can be irrigated . . . It is not right to speak about the area of the public domain in terms of acres that extend over the land, but in terms of acres that can be supplied with water . . . I tell you gentlemen that you are piling up a heritage of conflict and litigation over water rights for there is not sufficient water to supply the land.

His views were received with dismay. During the rest of his life Powell continued to press for the establishment of a federal bureau of irrigation. The Reclamation Bureau was created in 1902 in the last year of his life.

3.2.6.2 The Bureau of Reclamation

The Reclamation Act of 1902 established the Bureau of Reclamation, with the aim of opening up the American West through the development of irrigated agriculture. The Bureau quickly became popular in the western states and was inundated by requests for new projects. By 1903 six projects were under construction and by 1907 that figure had risen to 25. The first irrigation scheme to receive Bureau of Reclamation water was the small Newlands project in northern Nevada. Its first power project was the Roosevelt Dam near Phoenix, opened in 1909, and this led the way to rural electrification throughout the West.

What the Bureau of Reclamation brought to the West were large capital funds, modern engineering techniques and a bureaucratic organisation which was capable of constructing large-scale projects and advocating the building of new ones. This occurred at a time when the West was dominated by small towns and local government areas, and when the fierce individualism of the frontier still remained. In effect it meant that the Bureau of Reclamation was a large enough organisation from its inception to overcome any obstacles to its activities.

In the early years of the twentieth century the Bureau of

Reclamation concentrated its activities on single-purpose projects supplying irrigation water, which it did by a construction programme of dams and reservoirs. Up to the mid-1920s the Bureau was the pre-eminent organisation in the world involved with water storage and management. However, its own success meant that communities grew and urban demands for water increased accordingly. This necessitated the adoption of a multi-purpose approach to water resources development, culminating in what is still regarded by many as the flagship of the Bureau, the Hoover Dam on the Colorado River. This provided water for Las Vegas and Los Angeles, hydro-electricity, flood control and irrigation water for the lower Colorado. During the Depression of the 1930s the Hoover Dam provided a beacon of hope that engineering skills could once again generate prosperity. The 1930s also saw a broadening in approach of the Bureau's activities, as it became involved in solving the problems of arid land management and in helping farmers overcome the severe economic difficulties in which they found themselves.

During the early years of the twentieth century the land irrigated by the Bureau of Reclamation grew rapidly to around 2·2 million acres (0·89 million ha) in 1920. Further growth in the twenties and thirties, despite economic depression, meant that by the end of the Second World War the irrigated area stood at 4·16 million acres (1·68 million ha). It is perhaps somewhat surprising to learn that the really massive period of growth in the Bureau's history took place between 1945 and 1965. During this time almost 4 million irrigated acres (1·62 million ha) were added to the Bureau's holdings to produce a total of 8·01 million acres (3·24 million ha) (National Water Commission, 1973, p. 126). Since then a much slower period of increase has taken place so that by 1981 the irrigated area had only expanded to 10·1 million acres (4·09 million ha) (National Water Resources Association, undated, p. 13). Although the Bureau of Reclamation is, quite rightly, always thought of as the main irrigation organisation in the American West, it should not be assumed that the Bureau is responsible for most of the irrigated land in the 17 western states. Indeed, the reality is very different, for throughout most of its history the irrigated lands of the Bureau have only accounted for between a fifth and a half of all the irrigated areas. In the West, as elsewhere in the US, private enterprise still plays a dominant role in land development.

However, the Bureau of Reclamation has not been without its

critics, particularly with regard to the efficiency of irrigated land use and the cultivation of low value crops. Thus, for example, the National Water Commission noted in 1973 that about 23 per cent of Bureau land grew hay, which provided a gross annual return of only $106,53 per acre ($263·3 per ha) (National Water Commission, 1973, p. 126). A further 13 per cent of the land grew barley and corn with annual gross revenues of about $100 per acre ($247·1 per ha). At the other extreme only 7 per cent of land grew fruit and nuts with returns of $660 per acre ($1630·9 per ha) and only 9 per cent grew vegetables with returns of $600 per acre ($1482·6 per ha). This meant that in 1970 only about 16 per cent of Bureau land was used for high-value crops.

The National Water Commission also felt that a major weakness of the Bureau's work was that its activities were too heavily subsidised by the Federal government. It quoted the example that on some projects only 10 per cent of the construction costs for irrigation facilities was repaid. This was achieved by the absence of interest payments on capital lent for construction — by permitting income from power sales to be credited towards irrigation payments and by deciding that a large part of the total costs were allocated to non-reimbursable headings (National Water Commission, 1973, p. 128). The effects of this policy can be dramatic: for example, in the Manson scheme in the state of Washington, the National Water Commission were told that if a full-cost pricing policy for irrigation water was applied, annual water charges would increase from $32·5 per acre ($80·3 per ha) to $414 per acre $1203 per ha) (ibid., 1973, p. 130). Gross crop receipts from a nearby project were only $128 dollars per acre ($538·7 per ha), suggesting that the Manson scheme could never operate profitably on a full-cost reclamation basis.

The original policy of the 1902 Reclamation Act was that construction costs for irrigation facilities were to be repaid without interest over a 50-year period. With the enactment of the 1939 Reclamation Project Act it became permissible that revenues for hydro-electric power and other water uses could be used to repay the irrigation costs. In the early 1970s it was estimated that such revenues were accounting for about 60 per cent of the repayments made on irrigation facilities. As a result one of the recommendations of the National Water Commission with regard to irrigation water supply was as follows: 'All costs of new Federal irrigation facilities should be recovered from irrigators and other direct beneficiaries through contracting entities, with interest equal to prevailing yield rates for long-term US Treasury Bonds at the time of construction' (ibid., p. 497).

The 1902 Reclamation Act obtained funds for the construction of irrigation works by the sale of public lands in the West. A condition of the Act was that a farmer was limited to a total of 160 acres (64.75 ha) which he could own and still receive subsidised water from a Reclamation Project. Over the years this limitation proved too small, as increased mechanisation meant that efficient and economical farming could only be applied on larger acreages. Many farmers got round the regulation by leasing land in excess of the ownership limit.

The Reclamation Reform Act of 1982 addressed itself to many of the problems which had arisen over the years (US Department of the Interior, Bureau of Reclamation, 1985). A feature of the new law was two sets of provisions termed non-discretionary and discretionary. The non-discretionary provisions applied to all irrigated lands on reclamation projects. In contrast, the discretionary powers only applied to districts with new or amended contracts or to individuals and groups who decided to come under the new provisions.

In the districts where the discretionary powers applied, individuals and small companies were able to increase their ownership entitlement from 160 (64.75 ha) to 960 acres (388.5 ha) and still obtain subsidised irrigation water. Beyond this figure all water would be charged at the full cost rate. For large companies the acreage limitation was only increased to 640 acres (259 ha). Where the discretionary powers do not apply the 160 acres ownership limitation still has to be complied with. The Reclamation Reform Act of 1982, like the original 1902 Act, did not limit the amount of land which could be leased. It did, however, state that for each land holder, whether an individual or small company, only 960 acres (388.5 ha) of land could receive subsidised water. Beyond this figure full costs had to be charged.

In its first 80 years up to 1982 the Bureau of Reclamation expended $10.5 billion on project construction. Of this total about 84 per cent will eventually be repaid through the sale of water and power. The balance is regarded as non-reimbursable investment in flood control, wildlife management, salinity prevention and recreation. The detailed picture is more complex, however. In 1982 $7.3 billion had been invested in projects already completed and $3.2 billion spent on projects still under construction. By 1980 only $2.0 billion had been repaid to the Federal government, though it is hoped that a further $6·8 billion will be repaid in the future. The Bureau does claim, however, that the US government received revenues of $3·0 billion dollars in 1980 as an indirect result of its activities (National Water Resources Association, undated). Indeed, for the period 1941–82 it is claimed that the total Federal revenue generated from Bureau projects reached a total of $36 billion.

Today the Bureau of Reclamation lands extend to more than 10 million acres and produce more than half of the US's vegetable crop and one-quarter of its fruit and nut output. In an urban context, 19 million people are currently supplied with water from Bureau facilities, while 15 million people receive electricity from 49 Bureau power plants producing 47 billion kilowatts of electricity each year.

What, then, is the future role of the Bureau of Reclamation? To a great extent the Bureau has already fulfilled many of its original objectives. Irrigation development has occurred on a wide scale throughout the West and this, in turn, has led to the self-sustaining growth of the local communities. It would seem that in future irrigation is going to become less important to the economy and so the Bureau's role will have to change from that of an essentially pioneering organisation to one dealing with the preservation of the existing order. This changing role for the Bureau was effectively recognised in the Reclamation Reform Act of 1982, which, while firmly endorsing the earlier traditions of the Bureau, at the same time spelt out the need for a more market-orientated approach to irrigation development in the future.

3.2.6.3 Irrigation in the USA today

The USA is the only large country in the world for which detailed and largely accurate statistics on irrigation are available. Irrigated areas in the USA account for about 7 per cent of the world's total irrigated lands. About 23·5 million ha of land are irrigated in the USA (1980), representing about 9 per cent of the cultivated area of the country. As one would expect it is the drier regions of the country, especially to the west of the Mississippi River, where irrigation water withdrawals reach their highest values. In terms of the individual states California, Texas and Nebraska have the largest irrigated areas with figures of 3·9 million, 3·1 million, and 2·9 million ha respectively. Other states with large areas of irrigation are Colorado, Kansas and Montana.

The actual volumes of water withdrawn for irrigation purposes

145

Table 3.11: Irrigation water use by regions, USA, 1980 (million cubic m per annum)

Water resource regions	Ground water	Surface water	Reclaimed sewage	Total
New England	10·8	62·2	0·0	73·2
Mid-Atlantic	134·0	207·2	0·1	345·4
South Atlantic-Gulf	2,763·1	2,486·8	0·0	5,249·8
Great Lakes	248·7	165·8	41·5	469·7
Ohio	121·6	82·9	0·0	207·2
Tennessee	3·7	5·7	0·0	9·4
Upper Mississippi	483·5	40·1	0·0	525·0
Lower Mississippi	6,631·3	4,006·4	0·0	10,637·7
Souris-Red-Rainy	63·6	24·9	0·3	88·4
Missouri Basin	15,196·8	24,867·5	2·4	38,682·7
Arkansa-White-Red	11,604·8	3,315·7	20·7	15,196·8
Texas-Gulf	5,387·9	2,210·4	76·0	7,598·4
Rio Grande	2,210·4	3,730·1	0·0	5,940·6
Upper Colorado	111·9	10,223·3	0·1	10,361·4
Lower Colorado	5,388·0	5,111·6	8·6	10,499·6
Great Basin	1,381·5	6,769·5	5·1	8,151·0
Pacific North West	7,045·8	33,156·6	23·5	40,064·2
California	24,667·5	27,630·5	207·2	52,498·0
Alaska	0·0	0·0	0·0	0·0
Hawaii	635·5	621·7	0·0	0·0
Caribbean	193·4	248·7	0·0	428·3
Total	82,891·5	124,377·3	386·8	207,228·8

Source: Data from Solley *et al.*, 1983.

are enormous by any standards: in 1980 they totalled 207 cubic km (Table 3.11). About 60 per cent of all irrigation water was supplied from surface water sources, but wide variations are to be noted from one region to another. In general groundwater use tends to become more important in the drier zones of the south-west of the USA. The highest water withdrawals rates for irrigation are slightly in excess of 19,500 cubic m per ha per year in the basin of the upper Colorado. However, throughout the south-west figures of over 10,000 cubic m per year commonly occur (Figure 3.9). In the eastern part of the country withdrawal rates are much lower and are usually less than 3,500 cubic m per ha per year.

On any irrigation project the quantity of water withdrawn from a water source is normally considerably in excess of the amount consumed. Consumption in this sense is defined as water which does not make its way back rapidly to an adjacent stream course as

Figure 3.9 Irrigation water withdrawals from the major watershed units of the USA.
Figures in cubic metres per hectare per year

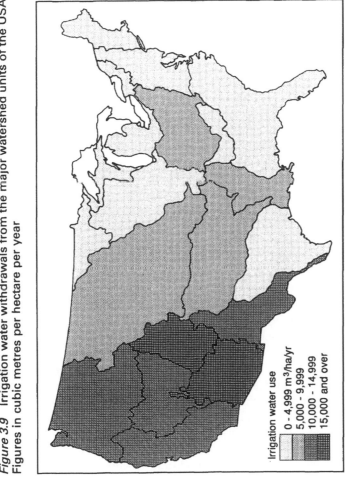

Irrigation water use

- 0 - 4,999 m³/ha/yr
- 5,000 - 9,999
- 10,000 - 14,999
- 15,000 and over

Source: Data from Solley W.B. *et al.*, 1983, *Estimated use of water in the United States in 1980.*
United States Geological Survey Circular, No. 1001.

a result of field drainage. The main consumptive loss on any irrigation scheme is undoubtedly evapotranspiration from the growing crop, though a certain proportion of the water which is applied will percolate below the root zone of the crop. Some of the water will eventually drain into drainage ditches and so it will return to stream and river systems. A certain amount of water will, however, percolate down to recharge groundwater reserves.

Of the water actually abstracted from water sources in the USA, a large proportion is consumed by either evaporation or percolation once the water has reached the fields. For the country as a whole the average loss in the USA is 55 per cent. In absolute terms consumptive losses reach their highest values in the lower Colorado basin, with figures in excess of 10,000 cubic m per ha per year. Throughout the south-west, owing to the high prevailing temperatures, consumptive losses are high everywhere.

Another type of water loss is conveyance loss. This term is used to describe the water which is lost between the point of abstraction from a river or well and the field in which the water is to be applied. The main cause of this loss is the percolation of water through the bed and banks of the unlined irrigation canals, though secondary losses through evaporation from canal water surfaces can be considerable in arid areas. Conveyance losses in the USA show wide variations, though overall the mean loss is 15 per cent. However, it should be remembered that in the USA many schemes employ modern irrigation techniques, such as concrete lined canals, which will greatly reduce percolation losses. In most of the developing countries conveyance losses are likely to be much greater, possibly accounting for up to one-third of all water withdrawn.

In the USA a number of theoretical models have been established to illustrate the relationships between water and land use, agricultural technology and environmental improvement. One of the most detailed of these models was that proposed by Heady and Nicol (1976). In this model, which focusses on irrigation use in the 17 western states of the USA, it was shown that changes in the West were closely linked with land use and levels of agricultural technology in other parts of the country, as well as on economic policies of food supply and control. It was pointed out, for example, that crop production in an area of rain-fed agriculture could often be a substituted for crops produced in an irrigated region.

The linear programming model used by Heady and Nicol

included information on 223 agricultural-producing regions, 51 water-supply regions and 27 consumer-demand regions, which were interdependent through transportation networks, water transfer systems and commodity supply and demand. The objective of the model was to minimise the cost of producing and transporting agricultural commodities from areas of production to those of consumption.

Table 3.12: Water use by agriculture in the 17 western US states by the year 2000 assuming different water prices

Water price (US$ per acre foot)	Irrigated land (million acres)	Total water withdrawals (million acre feet)	Consumptive use in agriculture (million acre feet)	Water released from agriculture (million acre feet)
7·50	27·2	155·9	68·0	–
15·00	22·6	139·3	56·7	11·4
22·50	17·2	120·3	42·6	25·5
30·00	12·2	105·0	31·9	36·2

Source: Heady and Nicol, 1976.

Using the model a series of possible futures for the USA was calculated, in which parameters such as population, water prices, land control policies and environmental goals were varied. In terms of the 17 western states of the USA it was shown that water prices would have a greater effect on conditions in the region in the year 2000 than any of the other variables (Table 3.12). As water prices increased the area of irrigated land declined markedly, so releasing large volumes of water for other beneficial uses. Under such conditions production from rain-fed agricultural areas becomes more profitable and a change in the relative distribution of production occurs.

This example illustrates what is often forgotten — namely, that in most large countries it is possible to substitute crop production in a non-irrigated area for production in an irrigated area. This is often beneficial for the country as a whole as irrigation water is generally highly subsidised by both direct and indirect payments from the central government. Such a policy has often meant that land has had to be withdrawn from cultivation in non-irrigated regions as it has been unable to compete with the subsidised irrigated production. This emphasises the need to study all aspects

149

of agricultural production and indeed the national economy, when decisions concerning irrigation are being contemplated. All too often in the past irrigation schemes have been viewed as a means of revitalising a local area with little regard to the effects such a project might have on the regional or even the national economy.

A feature of irrigated agriculture which is sometimes over-looked is the large amount of energy consumed by it. Thus, for example, in the USA, with one of the most advanced agricultural systems in the world, almost 7 per cent of all energy used 'on farm' is accounted for by irrigation (Steinhart and Steinhart, 1974). When it is remembered that less than one-tenth of the total cultivated area of the USA is irrigated, the very high energy costs/ ha associated with irrigation are clearly seen. The high costs associated with irrigation are also disturbing when one considers that the 'Green Revolution', which consists of the import of high-yielding crop varieties to developing countries, nearly always requires the introduction of irrigation and fertilisers into regions which had not formerly made use of them. Although spectacular initial gains may be made when these new varieties are introduced, the long-term viability of the system must give some cause for concern.

3.2.7 Conclusion

The expansion of irrigated agriculture has been proposed by some as a possible solution to growing food shortages in many parts of the world. Currently about 15 per cent of the world's agricultural land is irrigated, with almost three-quarters of this concentrated in south and south-east Asia, where two-thirds of the world's population already live. The irrigation systems of the humid areas of south-east Asia, which are almost exclusively dependent on rice and vegetable production, are amongst the most stable agricultural systems in the world. They are capable of supporting up to 2,000 persons per square km at a subsistence level. The stability of these systems is related to the fact that essential nutrients for plant growth are provided not from the underlying soil, but rather from the inflowing irrigation waters. Each annual cycle, therefore, sees a replacement of mineral nutrients utilised for crop growth in the preceding season. A feature of these systems is the high energy inputs which are required for seed bed preparation, planting and harvesting. Traditionally this energy has been supplied from human and animal sources, but over the last decade or so the use of fossil-fuel subsidies has increased through the use of fertilisers, small tractors and new rice varieties. Inevitably this has tied these once largely independent systems more closely into the global industrial economy.

There seems little doubt that an extension of irrigated agriculture in semi-arid regions could greatly increase yields. It does have to be recognised, though, that this is only likely to be achieved by a massive increase in the use of fossil-fuel subsidies. To double the yield per unit area Odum (1971) has claimed that a tenfold increase in inputs, such as machinery, fertilisers and pesticides, is required, based on evidence collected from both industrialised and developing countries. Whether there will be enough energy available at reasonable prices to permit both the intensification and spread of irrigated agriculture in drylands is, of course, another issue.

A particularly disturbing feature about many irrigation projects associated with large dams and complex canal networks, which have been established this century, has been the fact that the productivity of these schemes has rarely lived up to the planned values (Goldsmith and Hilyard, 1984). There have been many reasons for this, which makes it difficult to generalise, but a number of factors do appear to be of special significance. In many cases improper use of the newly available waters has often led to the related problems of salinisation of the soil and waterlogging. Perhaps somewhat surprisingly, most salinity problems are caused by excessive rather than insufficient water use. With inadequate drainage this leads to the build-up of high water tables and the evaporation of this water from the ground surface as the result of capillary action. Inevitably, this produces saline soils which are extremely difficult to reclaim (Thomas and Jakeman, 1985). In more severe cases of overwatering, the water table can rise to the surface to produce pools of standing water. When this water evaporates all the dissolved solids are deposited in the surface layers of the soil.

Salinisation and waterlogging are basically management and design problems which can be avoided by careful initial planning and adequate supervision once a project begins operation. All too often, however, in the developing countries there is a shortage of adequately trained farm managers and agricultural extension workers who can advise the individual farmers actually operating the irrigation systems. In many situations, therefore, it is manpower deficiencies, rather than technical problems, which can lead to the failure of a project.

3.3 MINERAL EXPLOITATION

3.3.1 Introduction

Extractive industries in drylands are examples of the intensive and destructive use of resources, usually carried out by large companies which possess the necessary political power to overcome the objections of indigenous groups. Extractive activity, by its very nature, is short-lived, with operations ceasing permanently when the ore being pursued is exhausted, or temporarily when economic conditions dictate that it cannot be extracted economically.

The basic activity consists of removing a resource from the ground surface or from beneath it, but there are different ways in which this can be accomplished. The simplest operation is extraction from the ground surface, which is known as open-cast or strip mining. This involves the use of earth-moving machinery to strip away superficial material so that the desired product can be reached. Coal, phosphates and iron ore are all obtained using these methods. These operations often cover large areas and usually leave behind significant holes in the ground and large spoil heaps.

Deep mining operations involve tunnelling into the ground, either vertically or at an angle into a cliff face. It requires sophisticated management and operational procedures as complex safety standards have to be maintained. The material being won in such conditions can vary from relatively low value but high volume products such as coal, to extremely valuable minerals like gold and diamonds. For both, the economic principle has to remain the same — namely, that the value of the product has to be greater than the cost of obtaining it. Large spoil heaps near the mineshaft are a characteristic feature of the method.

The third extraction technique involves drilling into the ground, but the product can be extracted without people going underground. This method is used primarily for the extraction of fluids, of which the most important are oil and natural gas. It should not be forgotten that in many parts of the arid zone, groundwater is also being mined. Compared with the other methods this operation is often associated with numerous extraction points scattered over large areas. The amount of spoil produced is very low.

Despite differences in approach the requirements of the mining industry do show remarkable similarities. They are all highly capital-intensive and employ machinery which has to be brought huge distances to the site. This machinery has to operate under extremely rigorous conditions. The technology levels of the operation are not usually high, though the machinery is both expensive and complex. All operations require a sizeable labour force for undertaking low-skilled activities such as truck driving, underground mining, pipeline and well construction, as well as a technical and management elite.

Around all mining sites the flora and fauna usually become impoverished. Local vegetation, if available, is exploited for timber for construction, as well as for cooking and heating purposes. The fauna is hunted for both food and for recreation. The impact of this activity follows a series of concentric zones around the mining camp with the intensity of effect dropping off away from the centre. The overall result can be so pronounced that even years after all mining and settlement has ceased in the region, the fauna and flora will not have recovered to their original state.

The remote nature of many sites means that they have to be self-sufficient with regard to a wide range of services and living accommodation. If the operation is large this can give rise to the company town where the mining organisation provides all the infrastructure necessary for the life of its work-force. If the town is very large it will also attract other service activities to it and a true urban system may come into being. Some of the best examples of company towns are found in South Africa where the black manual labour force is almost kept imprisoned in compounds, while the white technical and management personnel enjoy pleasant single-family dwellings nearby.

Certainly these settlements are centres of high economic activity in terms of income generation when compared with adjacent land use types. This will be beneficial for the labour force and to local service industries. However, most of the profits from the operation may be removed from the region where they are generated to the capital city of the country concerned, or even more likely nowadays to a foreign country where the headquarters of the multinational organisation undertaking the mining operation is located. Benefits will accrue to the country where mining is carried out in terms of taxes and mineral dues paid to the government and also in terms of infrastructure improvements to road and rail links.

For any mining activity to be a success the product has to be transported to the market. Many mineral products today enjoy an international market and movements of many thousands of kilometres are often involved. The degree to which this is a problem depends on the nature of the material itself. High value products such as diamonds can be moved relatively easily by a single courier. At the other extreme, bulky products such as iron ore, phosphates and oil require an efficient bulk transport system which may need to move in excess of a million tonnes each year.

3.3.2 Mining activity and environmental problems

In its simplest form mineral exploitation provides bulk materials for building purposes. These materials are low in cost, but extremely bulky and heavy. This necessitates that the nearest source to the demand is exploited. Around most of the larger towns on the Iranian plateau, alluvial silt, clay and loess were used for mud-brick production, or as economic conditions improved, for kiln-baked bricks. Elsewhere in the Middle East, quarries were established where outcrops of good-quality building stone were found near settlements. If such heavy materials had to be transported by land, they could only be moved a few kilometres economically, though rich landowners would often subsidise such movements from other business interests.

Today, in inland locations, either road or rail infrastructures often have to be constructed. These can cause severe environmental degradation, particularly as flood waters are often concentrated by culverts and elsewhere, waters are ponded back by embankments. The result can be serious erosion. When mineral workings are situated near the coast, the bulk transport of material can be most economically achieved by sea transport. To achieve this it is usually necessary to construct an offshore loading facility as natural deepwater harbours are rare.

The environmental effects of mineral exploitation are greatly dependent on the type of activity involved. In general, at the mining site itself the original environment is almost always totally destroyed as a result of mining activity. Around the mining site there are large spoil mounds, which are not aesthetically pleasing, but which may not have any other deleterious effect. The impact then drops away rapidly, except along the major communication lines. At about a kilometre or so beyond the boundary of the site, little tangible evidence of the operation is visible in the natural environment.

The effects of pollution can, however, reveal themselves in many ways. Depending on their nature, the results can be felt over large distances. A critical aspect is water pollution, which can affect a river system for many kilometres downstream from the point where mineral exploitation is taking place. It can be one or both of two types. The first is the generation of particulate material, both large- and fine-grained size, as a result of the operation itself. Heavy machinery quickly removes the vegetation cover, breaks down the soil structure and — if deep soil profiles exist — can lead to severe erosion. It is made considerably worse if the extraction activity is taking place in deeply weathered rocks or unconsolidated sediments.

Sediments of varying size range can often be generated as a result of the type of mining activity or through the processing of ore which takes place on the site. In the nineteenth century, for example, hydraulic jetting techniques for mining alluvial gold in the foothills of the Sierra Nevada in California led to the generation of huge waves of gravel which have moved slowly downstream along the major rivers over many decades. These have destroyed the local riverine ecosystem and often caused flooding problems. Elsewhere the washing and concentration of ores using grinding and water washing techniques can produce large quantities of fine-grained sediment. Silt- and clay-sized sediments move rapidly down a stream system, but inevitably in arid environments some material will be deposited along the stream bed. If this occurs in a perennial stream it can affect the flora and fauna.

Both strip- and deep-mining operations bring spoil material to the surface, including minerals that rapidly break down in the intense weathering environment in which they find themselves. One of the best examples here is iron pyrites which is often associated with coal deposits as well as with many ore minerals. It breaks down to produce sulphuric acid, which — when present in large quantities — can reduce the pH of streams to 3 or less. All life is killed under such conditions. What is even more worrying is that these acid conditions can release other elements from rocks and soils, such as aluminium, which are toxic to many forms of life. Once released these elements may pollute both stream and groundwater supplies. Drainage waters from deep mines with very high salt contents can be pumped into water courses, and so limit the uses to which river water can be put.

In many cases, sizeable quantities of the material being sought are left within spoil heaps on the ground surface. When these materials include ores of lead, zinc, cadmium, copper and other heavy metals, the danger of pollution to ground and surface waters resulting from spoil-heap drainage can be considerable. In most cases the effects are very localised, though the impact can be intense and cause severe disruption of the ecosystem.

Of the different types of mineral exploitation, the oil and gas industry has the smallest surface effect in terms of the mining activity itself. Wells are only a few centimetres in diameter and so the volumes of spoil produced are very small. The greatest impact with this type of activity is the infrastructure associated with oil development. Large housing complexes are often constructed which require fresh water provision and sewage disposal facilities. Even more important are the oil transmission pipeline systems and the storage and handling facilities which cover large areas.

Pollution from this industry seems to be most closely associated with spillages occurring at trans-shipment points where oil from pipelines is loaded into tankers. This, together with the flushing of the tanks of oil tankers, has led to the growing pollution of shorelines in the Persian Gulf, along which a large proportion of the oil of the world travels. The risks of pollution have also increased greatly over the last two decades with the development of off-shore oilfields. In the Persian Gulf one or two major leaks have already occurred, and it seems inevitable that there will be others in the future. Over the last eight years the Gulf area has been subjected to a new pollution hazard stemming from the war between Iraq and Iran. At various times both of these countries have regarded oil tankers as legitimate targets for attack. Although the more volatile constituents quickly disperse into the atmosphere, the heavier portions float on the water in large black masses, eventually to be deposited on beaches. When present in large quantities these petroleum products, although largely insoluble, can produce enough toxic dissolved chemicals to cause harm to near-shore aquatic environments.

3.3.3 Patterns of mineral exploitation

The pattern of mineral exploitation exhibits certain similarities which are independent of time and space. The discovery of precious minerals, such as gold, silver and diamonds, for example, always produces a boom of activity during which time individuals and small groups from around the world are attracted by the hopes of a quick fortune (Barnea, 1983). At this stage the level of equipment needed is low, and labour and capital is provided by the individuals themselves. This stage is usually short-lived and is associated with the working of the easily available deposits.

A second stage begins, often overlapping with the first, when increasingly difficult mining conditions require levels of capital investment which can only be provided by large groups that have the necessary financial backing. This is normally a time of fierce competition between interested mining groups, and the claims of the less successful operations are progressively taken over by the more successful ones. This process may continue until a third stage is reached, when a single company or an organisation will have a virtual monopoly over production at a single site or a group of sites. The pattern does not have to follow the one outlined above exactly, as the exhaustion of the ore can lead to the abandonment of mining activity during stage one. Equally, if mining activity is scattered over a sizeable area, it is likely that a number of companies will continue with mining activity.

With the bulky and less valuable minerals it is less easy to discern a pattern. However, discovery of the deposit is usually made some time before the mining activity commences. This is because the large volumes of material to be moved necessitate large capital expenditure in equipment, which takes time to organise. Perhaps even more important is the need to develop an efficient transport infrastructure to take the mineral directly to the market, or more usually, to a coastal location whence it can subsequently be shipped by sea. Given the large scale of such an operation, it is not unusual for it to take at least ten years to develop from initial discovery to the first production of the mineral itself. Equally likely is the fact that a large multinational company will have developed overall control over the mining activity, so causing a high proportion of the profits to be repatriated to the country in which the headquarters of the organisation are located.

In the developing world most governments do not have a strategy for environmental management decision-making with regard to mineral exploitation. In general, the government concerned is so pleased at the prospect of much-needed foreign exchange that it places few constraints on any proposed development. Indeed, the government may even be willing to finance certain infrastructure improvements as an inducement for production to begin. Worries about potential environmental pollution are not normally considered at this stage and if one is dealing with a very arid environment there is not likely to be much opposition to development from other land users.

The position is more complex in the developed world. In many cases the need to generate foreign exchange will be less strong an influence on government policy, though the need for the mineral itself may be vital. At the same time the government is likely to pursue a technocentric approach to resource development, believing that if ecological problems are caused as a result of mining activity, they can be overcome by the application of scientific knowledge. With this essentially optimistic approach to environmental matters it can be seen that on the whole mining development will receive governmental support. Ecological groups, such as Friends of the Earth and the Sierra Club, are likely to have philosophical as well as practical objections to some forms of mineral development, particularly in wilderness areas. Such groups will often organise protests which may be successful in preventing the development of new mines.

3.3.4 Case study — mining in Western Australia

Gold was discovered in the desert near Coolgardie in September 1892 and a gold-rush immediately followed (Morrell, 1968). Over the next four years about 100,000 men moved along the 480 km trail from the coast to the gold-fields, though not all of them stayed there. Most of the early finds were of alluvial gold in shallow soils, which proved easy to mine but were soon worked out. In June 1893 the Kalgoorlie gold-field was discovered some 30 km to the north-east of Coolgardie. This was to prove the most productive gold-field on the continent, though the richest part of the field (where the gold is found in a quartz dolerite greenstone rather than in vein quartz) was not opened up until a little later. The very large-scale prospecting activity associated with the gold discoveries meant that by 1895 the extent of the gold-bearing land in Western Australia was reasonably well known.

By 1894 Coolgardie had developed into a prosperous entertainment and supplies centre, but all supplies still had to be brought over land, chiefly by organised camel train. Costs were high and as a result food and commodity prices on the gold-fields rose to very high levels. The town was not without its problems. Typhoid was common because of the poor water supplies and fires often destroyed sections of the town. Little local water was available and so use had to be made of all the available sources. One of these was salt water in the nearby salt lakes. This water was purified using condensers, fired by whatever vegetation could be obtained. It meant that the local area was quickly denuded of suitable species. Other water was brought by animal trains from pools and wells wherever they might be found. In the early days of the gold-rush there was only sufficient water for drinking purposes.

About two years after the first discoveries most of the easily won alluvial gold had been worked out. The rich gold-bearing lodes could only be exploited by considerable capital investment. In Australia the financial markets had been undermined by bank collapses and stagnant economic activity, and so little capital was forthcoming for investment. A quite different situation prevailed in London where large amounts of money were searching for profitable investments. The year 1894 proved to be a boom one for gold shares, with large sums invested in both South African and Australian gold-mines. Much of this was speculative money with little assessment given to the risks involved. Not surprisingly losses were high.

By 1896 a railroad link had been completed from Perth to Coolgardie and Kalgoorlie. This permitted the efficient transport of supplies and machinery and led to an intensive period of mining activity. This revealed that the Coolgardie reefs were relatively shallow, and progressive decline followed. At Kalgoorlie the reefs proved richer and soon a number of large and deep mines began to dominate mining activity.

In 1896 it has been estimated that there were probably 65,000 people, mostly men, on the gold-fields. Coolgardie at this time had a population of about 6,000. The year 1896 marked the peak year of the boom in terms of both population and the volume of investment. Subsequently, both decreased, though the actual production of gold itself continued to rise. In 1896, 250,000 ounces were produced, but this more than doubled to 600,000 ounces in the following year, and peak production of 2 million ounces was reached in 1903. From now on it was the period of the big companies rather than the individual prospector. The most important developments were now of a technical nature in terms of better ways to crush the ore and to extract the gold.

159

Of the two main sites Kalgoorlie continued as the more prosperous centre and in 1899 had a population of about 35,000. By the early twentieth century, both settlements had all the trappings of urban centres found elsewhere on the continent. They were connected to Perth by railway and had telegraph communications with the rest of Australia and the world. In both towns there were public buildings, hospitals and schools as well as water supplies, electricity and telephones. A reliable source of water reached the gold-fields in 1903, when a 76 cm diameter pipeline, 560 km in length and costing £2·7 million (sterling), connected Kalgoorlie with a water source in the coastal hills near Perth. There was an electric tram service in Kalgoorlie, when such transport systems were still unusual in Australia. The town also had wide pavements and streets and a considerable level of retailing activity.

Kalgoorlie had therefore certainly developed well beyond the usual small centre associated with mining activity. What is even more remarkable is that this development had been achieved in less than a decade. By the early 1900s, Kalgoorlie was one of the most important regional centres in Australia. Today, almost 90 years after its heyday, it still retains a population of around 20,000 people.

One of the most important aspects of the Western Australian gold rush was the positive economic impact it had on the rest of the Australian economy. In 1897 money orders alone for £900,000 were mailed from Western Australia to the east. Dividends from profitable mines were high, attaining £2 million in 1899, 1903, 1904 and 1905. By the end of 1915, £25·5 million had been paid out as dividends, of which about one-half had been generated from four famous mines — Great Boulder, Ivanhoe, Horseshoe and Great Fingall (Blainey, 1981). However, of the thousands of companies which had been established to exploit the gold, less than one hundred ever paid dividends. Financial losses overall were enormous, particularly for London investors. By the early 1900s decline had set in on the gold-fields, though the larger mines continued to operate. By 1910 many sites were abandoned and even large settlements decayed back to the bush: the boom was over. Another flourish of activity took place in the 1930s when gold prices increased, but this did not match the scale of the 1890s boom.

The historical significance of the gold-fields is that they acted as a stimulus for development both in Western Australia itself and throughout the continent. The majority of the capital which made the mining possible came from London, and a very high proportion was lost without any return. However, in passing through the system the money did help to produce a sound infrastructure and a sophisticated service sector, which survived the decline in mining activity. The profits from gold certainly revitalised the Australian economy following a period of considerable economic depression.

Today mining activities in Australia tend to be on a very large scale and are associated with the production of minerals for distant markets, such as Japan. The post-Second World War period has seen a highly systematic and scientific approach to mineral prospecting in Australia, which has led to spectacular discoveries of nickel, iron ore and uranium. The 1960s saw the opening up of large iron ore deposits at Pilbura, Mount Newman and Mount Tom Price in the northern parts of Western Australia. However, these often very rich deposits could only be utilised when rail connections from the ore body to the coast had been completed. The trains on these lines have used the latest technology available and are capable of transporting up to 20,000 tonnes of ore in a single load. The isolation of the sites necessitated that the mining camps were self-sufficient, but even today, when high salaries are paid, there is still a rapid turnover of labour. The huge size of many of these operations has meant that they are often financed with foreign capital, and as a consequence many of the financial benefits flow out of Australia.

In Australia the crown has retained ownership of all minerals, and exploitation of sites is only possible following the granting of a licence. Initially only small areas were granted to individual miners, but in the twentieth century much larger licences covering many thousands of square kilometres have been awarded for iron ore and petroleum prospecting. Over the last few years disputes have arisen about mineral rights on aboriginal land, particularly with regard to the mining of uranium. Arguments have raged as to whether it should be permitted and as to how the benefits are to be shared.

3.3.5 Conclusion

Mining is a 'boom or bust' activity, whose location is limited by the availability of a mineral or rock which is economically attractive to exploit. In a historical context mining is a short-lived activity, but it is capable of generating enormous economic wealth. Just where this wealth ends up is a function of the economic system in which the mining is taking place. All too often in the past, and even at the present day, the venture capital necessary for initiating mining often comes from outside the country where the mining is taking place. Not surprisingly, therefore, most of the profits are shipped abroad, thereby minimising the impact on the local economy.

In other cases, however, mining acts as a catalyst to the development process. Mineral discovery attracts people to an area and almost inevitably leads to major improvements in local infrastructures. There are always employment opportunities for local labour and so at least some of the capital finds its way into the local economy. In most cases today the governments of countries in which mining is taking place are demanding fees or royalties which are linked in various ways to production. As a result the country is able to benefit by gaining capital which is so valuable for public investment. A danger here, though, is that the country can become overly dependent on a particular resource. If mining activity ceases (due to either exhaustion of the mineral or low prices), then the impact on the economy can be considerable. As far as dryland countries are concerned, mining should be looked upon as a useful short-term activity which is capable of providing significant wealth. However, it is unwise to regard it as a long-term component of the national economy.

4

Extensive Use of the Dryland Environment

4.1 RAIN-FED AGRICULTURE

4.1.1 Introduction

Rain-fed agriculture predominates throughout the humid tropics. As rainfall decreases towards the semi-arid zone, the chances of crop failure increase, until below about 250 mm per annum, successful cultivation is not really possible on a long-term basis. Along this same gradient the importance of irrigation increases. Rain-fed agriculture has the great advantage over irrigation that it requires no infrastructure to be built. It is therefore a much cheaper form of crop production and is utilised by farmers wherever possible. In the semi-arid zone there are great differences in crop production between the wetter and drier margins. In the wetter parts low yields or crop failure due to drought may occur only infrequently, perhaps less than one year in every ten, while in the drier parts crops might fail three or four years in every five.

The climate of the semi-arid zone in which rain-fed agriculture takes place is characterised by marked wet and dry seasons. On the high latitude side of the sub-tropical deserts, winter rainfall predominates, while on the equator side there is usually a summer rainfall maximum. The principle of cultivation is basically the same in both areas. During the rainy season recharge of the soil moisture reservoir takes place and then this reservoir is depleted as the crop grows (Arnon, 1972). In wet years there are adequate water reserves to permit the growth and maturation of the crop. However, in dry conditions the soil moisture reservoir is not fully topped up and so all the water is consumed before the crop reaches maturity, which naturally results in crop failure. In reality there is every extreme between crop failure and bumper crops. Unfortunately, the occurrence of the different types occurs on an entirely random basis, according to antecedent weather conditions. Although farmers do have an idea at the end of the rainy season as to what growth conditions are likely to be, they are completely powerless to do anything about them.

The key to understanding rain-fed agriculture is rainfall variability, which increases as the total rainfall amount decreases. In the wetter part of the semi-arid zone — such as in parts of Turkey, Jordan, Iraq and Iran — successful rain-fed agriculture has been carried out at a number of sites, probably almost continuously for 10,000 years. This has been based on cereal cultivation, chiefly wheat and barley, with many other crops, including vines, olives, melons and vegetables, grown as well. Pastoral activity, especially with sheep and goats, has been closely associated with cultivation. The village community has been the major social unit, with certain village sites being occupied over periods of hundreds of years.

The traditional societies which developed were based entirely on solar energy. Solar energy caused vegetation growth which in turn provided food for animals and humans alike. The animals, especially oxen, donkeys and mules, provided the motive power for operations such as ploughing and the movement of goods. Crop residues were used for animal feeds, as fertilisers or as heat sources. Animal dung was also used as a fertiliser or as a fuel. Wood for construction would be obtained from nearby woodlands, usually in uplands, and these same areas provided firewood for cooking and heating. As a result the villages were largely self-sufficient in most of the necessities of life.

The main bulk of the food produced was used for subsistence in the local area, though small food surpluses would be moved by animals to nearby market centres for sale. These markets also provided goods such as metal cooking utensils, which could not be produced at the village level. Until the last hundred years or so, most of these village settlements were located in the wetter part of the semi-arid zone where crop production was more assured. However, with increasing population pressures the limits of cultivation have been pushed further and further into the drier margins of this zone. The result has been that low yields and even crop failures have become commonplace, with consequent major impacts on the stability of village life. This cultivation of what are essentially marginal lands has also produced a series of environmental problems. The ploughing of land on the desert margins has produced ideal conditions for wind erosion, and over large areas top soil has been severely depleted. Elsewhere, particularly in foothill zones, ploughing and deforestation have led to severe sheet and gully erosion and subsequent land degradation. In many countries conservation programmes are now trying to alleviate these forms of desertification.

The post-Second World War period, besides witnessing a large expansion in the cultivated area, has also seen a change-over from a solar-powered economy to one increasingly dependent on fossil-fuel subsidies. Two- and four-wheeled tractors have rapidly replaced animal power for agricultural operations like ploughing, except in the remotest areas. In transport the impact of the internal combustion engine is even more impressive, as it has virtually replaced all other modes of transport. The net result of this change-over is that these traditional societies have now become dependent on the world economy for the supply of high technology machines such as tractors, cars and buses. As a result the independence and self-reliance of the traditional village community has been severely undermined. Perhaps even more important is the fact that these countries have now developed large foreign debts in paying for these goods and are having difficulties in maintaining the interest payments.

Today, rain-fed agriculture ranges between two extremes. On the one hand there is the traditional system, which depends largely on one crop (usually wheat or barley), but which is supplemented by a wide range of other crops, including sorghum, vines, olives, fruits and vegetables. The range of crops in any area mainly depends upon local climatic conditions. Cultivation is for purposes of local subsistence consumption at the village level, though with the growth of large urban centres in the post-Second World War period, many villages situated near these centres are also producing fruit and vegetables for these markets. In the remoter regions traditional cultivation techniques may still predominate, though where suitable ground conditions prevail mechanisation is being steadily introduced.

At the other extreme there is the extensive cultivation of one crop (often wheat) for distant markets. This commercial mono-culture developed mainly in the lands of the New World where population pressures on the land have been low. In the past monoculture was practised exclusively, but today in many areas other activities are undertaken as well in an attempt to lessen the

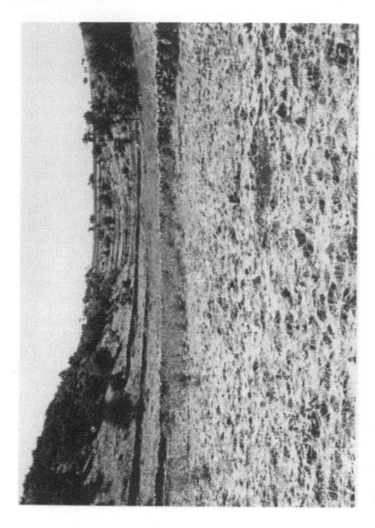

Plate 4.1 Gully plugs to reduce downstream movement of sediment in the Wadi Ziqlab in the northern highlands of Jordan.

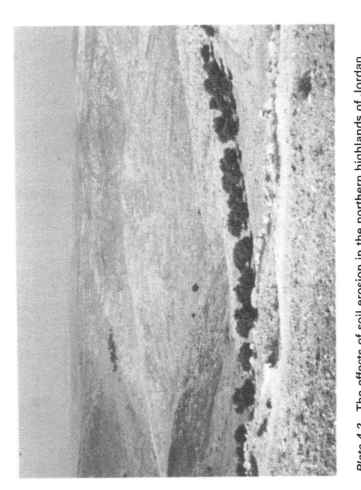

Plate 4.2 The effects of soil erosion in the northern highlands of Jordan. Almost all the soil has been removed by sheet erosion leaving bare limestone pavements. The occasional olive grove still manages to survive.

effects of crop failure. This commercial type of rain-fed agriculture often receives large subsidies from the central government, either directly in the form of cash payments or else in the form of infrastructural provision.

4.1.2 Case study — northern Jordan

The topography of the Northern Highlands of Jordan comprises a steep escarpment overlooking the Dead Sea lowlands with a dip slope falling away gently eastwards towards the Syrian desert. These highlands rise in places to over 1,000 m. To the west, slopes fall steeply to the Rift Valley, whose floor is 200 m below sea level (Figure 4.1). The highlands are deeply dissected by major streams trenching the escarpment and draining down into the Dead Sea lowlands. In the highest parts of the uplands annual precipitation is up to 600 mm. Although precipitation falls during eight months of the year, four months (December to March) normally provide more than 80 per cent of the total. Eastwards over a belt about 30 km wide and some 90 km long, precipitation falls to less than 200 mm per annum. To the west, down into the Rift Valley, an even steeper precipitation gradient occurs, with the 200 mm isohyet being reached in less than 15 km.

Rain-fed agriculture is practised wherever possible throughout the highlands and on to the dip-slope plateau. In the dissected uplands cultivation occurs on slopes of up to 20° or more. Field sizes are generally small and the soils are shallow and stony. On the plateau, undulating terrain prevails and the brown soils are much deeper. Here precipitation quickly declines eastwards towards the desert. Since the Second World War the lands around Irbid have been developed by mechanised farming for wheat production. Prior to the availability of tractors, the ploughing of these soils on an extensive scale was not a feasible proposition. In their desire to increase wheat production, the farmers have undoubtedly taken the cultivated land in this region beyond the limit for sustainable long-term crop production. As a result wind erosion of the dry steppe soils has been considerable and crop yields have fluctuated sharply from year to year.

In the highlands the traditional method of cultivation used oxen, mules or donkeys for ploughing. In general a two-year rotation was practised, with cereals followed by fallow or winter crops. Deep ploughing to open up the soil and make ridges so as to retard

Figure 4.1 The northern highlands of Jordan

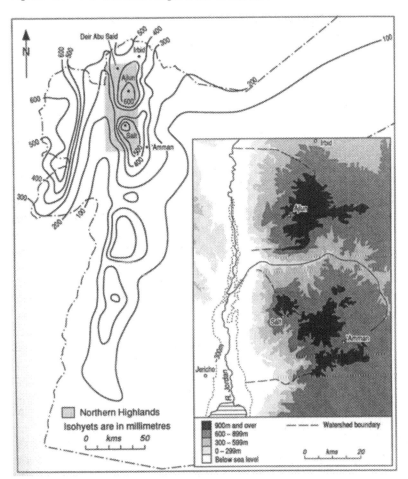

runoff and increase percolation took place before the winter rains came. To achieve this end the land was ploughed parallel with the contours. A single-prong plough was used which penetrated the soil to a depth of 20–30 cm. Sometimes the land was left fallow after deep ploughing, but if it was to be used for cereal production it would usually be followed by shallow ploughing to depths of 10–15 cm (to remove weeds and to prepare the ground for planting and for turning in the seed).

A land use survey of the Wadi Ziqlab, one of the major streams dissecting the escarpment of the highlands, was carried out in the mid-1960s. This revealed that about one-third of the total area of the watershed was used for annual crops, though at any one time a sizeable proportion would be fallow. A further 6 per cent was under orchards, and the rest was used for rough grazing. From the 1950s onwards modernisation of agriculture was proceeding quickly, and the government was constructing a network of hard-topped roads throughout the region.

In a survey of five villages in the Wadi Ziqlab between 1955 and 1965, it was found that about 80 per cent of the field crops were under either wheat or barley. Wheat was by far the most important crop, with barley only grown on thin soils and where low rainfall predominated. A wide range of other crops was recorded, including lentils and chickpeas, water melons, sesame, cucumbers, tomatoes and tobacco. Vegetables were also grown around the villages. Olives were the most important of the orchard crops, though vines were widely grown. A few fig trees were also cultivated. Over the survey period the area devoted to cereals and the main winter crop, the pulses, did not change significantly. However, there was a marked rise in the cultivation of tomatoes, which proved an extremely profitable crop for the farmers in the growing urban centres. Olives also increased in area by 17 per cent.

A record of precipitation and crop yields over a ten-year period is shown for one of the villages of the Wadi Ziqlab, Dair Abu Said (Figure 4.2). This reveals a relatively close correlation between total precipitation and yields for cereals, but much less corres-pondence for other crops. With these it is often the timing of the precipitation which is more crucial than the actual amount.

In northern Jordan one of the greatest problems has been an increase in population numbers since the 1930s, which has put tremendous pressure on the available land and soil resources. This semi-arid region already has what might be described as a fragile environment and so the impacts from excessive environmental pressures are considerable. In Jordan the main problems have been deforestation of the uplands for constructional timber and for fire-wood, the widespread cultivation of cereals leaving a vegetation-free surface during the rainy season, and the overgrazing of pasture by sheep and goats. The impact of these activities has been greatest around individual villages and falls off with distance, except along roadways. However, in the Northern Highlands the villages are so close that many areas receive pressures from more than one direction.

Figure 4.2 Relationships between precipitation and crop yields at Dair Abu Said in the northern highlands of Jordan

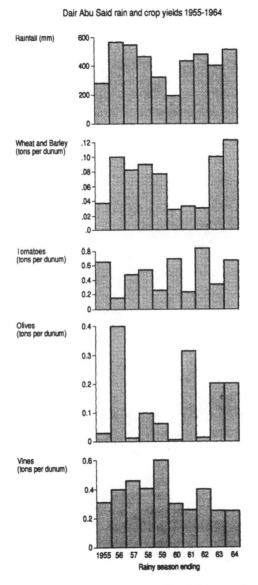

Dair Abu Said rain and crop yields 1955-1964

The net effect of these activities has been soil erosion on a large scale. This erosion (mainly sheet erosion) has been caused by water runoff during the rainy season. Evidence suggests that the

171

Figure 4.3 Slope gradients in the Wadi Ziqlab, northern highlands of Jordan. (Top) less than 15 per cent; (middle) 15 to 40 per cent; (bottom) over 40 per cent

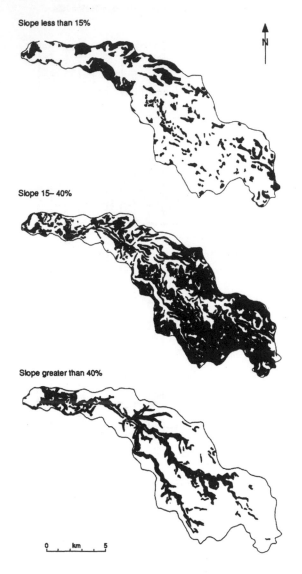

Slope less than 15%

Slope 15– 40%

Slope greater than 40%

0   km   5

Source: Redrawn from Beaumont P. & Atkinson K., 1969, 'Soil erosion and conservation in Northern Jordan', *Journal of Soil and Water Conservation*, v. 24, pp. 144–7. Reproduced by permission of the Soil Conservation Society of America.

main erosion takes place when rainfall intensities exceed 15 mm per 24 hours. Bearing in mind that on average there are only 50 rain days per year, and that rainfall intensities of more than 15 mm per day are restricted to about 20 per cent of the rain days, this means that major soil erosion only takes place on about ten days each year.

Another point of great significance is that the streams trenching the Northern Highlands have enormous erosion potential falling as they do to a base level well below sea level. This has meant that they have deeply incised their channels with the result that little of the land within their watersheds is gently sloping. Indeed, in many river basins, such as that of the Wadi Ziqlab, most of the area possesses slopes of between 15 and 40 per cent (Figure 4.3) (Beaumont and Atkinson, 1969). On these slopes, many of which are cultivated, soil erosion rates are very high indeed. A land capability map for the Wadi Ziqlab, constructed using the methods employed by the US Department of Agriculture, reveals that almost all the land has severe limitations which should preclude its use for cultivation (Figure 4.4). Yet despite this there are nine villages within the catchment which are dependent for their livelihood on dryland farming.

Back in the 1960s the government of Jordan realised the erosional problems which it was facing in its dryland farming areas, and in conjunction with the United Nations, Jordan instituted conservation measures to overcome them. These included gradoni terracing and tree planting on the steeper slopes; the construction of gully plugs on valley floors to promote sediment deposition; contour walling to reduce the speed of water runoff; and the strip planting of crops to minimise areas of bare ground. Although the techniques employed have often been successful in lessening the rate of erosion, they have not prevented it completely. The other difficulty is that the maintenance of these conservation measures is expensive in both time and labour, so that the local farmers have found problems with the upkeep of the original structures put in by the government.

The increasing population pressures, caused by natural increase as well as the influx of Palestinian refugees following the Israeli invasion and occupation of the West Bank in 1967, have meant that over the last 20 years environmental pressures have become even greater. Nowhere is this better seen than in the steppe areas of the plateau to the east of the dissected highlands. Here, the increasing availability of machinery has led to the ploughing of the

173

Figure 4.4 Land capability classes in the Wadi Ziqlab in the northern highlands of Jordan

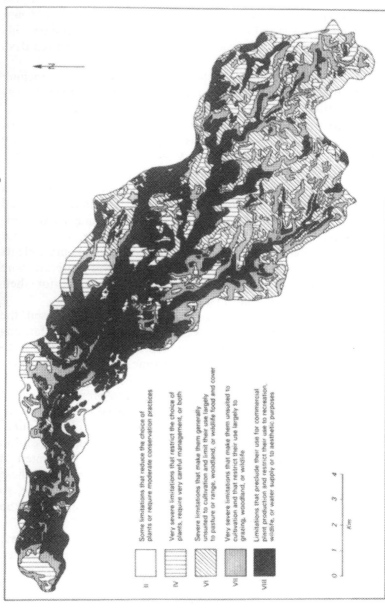

Some limitations that reduce the choice of plants or require moderate conservation practices

Very severe limitations that restrict the choice of plants, require very careful management, or both

Severe limitations that make them generally unsuited to cultivation and limit their use largely to pasture or range, woodland, or wildlife food and cover

Very severe limitations that make them unsuited to cultivation and that restrict their use largely to grazing, woodland, or wildlife

Limitations that preclude their use for commercial plant production and restrict their use to recreation, wildlife, or water supply or to aesthetic purposes

III

IV

VII

VIII

VIII

0   1   2   3   4
          Km

Source: Redrawn from Beaumont P. & Atkinson K., 1969, 'Soil erosion and conservation in Northern Jordan', *Journal of Soil and Water Conservation*, v. 24, pp. 144–7. Reproduced by permission of the Soil Conservation Society of America.

deep brown soils for wheat cultivation. This is an environment which is already on the very margin of dryland farming, and now soil erosion by the wind has become a serious problem.

4.1.3 Case study — Australian wheat production

In Australia, the semi-arid zone where wheat can be grown without irrigation is found in a north-west to south-east belt in Western Australia and a west to east zone from South Australia into New South Wales (Figure 4.5). In both these areas precipitation is at least 250 mm per annum. In some parts, owing to the high evapotranspiration rates, rainfall has to be well over 300 mm per annum to make successful wheat cultivation possible. Significant wheat production began in Australia around the mid part of the nineteenth century. An important factor which stimulated this production was the repeal of the Corn Laws in Britain in 1846, which permitted overseas grain to be used to supply the British market. This occurred at a time of rapidly growing population in Britain and meant that the demand for cereals also increased markedly. By the 1870s telegraph communication between London and Australia had been established, while the development of steam power for shipping, together with the opening of the Suez Canal in 1869, meant that an efficient bulk transport service could be maintained between Australia and Western Europe.

Thus, from the mid part of the nineteenth century Australian farmers had the incentive to increase wheat production for the British market and also the means of transporting it cheaply. This increase in production was achieved by an expansion of the cultivated area into the semi-arid lands of the Australian interior. The earliest development of the red-brown semi-arid soils took place near Adelaide, where these soils reach the coastal zone. In the 1840s South Australia was exporting wheat to the other Australian states, and thereafter wheat cultivation spread rapidly during the rest of the nineteenth century. By 1900 the state of Victoria had overtaken South Australia as the state with the largest wheat acreage, only to be surpassed a decade later by New South Wales. The development of the eastern part of the interior wheat belt was dependent on an efficient and cheap transport infrastructure. This was provided by the newly evolving railway networks which were constructed between the 1860s and 1930s, largely from public funds. In the southern and western states the

Figure 4.5 Expansion of the dry-farming grain belt of southern Australia

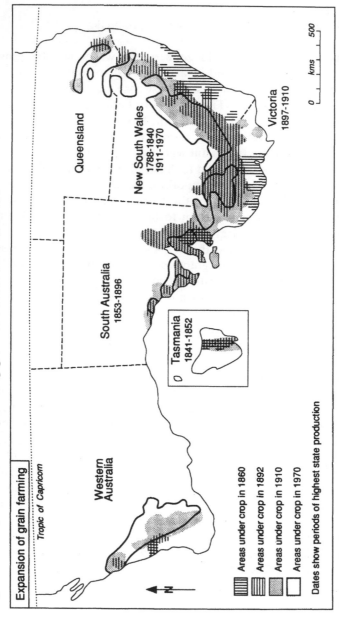

Source: Redrawn from Heathcote R.L., 1975, *Australia*, Longman, London. Reproduced by permission of the Longman Group.

systems were centred on the state capitals and had specially constructed freight rates to permit even the remotest farmer to send wheat via the state system. It did, however, mean that the shortest route was not necessarily the cheapest. In Western Australia the gold mining boom of the 1890s encouraged an increase in the wheat acreage, although the major expansion in this state did not occur until the 1950s (Heathcote, 1975).

In the initial period of wheat-farming expansion in the latter part of the nineteenth century, the land first utilised was grassland, which could be ploughed with little or no preparation. However, by the 1880s the wheat land frontier had moved into the semi-arid scrubland, where dense mallee and acacia shrubs were found. Thereafter, the pace of expansion slowed down as this land proved extremely difficult to clear, both in terms of labour expended and wear and tear on machinery. In the post-First World War period, the advance of wheat farming into the semi-arid zone was promoted by government policies (such as the Soldier Settlement Schemes), as well as technological advances in breeds of wheat and the increasing use of artificial fertilisers. The policy of Imperial Preference also provided a protected market for Australian agriculture, which lasted until Britain's entry into the Common Market in the early 1970s. It was at this time in south-east Australia that the spread of wheat farming reached the inland limit of cultivation in terms of available moisture and the length of the growing season. A much later expansion of wheat lands occurred in Western Australia in the 1950s and 1960s, with the opening up of sandy soils through the use of fertilisers and the correction of trace-mineral deficiencies.

The 1930s saw the first major check on the expansion of wheat production in Australia, as a result of drought and world economic depression. By the late 1920s expansion of wheat cultivation had occurred on to lands where dry farming was not really viable, and as a result drought had a devastating effect on these lands. Superimposed on this was a world depression which led to a reduction in wheat prices and put even greater pressure on the wheat farmers. The result was a reduction in the wheat acreage from a peak of 7·28 million ha as the government provided incentives for the area to be reduced. The least successful farmers went out of business and their land was incorporated into other farms to produce larger and more economic units through government subsidies, via the Marginal Lands Scheme. At the same time the government encouraged farmers to diversify their output by a mixture of cereal and livestock farming in an effort to lessen the impact which environmental and economic factors had on a single-crop economy.

From the 1930s, therefore, cereal farms increased in size, but at the same time their farming activities became more varied. Much greater emphasis was placed on sheep rearing to lessen the dependence on wheat. This process of amalgamation of farm units continued, and by the late 1960s the average wheat and sheep farm was about 1,000 ha in size. Large farms of over 2,000 ha were found in Western Australia, while the smallest (under 500 ha) were located in South Australia where extensive wheat farming first began.

Wheat production in Australia is an example of extensive rain-fed agriculture which has always been geared to a far-away market. As such it has always been commercially orientated. In its earliest days (during the 1840s) it was the urban areas of Sydney and Melbourne which provided the demand, but during the latter part of the nineteenth and early twentieth century it was the British market which stimulated production the most. In the post-Second World War period new demand areas in south-east Asia, and particularly in Japan, have provided the continued spur for farming activity at a time when the British market was beginning to be supplied from other sources.

This commercial wheat production has always been dependent upon an efficient transport infrastructure to move what is essentially a bulky and low-value product. Indeed, development of many of the interior areas far from the coast had to wait until the construction of railway networks in the latter part of the nineteenth century. Much of this network was heavily subsidised by the government as a conscious policy decision to open up the interior and the freight rates were drawn up in such a manner as to make it economical for even the remotest farmer to ship grain to a port. These farmers were not, therefore, competing with producers from elsewhere in a true free-market economy: instead, wheat farming was almost regarded as a government tool of land development, with a view to encouraging the small farmer and so providing a denser pattern of rural settlement in the semi-arid zone.

4.2 PASTORAL SYSTEMS

4.2.1 Introduction

The aim of pastoral systems is to use animals to harvest limited amounts of vegetation scattered over large distances which cannot easily be gathered by any other method. In energy terms it is very inefficient as only a very small proportion of incoming solar radiation is converted into usable material, yet without the method no benefit at all would accrue. The low energy efficiency does, however, mean that only small human populations can be supported by this method of husbandry, even though very large areas of land may be involved. Meat, hides, wool, cheese and milk are the main products of the system. There are really two types of pastoral farming which for simplicity can be termed traditional and commercial, although each of these can in turn be subdivided. Some overlap also occurs between the two.

4.2.2 Traditional pastoralism

In drylands traditional pastoral activity can either be part of a mixed farming system or be carried out separately as nomadism. In the wetter semi-arid regions, pastoralism is used to supplement dryland farming by harvesting vegetation from difficult upland terrains or from areas distant from the villages. It is important to realise that in many parts of the drylands of the Middle East and the Maghreb, diverse environments are found within relatively short distances of one another. In some cases the animals might be kept in or around the villages all the year round; in others, if high mountains are close by, the flocks are sent to graze upland pastures during the summer months. The relative importance of pastoralism in these areas depends largely upon climate conditions. The drier it becomes, the greater role pastoralism plays in the local economy. The major animals used are sheep and goats.

At the other extreme there is true nomadism where the people depend largely on their animals for their livelihood. The flock represents the major capital asset of the people and so there is a tendency to keep as many animals as possible. The main food source is milk, cheese and blood, though old and infirm beasts and excess young males are slaughtered for meat. Traditional nomadic

179

societies were best developed on the margins of drylands through-out Africa, the Middle East and Central Asia. By their very nature such peoples have been difficult to enumerate and estimates of their numbers have always been imprecise. Figures quoted by Heathcote (1983) suggest that there may have been as many as 20 million nomads in the 1950s and 1960s. Of these perhaps 11 million were located in north Africa and the Middle East. Our knowledge of the size of the nomadic population groups prior to the twentieth century is virtually non-existent, making it difficult to assess the long-term impact which these people and their animals have had on the environment. However, unlike the Middle East — where the numbers of nomads have declined drastically over the past 50 years — the situation in Africa is very different. In the Sahel and parts of the Sudan, both population and animal numbers seem to have increased significantly in the period since the Second World War, causing severe overgrazing and in extreme cases, desertification.

With nomadism the various groups follow a regular pattern of movement from one grazing ground to another at different times of year, so that they can always be where biological productivity is at its maximum. Nomadic groups were never fully self-sufficient and depended on contact with sedentary agricultural groups for cereals (such as wheat and rice) to supplement their diets of milk, cheese, blood and meat. These cereals and other requirements were often obtained through trade of skins, wool, cheese and carpets. However, nomads have often taken what they needed through raiding parties on villages and in some cases have controlled village communities by force of arms. In many parts of the world nomadic tribes have exacted tribute from any caravan trade which has crossed their territory. This was certainly the case with the Tuaregs in the western Sahara. When tribute was paid, safe passage of the caravan through the tribal grazing grounds was given. In more isolated regions, such as in the highlands of the central Sahara, it was not unusual for nomadic peoples to plant cereals along the floors of the major wadis draining the highlands. The soils would be watered by any flows coming down the wadi and in wet years would provide adequate moisture for crop growth. When these crops failed, the nomads would be forced to obtain their needs by trade or raids.

In the Arabian Peninsula the Bedouin and their camels used to undertake a grazing strategy which covered large distances in semi-arid and arid regions with relatively small altitudinal variations. In contrast, in Khuzestan and Iran the Bakhtiari and other tribes moved from winter quarters on the lowlands of Khuzestan close to sea level to summer pastures in the Zagros Mountains at heights in excess of 3,000 m. Johnson (1969) referred to such movements as horizontal and vertical respectively. However, it does tend to suggest a conformity to type which in the majority of cases does not occur.

When people and animals are moved into mountain areas in large numbers, considerable congestion can occur along certain key valley routes and over strategic passes. In the Zagros Mountains of Iran so many groups were present that a complex sequence of movement evolved. Different tribes with different animals and even different types of the same animal meant that the various groups could follow each other along certain key routes over periods of many weeks, without all the vegetation being depleted. This was because the different animals preferred to graze differing vegetation types and the newer vegetation of later flowering plants kept becoming available as the migration season progressed. The complexity of these movements reveals a level of planning and co-ordination between widely scattered groups which was quite remarkable in a society which lacked any form of modern communication system. Such movements prevailed in the Zagros Mountains up to the 1920s, but with the coming to power in Iran of Reza Shah they began to be broken up as he pursued a policy of nomad sedentarisation in an attempt to destroy their political power. Though this sedentarisation was initially carried out for political reasons, in more recent times it has been the attractions of urban life and high wage-levels which have led to a further breakdown of nomadic activity (Birks, 1981). Young people in particular are no longer willing to endure the hardships of a life which requires frequent movements and few personal possessions.

The decline in nomadic activity in the Middle East since the 1920s has meant that upland and more remote pastures are no longer so fully utilised for animal production as they were previously. In some parts of the semi-arid zone the improvement in communications, especially the provision of hard-topped roads, has allowed the movement of animals by trucks into areas which were formerly only accessible to herds following long journeys on foot. This has permitted certain of the isolated pastures to be used, but there are still many areas which have now to be left ungrazed.

The wealth of the nomad was judged by the number of animals

possessed and so in nearly all nomadic societies the objective was to increase animal numbers whenever possible. Many tribes also believed that more animals provided extra insurance against deaths in times of drought. Given the movement of people and animals from place to place, relatively little attention seems to have been paid to concepts of carrying capacity for various grazing grounds. A possible reason for this is that the considerable climatic variability experienced by these areas on the desert margin meant that it was never really possible to talk about a 'normal' year. Because of this the nomadic group would always want to make maximum use of pastures available in a wet year, and to minimise the potential loss of animals by starvation during dry years. This could be achieved by tribal knowledge of alternative grazing resources.

4.2.3 Commercial pastoralism

The modern form of pastoralism is ranching. This developed during the nineteenth century in the arid and semi-arid lands of North and South America, South Africa, and Australia. In its earliest days it was a form of pioneering settlement, but it soon developed into a major economic enterprise. Unlike traditional pastoralism ranching depended upon the production of goods, the most important of which was meat, for distant markets. It was centred on a fixed base, the ranch, which acted as the hub of all productive activity. Since the late nineteenth century the grazing areas have become limited in extent and bounded by fences. In the early days of ranching, in any one region it was the better watered grasslands which were first occupied by European settlers. However, with the growing success of the industry, drier range-lands were also taken over. During the twentieth century the expansion of dryland farming into the drier parts of the semi-arid zone has squeezed pastoral activities into regions in which little economic activity was previously possible. In terms of resource use there is little doubt that ranching was an exploitative form of land use. Over the years grazing pressures have reduced the more palatable grass species and these have been replaced by woody species ill-suited for grazing. In a number of areas it would seem that the current grazing capacities of rangelands are only about one half of what they were a century ago during the heyday of ranching. Despite this, actual animal numbers have often increased substantially since the nineteenth century, as much larger and often drier regions are now being used for animal production.

The pattern of development of commercial pastoralism has followed similar lines in different areas. Most of these developments took place from 1840 onwards as European peoples poured into North America, Australia, and South Africa as a result of growing population pressures. Thus, for example, the USA alone received 27·1 million immigrants between 1840 and 1910. With this level of immigration, the spread into the semi-arid lands took place rapidly so that by the early years of the twentieth century all suitable lands had been colonised. These lands were often areas of considerable biological productivity, as witnessed by the huge herds of bison on the Great Plains of the USA and of deer and antelope on the veldt of South Africa. In a remarkably short time these animals were displaced to make way for introduced animals such as cattle and sheep. By far the most spectacular example is provided by the bison herds which were reduced from probably in excess of 11 million head to only a few dozen between 1860 and 1885. A similar fate overtook many of the indigenous peoples, who were killed, subjugated or else shipped to other areas. The Red Indians of North America are a good example, though similar clashes took place between the Boers and the Bantu in South Africa, and the white Australians and the aborigines in Australia. Once subjugated these peoples often provided cheap manual labour for the newly emerging pastoral activities.

In the early days commercial pastoralism was relatively unsophisticated. Animals were grazed on open ranges and required a large labour force to keep them together. Water was provided for the animals from natural waterholes and by the damming of small creeks to catch flood flows. Marketing of the animals was always a problem because of the remoteness of the grazing lands. It usually involved animal drives of many hundreds of kilometres to the nearest railhead or port. Inevitably these drives resulted in the loss of many animals and a deterioration in their condition, which depressed the prices obtained for them.

From about 1860 onwards a number of changes occurred which greatly improved conditions for commercial pastoralism. It is difficult to assess them in terms of individual importance, but together they revolutionised pastoral techniques and methods. Some of the most important were transport-related. Although railways had been invented early in the nineteenth century, it was only from 1860 onwards that they began to reach into the semi-arid regions, rendering the cattle drives outdated. By the early years of the twentieth century all the major pastoral regions of the world possessed adequate rail communications to markets or to ports. At sea this same period saw the development of steam ships, which provided a reliable and cheap method of bulk transport. The value of this was made even more significant with the introduction of refrigeration in the 1870s, for it meant that animals no longer had to be kept alive to reach distant markets in good condition. Finally, in the same decade there was the extension of the telegraph network to all parts of the world, which enabled the producing and market areas to keep in contact with each other and so be better able to plan deliveries to satisfy market conditions.

In the semi-arid lands themselves three inventions were to revolutionise conditions. The manufacture and distribution of the Colt revolver and other personal firearms provided the pastoralists with a fire power that allowed them to overcome hostile natives who often outnumbered them. Barbed wire meant that the ranges could be fenced and the animals confined to chosen grazing areas, and this greatly reduced the demand for labour. Finally, the invention of the wind pump in the 1880s meant that water could be provided for the animals from aquifers which underlay the grazing lands. This access to permanent water supplies permitted animal numbers to grow rapidly as new grazing lands were able to be opened up. In future it would be food and not water which was the limiting factor in production.

In all countries the low productivity per unit area meant that the land had to be available cheaply if profitable production was to take place. In different countries land occupation tended to follow similar lines. Initially the land would be settled by people with no title to it. This would usually be followed by some form of government recognition of this occupation and the regulation of it through a system of low-cost loans or licences. In the USA the Federal lands of the West were occupied by farmers who used the Homestead Act of 1862 to gain control of watering points. This Act permitted the acquisition of 64·75 ha (160 acres) of land. Federal land around this was then used for grazing purposes without cost. Such a situation continued until 1934, when the Taylor Grazing Act introduced grazing fees and the control of stock numbers in certain areas. A similar pattern occurred in Australia with initial illegal grazing being followed by licences from 1835 and pastoral leases from 1847 (Heathcote, 1975).

The heyday of commercial pastoralism was probably reached

between 1880 and 1920, depending on which region of the world is being studied. Since then these areas have experienced a relative decline in profitability. This has been partly because long-continued grazing has changed the vegetation cover in many regions. Grasses and other more palatable species have been reduced in importance and replaced by woody species. This has meant that the ecosystem is not as efficient as it used to be at providing food in times of stress such as drought. The result is that animal numbers now reveal much greater fluctuations under these conditions as the resilience of the system is reduced. For the farmer, therefore, risks are much greater and sustained profitability reduced.

Equally important has been the pressure on the better-watered lands from arable farmers. This has gone on for a long time, starting from the 1870s on the High Plains of the USA and in South Africa. However, it has increased in intensity in the twentieth century, often as a result of government pressure. The 'Closer Settlement' and 'Soldier Settlement' policies in Australia, for example, have meant that better-watered pastoral lands have been acquired and reallocated for arable farming. The net result is that pastoral activity has been driven to ever more marginal lands where the vagaries of climate have made sustained animal production an extremely difficult task.

Ranching in the 1980s still shows many similarities to the systems developed in the early years of the twentieth century. The animals remain relatively unchanged but the apparatus of husbandry has greatly altered through technological evolution. Light aircraft and helicopters are now used for observations on the range and even in some cases for round-ups. Four-wheel drive vehicles and motor cycles have replaced horses as means of communication on the ground, and all personnel are able to communicate with the ranch headquarters by radio. Animal movements are carried out by large trucks, often pulling many trailers.

The development of feed-lots has changed part of the production process in some localities. These are areas where tens of thousands of cattle are kept in relatively small non-grazing situations. In effect they are intensive meat production units. They have proved remarkably successful and in the USA now form an important part of the meat production chain. The normal procedure is to move animals from the semi-arid and arid ranges to the feed-lots, where they can be fattened prior to slaughter.

An interesting change in the location of the markets of

185

commercial pastoralism has also taken place. Since 1945 new markets have developed in both south-east Asia and the Middle East and these are now taking a sizeable proportion of the meat output of Australia. The Middle East market is especially interesting. As it is an Islamic region, the animals have to be slaughtered using special ritual methods. This has necessitated the shipping of large numbers of live animals from Australia on a scale which has probably not been seen since the late nineteenth century, when refrigerated transport was developed.

4.2.4 Case study — South Africa

Some of the earliest extensive pastoral farming by people of European descent occurred in South Africa. The western third of South Africa receives less than 350 mm of precipitation each year, and here pastoralism is the dominant land use activity (Figure 4.6). The first European settlers were the Dutch, but in 1806 (as a result of the Napoleonic Wars) Cape Colony was occupied by the British and the Dutch connection, which dated back to 1652, was ended (Guelke, 1976). In the early nineteenth century it is estimated that there were some 25,000 whites in the colony, almost all of Dutch origin (Adam and Giliomee, 1979; Elphick and Giliomee, 1979). British colonists only began to arrive in large numbers in 1820. In the eighteenth century the farmers around Cape Town used to move their cattle and sheep on a seasonal basis over the mountains to the dry interior. However, as time progressed the penetration into the interior became so great that the seasonal movement of animals was stopped and a true pastoral society developed in the interior. Even at this early date a small element of commercial agriculture occurred with the transport and sale of hides, skins and tallow to the settled areas of the coastal fringe (Harris and Guelke, 1977). Most farming was of a subsistence nature.

At this time the Dutch East India Company, which exercised its rule over the colony, wished to control the development of pastoral farming. To achieve this end it issued a series of regulations between 1708 and 1720 concerned with the organisation and taxation of pastoral farming. With these regulations each farmer was permitted to use a circular area of about 2,500 ha, for which a small rent was paid (Christopher, 1982). With the advent of British administration in the early nineteenth century, the major weakness of the Dutch system — namely, the lack of security of tenure

Figure 4.6 South Africa: relief and rainfall

— was abolished. It was replaced in 1814 with the concept of perpetual leases for farms with standard areas of 2,500 ha, though in 1829 the government recognised that this land unit was too small for the more arid regions. The idea of a 2,500 ha standard farm unit was adopted when parts of the Orange Free State, Transvaal and Natal were opened for settlement during the Great Trek of the 1830s and 1840s. Originally the farm unit had been a circle with a radius of half an hour's ride on horseback. This concept was now changed in favour of a square unit with dimensions of one hour's ride in each direction, which increased the size of the unit to 3,200 ha. Such standard farms continued to be granted until the 1890s (Christopher, 1976). By the 1890s, on the margins of the Kalahari desert, standard-size land grants of 12,500 ha were being made, in order to cope with the prevailing semi-arid conditions. In Cape Colony itself standard-size land units were no

187

longer granted after 1829 and a much more flexible policy was pursued. Indeed, some larger pastoral farms of up to 50,000 ha were created. However, the main aim of government policy was for rural settlement to be dominated by small family farms. Despite this objective British land investment companies and private speculators did manage, in the late nineteenth and early twentieth centuries, to establish large holdings. In some cases, though, government pressure resulted in the break up of these properties for individual family farms.

South Africa is unusual in comparison with other countries developed by Europeans in the nineteenth century, in that almost all the land in the possession of the government was transferred to private ownership. One of the most important effects has been to raise land prices to levels which are far in excess of their intrinsic agricultural worth (in terms of what can be produced from them). In some cases it has meant that people have developed areas which were not suitable for agricultural production, while the practice of dividing the land up between one's offspring has often meant that farms have become too small to be economic. The net effect of this has been to subject the land to very severe pressures in an attempt to obtain maximum productivity. Such overgrazing has led in many cases to serious land degradation, yet despite this, pastoral farming is still highly regarded as a way of life (Christopher, 1982). In order to control the breakup of farms the government introduced the Subdivision of Agricultural Land Act in 1970, which made all further division of land subject to government approval.

The basis of pastoral activity in the drier parts of South Africa has been sheep production with some cattle raising as well in the more favourable areas. The original farms were largely self-sufficient, using the animals as a means of exchange with the blacks or local merchants. However, during the nineteenth century, conditions changed rapidly as the effects of the industrial revolution in the UK led to a demand for animal products and wool in particular. This occurred at a time when transportation systems for both goods and information were also improving. The result was the transformation of farms from a semi-subsistence basis to one geared to a commercial external market. However, many of the more isolated holdings did not become fully commercial in their attitudes until well into the twentieth century. The demand for wool grew rapidly in the period after 1830 and in the next 20 years wool exports increased 200-fold (ibid.). By 1855 there were already 6 million sheep and by 1910 almost 22 million.

Numbers peaked in the early 1930s but then dropped significantly as a result of severe droughts and economic depression. In the early 1980s there were almost 32 million sheep in South Africa.

Like pastoral farmers everywhere the farmers of the western part of South Africa place very great pressures on their land:

> This is reflected in the fairly general indifferent attitude to the natural veld and the over-optimism shown in the tendency to base the carrying capacity of the veld on the exceptionally favourable years, with the result that the veld is mostly overloaded or too heavy demands made on it, thus weakening the vegetation and exposing the soil to wind and water erosion. (South Africa, 1970, p. 117.)

Almost all the small farms are overstocked in an attempt to become economic and the problem has become worse in drought by the abandonment of 'trekking' (whereby the animals were moved to other areas when grazing was still available). As a result the degradation of vegetation has been very severe and soil erosion a serious problem. To address this problem the government established a stock reduction scheme in the 1970s so that the farmers were paid to reduce their stock numbers to what was considered the true carrying capacity of the land. Some improvements have been noted, but these have been due at least in part to a sequence of wet years, and sheep numbers were stabilised between 31 and 33 million, which is well below the 38 million peak of the late 1950s and early 1960s.

The cattle industry in South Africa, unlike the sheep industry, has always tended to be geared to the internal rather than the overseas market. The biggest change which has occurred this century has been in terms of transport, with the introduction of the motor car and the lorry: prior to this cattle were the most important motive power. This transport revolution led to a decline in cattle numbers in the drier western parts of the country where cattle were largely kept for transport purposes. The most important extensive cattle grazing areas are found in the northern part of Cape Province and northern Transvaal, where rainfall is normally less than 400 mm each year. Many holdings are still uneconomic, though a policy of consolidation has improved the situation in recent years. In these areas carrying capacities are low, with one head of cattle for every 6 to 25 ha. In the north of Cape

Province the availability of water is a critical factor: if it is not available, then cattle rearing cannot take place.

4.2.5 Conclusion

Pastoral activity, whether traditional or commercial, can be regarded as an opportunistic form of land use, in which the annual variability of grazing resources is always an unknown factor. This has necessitated the development of management strategies which minimise the impact of harsh conditions, such as drought, yet at the same time are able to benefit from years when high rainfall makes vegetation plentiful.

With the traditional pastoralist, herd size is the most important variable in coping with the harsh environment. In times of plenty animal numbers are allowed to increase, while in drought the lack of both water and pasture causes animals to die. In commercial systems a twofold strategy is employed. Well-drilling techniques have meant that it is now rare for animals to die owing to lack of water, and so the availability of feed is the critical factor. In times of stress (such as during drought) the rancher can react in a number of different ways, having a much greater flexibility of response than his or her traditional counterpart. The simplest response is to do nothing and let the animals die, but this is wasteful of capital and something to be avoided. Given the fact that water is almost always available, the rancher can take an alternative approach whereby the animals are permitted to lose condition as a result of inadequate food resources being available, but sufficient food is provided to keep them alive. This is an extremely expensive operation on large ranches and is only worthwhile for short periods. A more useful approach is to reduce animal numbers by transporting them to areas outside the arid zone where feed is available. This can be done on a permanent basis when the animals are sold for slaughter, or as a temporary measure with the idea of moving the animals back again when conditions permit. Perhaps somewhat surprisingly, although attempts have been made on numerous occasions to improve the long-term carrying capacity of different ranges, the results have often been disappointing. This has been largely due to the fact that the pastoralists have often tried to ensure minimum income fluctuations from what is a very variable environment. Overgrazing and the exploitation of vegetation resources are still common even today.

It is difficult to see how this can ever change as long as commercial pastoralism continues with the profit motive as the driving force. In these very variable marginal pastoral environments it is essential to develop a management strategy whereby the available and sometimes very rich vegetation resources can be cropped and yet, when drought occurs, animal numbers need to be drastically reduced. As yet no satisfactory policy to achieve this has been evolved.

Part Two

Regional Resource Management — Case Studies

5

The Sahara and Central Australia: Pastoralism under Different Management Systems

This chapter sets out to examine two very different pastoral systems. Both achieve their livelihood by the exploitation of semi-arid rangelands on the margins of huge central desert areas in Africa and Australia respectively. However, the social and economic systems under which they operate are vastly different. In Africa a traditional subsistence system still predominates, which, until the last ten years or so, has been surprisingly insulated from the world economy. In contrast, in Australia — except for the very earliest days — the commercial production of animals has always been the driving force behind pastoralism.

5.1 THE SAHARA

5.1.1 Introduction

The Sahara is the largest expanse of desert in the world, stretching unbroken from the Atlantic Ocean to the Red Sea (Cloudsley-Thompson, 1984). In Egypt and eastern Libya, desert conditions (where precipitation totals are less than 200 mm per annum) reach the shores of the Mediterranean, but in the north-west the Atlas Mountains of Morocco and Algeria produce more humid environments. On its southern margin the desert grades slowly into steppe desert and shrub savannah from about 17° North. In the extreme south-east the Sahara terminates against the highlands of Ethiopia, where summer monsoon conditions prevail.

By nature of its very size the Sahara has always been intriguing to the modern world, but since the 1960s its southern margin has been the setting for one of the world's worst famines. A series of major droughts commenced in west Africa during 1969–73 and have occurred sporadically since then along the transitional belt from desert to savannah, known as the Sahel. With the most recent episode of famine, dating from 1983, the location has shifted eastwards to Chad, Sudan and Ethopia. This famine has now been taken up by the Western media and has gained extensive television coverage. The result has been a fund-raising bonanza by pop stars and others which is unprecedented in terms of aid to the Third World. In contrast, the earlier famines of the 1970s received relatively little media attention, despite the fact that large numbers of people also perished at that time.

5.1.2 The environment

Within the desert itself three major highland areas provide some respite from the intense aridity which surrounds them. These are the Ahagger Massif of southern Algeria (2,997 m), the Tibesti Uplands of northern Chad (3,263 m), and the Marra Mountains of western Sudan (3,085 m). Major rivers are scarce. In the east, the Nile, fed largely by monsoonal rains over Ethiopia and east Africa, flows north across some of the most arid parts of the Sahara. On the southern margins of the desert, where precipitation begins to increase, four major rivers drain the semi-desert and savannah zones — the Senegal, the Niger, the Benue and the Bahr al Ghazal. Given the semi-desert nature of much of these catchments, it is not surprising that these rivers reveal wide variations in discharge from year to year in response to precipitation fluctuations.

The driest part of the Sahara is an east to west band between 18 and 30° North, where annual precipitation totals are less than 50 mm (Figure 5.1). Here rainfall is extremely unreliable, and it is not uncommon for a year to go by with no precipitation at all being recorded. On the margins of the Sahara, both to the north and the south, precipitation gradients are quite steep and the change from less than 50 mm to over 400 mm per annum occurs over a distance of less than 500 km. These gradients change their spatial locations each year, as well as their steepness, in response to varying climatic conditions.

It is now generally accepted that the Saharan region has been subjected to considerable climatic change over the last 20,000

Figure 5.1 The Sahara desert: mean annual precipitation and seasonal variations

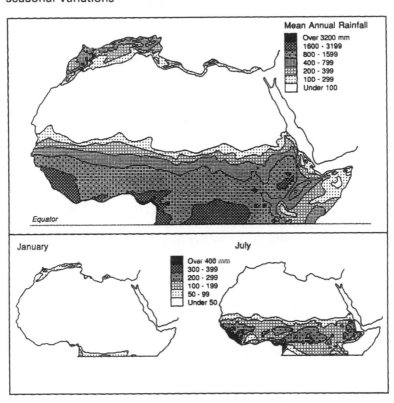

Source: Redrawn from Thompson B.W., 1970, *The Climate of Africa*, Oxford University Press. Reproduced by permission of Oxford University Press.

years. The nature of this change is known on a broad scale, though local differences do occur owing to the presence of highland massifs. It has been shown that tropical Africa, and in particular the arid and semi-arid zones, experienced a drier climate than the present between 20,000 and 15,000 years BP and a much wetter regime between 12,000 and 7,000 years BP (Grove, 1973; Street and Grove, 1979). More detailed work for the Sudan identifies five main periods of climatic oscillation (Wickens, 1975). Between 20,000 and 15,000 years BP the climate was much drier than at present and the vegetation zones are believed to have been displaced about 450 km to the south. From 12,000 to 7,000 years BP there was a very wet period with a northwards shift of the climatic belts of 400 km relative to the present day. This was followed by a short dry period of about 1,000 years duration. Another wet episode ensued between 6,000 and 3,000 years ago, with the vegetation zones some 250 km to the north of today's equivalents. Finally, over the last 3,000 years there has been a gradual falling off in rainfall to current values. In effect what has happened over the last 20,000 years is that individual isohyets have moved over a north to south range of 850 km, with some of these movements taking place rapidly over a few hundred years. In contrast, over the last 2,000 years relatively little change seems to have been registered.

Confirmation of a wet period in the Sahara between 6,000 and 4,000 years ago is provided by archaeological evidence (Milburn, 1984). This suggests that cattle pastoralism was common throughout the area during this period. However, by about 3,800 years BP the Saharan plains were abandoned as a result of dessication, with the people retreating to the wetter highlands, the Sahel or the Nile valley (ibid.).

5.1.3 Traditional land use

The type of land use practised on the desert margins and in the desert itself depends on the life-styles developed by the peoples occupying the area. In this section attention is focused on pastoral activities, though (as will be seen later) the interaction between the pastoralists and the cultivators is of growing importance. Within the Sahara desert itself three major groups of people can be identified (Norris, 1984). These are, from east to west, the Tubu, the Tuareg and the Moors.

The Tubu people are found in the eastern Sahara, centred on the Tibesti plateau, though many live to the south of the desert fringe in better-watered regions. Living quarters in the desert consist of huts made of palm fronds or occasionally of stone. Dates are their staple crop in the more arid areas, but they also possess herds of camels, sheep and goats. A wider range of crops is grown in better-watered areas. Given the variety of environmental conditions, it is not surprising that different life-styles are followed by these peoples, depending on where they live. Some groups are entirely nomadic, but the majority are semi-nomadic with date groves and gardens in the oases in the Tibesti Uplands and with traditional pastures in the desert.

The central part of the Sahara is the home of the Tuareg, with territory from the Ahaggar Massif in the north to the Nigerian border in the south (Norris, 1975). Currently they probably number about 300,000, with the majority of them living in and around the Ahaggar and Air Uplands. Their major economic activity was camel herding and trade by camel caravans with both north Africa and Nigeria. Small-scale cultivation was also practised in upland valleys.

Occupying the western Sahara and forming a group of about one million people are the Moors. They range down the Atlantic coast from Morocco to Senegal and inland to the Adrar of the Iforas in Mali. In the southern and better-watered parts cattle and sheep are the dominant livestock, while in the drier north, camel herds take over. Dates are common in the uplands of the Adrar and wherever irrigation water is available crops are grown, including wheat, barley, millet, melons and vegetables. In the inland areas the caravan trade used to form a vital part of the economy. On their travels, nomad groups roam northwards into Morocco and on the southern boundary of the Sahara down to Senegal and Upper Volta.

Despite the prevailing aridity the Sahara is not totally devoid of water, for throughout its extent there are numerous places where groundwater seeps to the surface or is located only a few metres below it. In such situations oasis communities have been developed which have provided bases for the indigenous peoples, who were dependent on a mix of irrigated agriculture, pastoralism and trade (Allan, 1984). Wherever water is available in quantity, crops are grown by irrigation. However, where water has to be lifted into irrigation canals by animal or human power the area which can be watered is limited. An advantage is that these traditional water-lifting techniques ensure that groundwater systems are protected from excessive use.

Oasis life was a strange mixture of activities. Irrigated agriculture called for the intensive cultivation of small gardens and plots to produce high yields from crops such as dates, wheat, barley, millet and vegetables. Given the high temperatures, both winter and summer crops could be cultivated. Wheat and barley were grown in winter and millet and other tropical crops during the hot summer period. Despite the intensity of cultivation, crops alone were rarely able to sustain the prosperity of the oasis settlement and recourse had to be made to other economic activities (Allan, 1981).

Pastoral activities, although often oasis-based, were extremely extensive on account of the very low biological productivity per unit area of the desert. Some of the pastoralists were true nomads following available pastures in an annual cycle, while other groups operated from desert oases or village communities on the desert margins (Johnson, 1969). Even in the central Sahara the nomads were able to raise crops in the valleys of the Ahaggar, Tibesti and Air Highlands during most years by employing dry-farming methods.

An essential part of oasis life was trade, with almost all goods being carried by camel caravan. Given the relatively small carrying capacity of the camel, goods had to be of high value and low bulk to be worth transporting the many hundreds of kilometres which were nearly always involved. Items included slaves from the Sudan, gold from Ghana, ostrich feathers, copper and perhaps the most important of all, salt. Rock salt was found near oasis settlements such as Idjil in Mauretania, Taoudenni and Taghaza in Mali, Amadror in the Ahaggar, Bilma in Niger and Tamanrasset in Algeria. Most of this salt was carried southwards to west and central Africa, though some was also taken northwards. The camel caravans moved from oasis to oasis across the Sahara. These oases provided the necessary water and provisions as well as political stability, which enabled the caravans to operate in safety. The nomads controlled the routes between the oases and often extracted a tribute from the caravans to guarantee safe passage in the form of cereals or other necessities. The oases therefore provided the necessary links in the trade chain between north and west Africa and at the same time formed the bases for the nomads and their animals when they migrated from the better-watered areas of the Sahel and the Atlas Mountain fringe to the meagre pastures of the Sahara itself. Over the years a complex way of life developed in which all groups within the Sahara itself and the peripheral areas became linked together in the same economic system.

Along the southern margin of the Sahara is the Sahel, a region dominated by a summer precipitation regime (Figure 5.1). From this area the Moors and the Tuareg would move northwards with their cattle in summer to crop the newly emergent vegetation of the desert. Later in the year, when the winter dry season approached, they would move southwards again into the Sahel of Niger and Mali for their animals to eat the cereal stubble left by the settled farmers following the harvest period and to obtain water from wells and water holes.

Given the different rainfall patterns on the northern and southern margins of the Sahara, it is not surprising that the temporal patterns of utilisation of the desert vegetation were different. In north Africa, in Algeria, Tunisia and Libya, the Bedouin practised semi-nomadism, with the year divided into two almost equal parts. During summer and autumn the Bedouin would spend their time in villages tending their crops, chiefly cereals and dates. Then in winter and spring they would travel into the desert with their herds of sheep and camels (Bedoian, 1978).

In this northern area, where winter precipitation predominates, the pastoralists always moved into the desert in winter and spring. This meant that as these north African groups were moving southwards, the people of the Sahel, where summer rainfall occurs, would be retreating from the desert into better-watered regions. In summer the opposite situation would prevail, with the southern tribes advancing into the desert as their northern counterparts were withdrawing from it. Winter was therefore a time of dominantly southern movement of people and their animals, while the summer months recorded a northward drift. In effect a tidal flow of animals into and out of the desert pastures was maintained on an annual basis. The extent of the penetration into the desert and the actual patterns of movement followed by the herds depended largely on the amount of precipitation which had fallen during the rainy season. In an exceptionally wet year pasture areas may be opened up which for years previously had not been able to sustain large animal numbers. In contrast, in dry years even the normally reliable pastures would provide insufficient growth to sustain the herds and so large numbers of animals died.

Besides the nomads and the semi-nomads of the desert itself, the cultivators of the Sahel and the Maghreb have also used the vegetation on the margins of the Sahara for grazing their animals during the period of maximum biological productivity. Although their movements have been much more restricted than the desert dwellers, they have been significant users of the desert fringe and have often been in competition with the other pastoralists for vegetation during times of drought.

Significant changes to the Saharan economy occurred in the nineteenth century as a result of the actions of colonial powers such as France and England. Trade was one of the first areas to suffer. In the last decades of the nineteenth century, French influence began to be felt in north Africa. It became French policy to gain control of the oasis settlements through military force and

201

the resulting political instability led to a disruption of trans-Saharan trade. Along the west African coast British colonial activity was opening up countries for the first time to sea-borne trade. This too caused a decline in the caravan trade as more goods were diverted to ships. The colonial powers also frowned on one of the main commodities of trans-Saharan trade (slaves) and did everything in their power to discourage such activities.

Thus, by the end of the nineteenth century the caravan trade (one of the essential parts of oasis life) was beginning to decline. This, together with the political unrest associated with colonial occupation, meant that the fragile economies of the desert peoples were being disrupted. As a result these people were forced to turn to their existing activities of pastoralism and irrigated agriculture to make up for the loss of income caused by a breakdown in Saharan trade.

5.1.4 Case study — Niger

5.1.4.1 Introduction

Although it is possible to generalise about conditions over the whole of the Sahara, it is only by detailed study of individual parts of the region that insights can be gained as to the importance of local variations. In the Sahel region of West Africa a case study of desertification in the Eghazar and Azawak region has been prepared by the government of Niger for UNESCO (Mabbutt and Floret, 1980). The area studied is situated to the south-west of the Air Massif in the central part of Niger, between 15° and 18° North in the northern part of the Sahel (Figure 5.2). Topographically the region slopes down to the south-west from the Air Uplands towards the valley of the Niger, and it consists of a series of plateaux and dry valleys. In Niger the Sahel is divided into two parts: the northern zone with rainfall between 100 and 350 mm has nomadism as the dominant land use, while in the southern zone, where rainfall varies from 350 to 550 mm, rain-fed agriculture predominates. The study area is located entirely in the pastoral zone between the regional centres of Agadez and Tahoua. The majority of the people are of Tuareg origin, though groups of Foulani peoples have moved north into the area since the 1940s. The total population of the region is about 115,000, about 80 per cent of which are Tuareg. Some 9,000 inhabitants are classed as

Figure 5.2   The Eghazer and Azawak regions of Niger

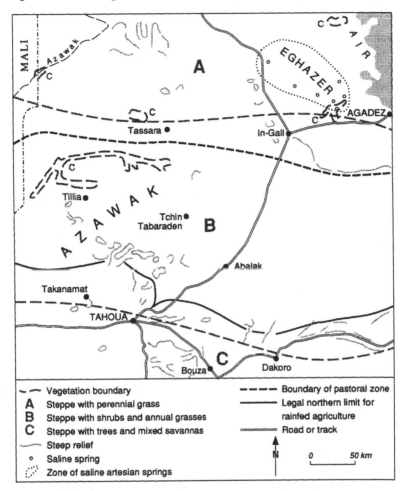

~·— Vegetation boundary
A   Steppe with perennial grass
B   Steppe with shrubs and annual grasses
C   Steppe with trees and mixed savannas
~~~ Steep relief
○   Saline spring
⋰   Zone of saline artesian springs

‒ ‒ ‒ ‒ Boundary of pastoral zone
——— Legal northern limit for
rainfed agriculture
═══ Road or track

0   50 km

Source: Redrawn from Mabbutt J.A. & Floret C., 1980, *Case Studies on Desertification*, UNESCO, Paris. Reproduced by permission of UNESCO.

urban dwellers. Population growth rates are high, with numbers expected to double in 25 to 30 years.

5.1.4.2 Land use evolution

Prior to the twentieth century there seems to have been a southerly movement of Tuareg peoples into the Sahel, though the actual times of these movements are not known with any precision. Towards the end of the nineteenth century, just before the colonisation of the region, the Tuareg confederations were at war with each other and political instability prevailed. In 1901, a French military occupation of Tahoua took place and as a result the nomads were confined to the northern part of the Sahel to limit their influence on the agriculturalists. A major Tuareg revolt occurred in 1917, which was crushed, and afterwards the power of the chiefs was severely curtailed. Prior to the colonial period, new groups had only been allowed into the area with the permission of the local chiefs. Under colonial rule such checks no longer existed, and with the build-up of population pressure in the south, groups of Wodaabe Foulani cultivators began to move northwards into the Sahel region. This northwards spread of arable farming took place in formerly grazing areas, so forcing the pastoralists to move into drier zones already peopled by Tuareg groups. This north-ward movement of the Foulani into a zone occupied much earlier by the southward-moving Tuareg meant that increasing pressures on available vegetation resources occurred in the period after 1940.

The colonial period also brought other changes to the pastoral-ists' way of life. Political stability meant that the herds could wander over larger areas than had previously been possible. The practice of slavery was reduced, and this had important ramifi-cations as the slaves were responsible for guarding and watering the herds. As slavery diminished the small animals, such as sheep and goats (which were highly labour-intensive) were replaced by larger ones, such as cattle.

5.1.4.3 The environment and its resources

The rainfall in the northern part of the Sahel occurs as a result of the summer monsoon and varies from 100 to 350 mm. The rainfall amount decreases from south to north, as does the length of the rainy season. The rains begin with the arrival of the Inter-Tropical Front, which marks the contact between humid maritime air in the south and dry desert air to the north. When these two air masses interact they produce considerable turbulence, leading to violent convective storms. The Inter-Tropical Front normally begins to affect the southerly part of the region from the end of May and reaches the north by about July. It begins to retreat from the north by mid-September and from the south by early October.

Agadez is situated on the northern limit of the Sahel region with

Table 5.1: Mean monthly rainfall and number of rain-days,
Tahoua and Agadez, 1921–54

	Tahoua		Agadez	
Month	Rainfall (mm)	Number of rainy days	Rainfall (mm)	Number of rainy days
January	–	–	0·1	0·1
February	–	–	–	–
March	0·2	0·2	–	0·1
April	2·8	0·5	1·2	0·1
May	17·4	2·8	6·1	1·5
June	48·0	6·0	7·3	2·3
July	110·1	9·5	43·1	6·2
August	140·2	11·6	90·3	9·7
September	53·3	8·2	15·7	2·5
October	12·7	1·2	0·3	0·1
November	–	–	–	–
December	–	–	–	–

Source: Government of Niger, 1980, p. 120.

an annual rainfall of 164·1 mm (Table 5.1). The other major
regional centre, Tahoua, some 300 km to the south-west, falls
within the arable part of the Sahel and receives 384·7 mm. Both
stations do, however, reveal considerable variations over time.
The importance of rainfall variability in such an arid climate
should not be overlooked. At Agadez, in the drier northern zone
of the Sahel, there is a 60 per cent chance that rainfall will be
between 100 and 200 mm, but a 14 per cent chance that it will be
less than 100 mm. In the major drought, between 1968 and 1973,
the 100 mm isohyet moved 200 km to the south and the 350 mm
isohyet 150 km (Government of Niger, 1980). Earlier droughts
have been known, but quantitative information about them is
scarce. It is known that there was a major drought throughout the
Sahel between 1910 and 1915, when heavy livestock losses
occurred. Drought also affected the region from 1940–3, but this
seems to have been much less severe than those of 1910–15 and
1968–73. The early years of the 1930s are remembered as a time of
famine, but this was the result of locusts destroying the millet crop
rather than rainfall deficiency.

For the pastoralists who occupy the Sahel the rainy season is a
time of abundance of both vegetation and water, and so their
animals can wander far from wells and other water sources. The

rainy season diminishes in both time and rainfall amounts from south to north. In the south of the Sahel it begins normally at the end of May and lasts until early October. On the northern limits of the region the rains do not arrive until July and have usually ended by mid-September. After the end of the rainy season, in the autumn, pasture is still plentiful and water is available in large but shallow surface pools. During the cool, dry winter season pasture resources quickly become depleted, with only dry vegetation remaining. Surface pools have dried up by this time and the animals have to be watered from wells to survive. The critical period, though, is the spring-time in May and June before the rains arrive. At this time all pastures are close to total exhaustion and many of the shallow wells are dry. The animals are exceedingly weak, and even the late arrival of the rains can bring heavy losses. In contrast, the arable farmers of the southern part of the Sahel have their time of shortage during the rainy season and their harvest in the middle of the dry season.

The timing of the rainfall, as well as the amount, is critical to vegetation growth, especially of annual species. Heavy rains at the beginning of the season, followed by dry weather, can mean that seeds will germinate and then die. Without adequate seed production, vegetation in the following year may be sparse. Under such conditions what appears to be ample rainfall can be recorded and yet the effects on the vegetation can be disastrous. At Agadez in 1968, for example, annual rainfall was 165 mm, yet 50 mm fell in late April, to be followed by a month of dry weather in which the vegetation withered and died (ibid.). For successful vegetation development, what is needed is sufficient moisture to enter the soil so that the plants are able to develop without a check to their growth. This is normally achieved by a series of heavy and well-spaced showers throughout the mid-part of the rainy season.

From the pastoralists' point of view, the vegetation resources in the Sahel can be divided into two major zones. In the north, on the margin of the Sahara, are grasslands, whose productivity varies with rainfall. During the rainy season the carrying capacity of these lands is high, but only lasts for a short time. To the south the vegetation changes to a mixture of shrubs and grasses as annual rainfall increases. Carrying capacity here is less than a fifth of that of the grasslands to the north at the time of maximum productivity, but this region has the advantage that shrubs can provide forage when the grasses are parched. It can therefore be grazed all the year round. During the rainy season groups of people migrate northwards into the grasslands along varying routes and at different times. Most come from the southern part of the Sahel, though some do migrate from as far as northern Nigeria. At this time grazing pressures on the Sahel grasslands can increase by tenfold.

The species of animals used by the pastoralists show a certain variation, depending on climatic conditions. Camels do best in areas between 100 and 200 mm, and are the prized animal of the Tuareg. Cattle need 200–400 mm of rainfall to thrive. Two main types are found: the Tuareg keep the Azawak zebu, whereas the Foulani herd the much larger Bororo zebu. Sheep and goats are found throughout the Sahel, irrespective of rainfall conditions. The diet of the nomads is a mixture of animal products and cereals. Milk products form the chief food during the rainy season, but as the dry season progresses millet becomes more important. Millet and other foodstuffs are bought when surplus animals are sold at markets in the south. When the millet harvest is poor and prices high, the nomads have to sell more animals to survive and as a result their herds may be depleted.

5.1.4.4 Modernisation of land use

Since 1950 the cattle population in the Sahel has increased rapidly. This has been the result of the abolition of slavery and the change from small to large animals, which are easier to herd, together with the arrival of the Foulani people and their cattle herds. Cattle mortality has also been greatly reduced by the introduction of a veterinary programme which has almost eliminated diseases such as rinderpest and pleuroneumonia. The sharp increase in cattle numbers caused the government of Niger to introduce a pastoral policy for the Sahel in the early 1960s. One of the first projects was the construction of a number of deep wells to provide water for the cattle during the dry season. It was expected that the wells would be utilised for no more than eight months each year. The idea was that the wells would not be opened until the pools and shallow wells were exhausted from about February onwards. Around each well a forage area of 8 km radius was delimited, which had a carrying capacity of about 5,000 cattle (Figure 5.3).

In 1961 and 1962 laws were enacted which defined a precise northern limit for cultivation, in an attempt to stop arable farmers moving into the pastoral zone. A pastoral modernisation zone in the Sahel to the north of the arable farming limit was also created. This legislation marked an important change in the management

207

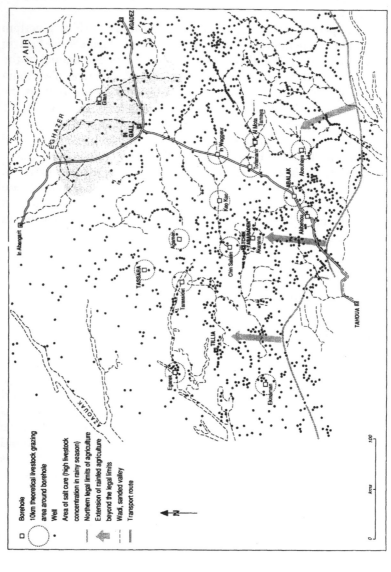

Figure 5.3 Boreholes, wells and the northern limit of rainfed agriculture in Niger

Source: Redrawn from Mabbutt J.A. & Floret C., 1980, *Case Studies on Desertification*, UNESCO, Paris. Reproduced by permission of UNESCO.

of the Sahel as it was the first time that an outside agency (the Niger government) had developed a management strategy for the nomads. Previously it had been the individual groups of people within the region who had been the chief decision-makers with regard to resource management.

During the 1960s, 20 pumping stations were established, though some were later abandoned. Originally it had been planned to allocate certain tribal groups to particular wells, and to charge the nomads a watering fee for their animals. However, both these schemes proved unworkable. The greatest difficulty with the operation of the deep wells was that much higher livestock numbers than predicted around each of the boreholes led to severe overgrazing. At the Abalak pumping station in 1961, for example, it was estimated that daily use was 7,000–8,000 cattle, 8,000–10,000 sheep and goats, and 1,500–2,000 camels. This produced an average of 18,000 head. When it is realised that cattle, sheep and goats are only watered every other day and camels perhaps only every four or five days, it can be seen that the actual number of animals present may have been approaching 36,000 head. This figure is obviously well in excess of the original estimates. Similar work at other pumping stations showed that actual use was between two and four times the planned values. The only effective control the government had over grazing was in terms of the dates of opening and closing of the pumping stations. Stations were not opened until all the pools had dried up, and were closed as soon as the first rains arrived. The basic problem, though, was that once the stations were open there were enormous concentrations of animals around them, and self-management of the rangelands was no longer possible.

5.1.4.5 The Sahel drought, 1968–73, and its aftermath

The Sahel drought began after a long period of average or even above average rainfall conditions stretching back to the 1940s. During this time animal numbers, particularly of cattle, had been increasing steadily and by 1968 it was estimated that there were about 1·3 million animals in the Tchin Tabaraden and Agadez districts (Figure 5.4). It is interesting to note that some pastoral workers at the time did not feel that stocking rates were excessive if full use was made of the rangelands (Government of Niger, 1980). However, the problem was that during the dry season, animals had become highly concentrated around the deep wells.

The drought can be said to have begun in 1968. This was a year

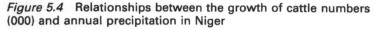

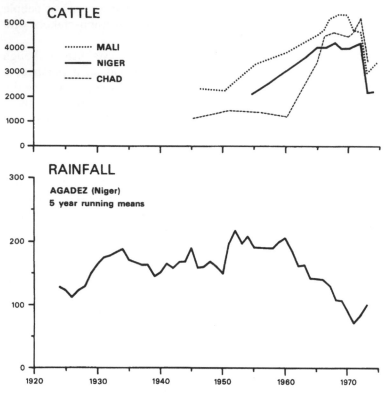

Figure 5.4 Relationships between the growth of cattle numbers (000) and annual precipitation in Niger

Source: Redrawn from Warren A., 1984, 'The problems of desertification', in J.L. Cloudsley-Thompson, *The Sahara Desert*, Pergamon Press, Oxford. Reproduced with permission from Pergamon Books Ltd.

of slightly above average rainfall, but almost one-third of the total fell in late April and was followed by a long dry spell during which many of the annual species died. The poor pastures left meant that many animals perished. In 1969 the annual rainfall at Agadez fell to about one-half of the long-term average, and in 1970 it was only a quarter. In the following three years precipitation totals continued at about one-half normal values, and as the drought built up the pressure on the depleted pasture resources grew. Despite water being available from the deep wells, animals died because of lack of food. The most obvious way to lessen the effects of the drought was to move the herds of animals to wetter areas where better pastures could be found. However, the responses of the local peoples to the drought varied considerably. The Tuareg peoples, with their well-developed social structures and their camels and Azawak zebu cattle, decided that the risks of a long southward migration were greater than those of staying where they were. In contrast, the Wodaabe Foulani, whose cattle were better adapted to long movements and who possessed only loose social structùres, almost all moved to the south.

As a result of the drought a number of the pastoralists lost all their animals. For them to survive the government established camps near Agadez where they could be fed with cereals and powdered milk. After the drought some became entrepreneurs, opening small shops, while others experimented with agriculture. This agriculture was of two types. In the south new rain-fed agriculture was established at the northern limit of dryland farming, with crops of millet and sorghum being grown. The northern spread of cultivation had been limited by legislation in the early 1960s, but after the drought it quickly advanced 100 km to the north as the government felt unable to do anything about it in a time of food crisis. Irrigated gardens were established further north, particularly around boreholes where surplus water was available. Wheat and tomatoes were grown in winter and millet, sorghum and maize in summer. Unfortunately the plots were often too small to sustain the family groups and extra food had to be obtained from elsewhere. Both types of agriculture had a similar effect — namely, that of limiting the amount of land available to the pastoralists. The vast majority of the people who had lost everything settled on the periphery of the major towns or in newly established centres along routes. Many of these people became totally dependent on government handouts.

Table 5.2: Estimates of livestock numbers and losses in Tchin Tabaraden and Agadez, 1968–74

	Agadez			Tchin Tabaraden		
	1968	1974	% loss	1968	1974	% loss
Cattle	120,000	15,000	88	300,000	175,000	42
Sheep	100,000	20,000	80	130,00	101,000	22
Goats	200,000	60,000	70	325,000	202,660	38
Camels	100,000	55,000	45	90,000	?	?

Source: Mabbutt and Floret, 1980 ('Desertification in the Eghazar and Azawak region', pp. 115–46).

By the time the drought ended in 1974 the animal losses had been enormous, particularly in the northern areas around Agadez (Table 5.2). Here, sheep and cattle losses were over 80 per cent. Further south, in the Tchin Tabaraden district, the losses were only about one-half of those near Agadez — a good indicator of where the impact of drought had been greatest. With the higher rainfall pastures improved, with the result that the Foulani people returned and herd numbers began to increase once more. This was further encouraged in the mid-1970s by government attempts to further build up livestock numbers. Animals were supplied and loans were made available, and breeding centres were also established.

The drought had an important impact on the vegetation itself,

Figure 5.5 Niger: areas susceptible to desertification

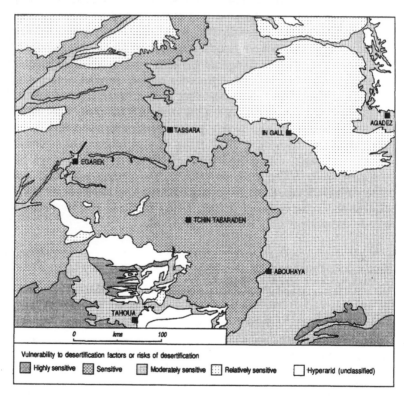

Source: Redrawn from Mabbutt J.A. & Floret C., 1980, *Case Studies on Desertification*, UNESCO, Paris. Reproduced by permission of UNESCO.

particularly around the watering points where the animals were concentrated (Warren, 1984). It is, however, interesting to note that the animals could only graze to a range from 8 to 12 km from the wells, and that beyond this the deterioration of vegetation was not so great. For grasses the critical feature is the spreading of seed. Provided that this has occurred, then it does not matter too much if almost all the vegetation is eaten away. The key to preserving the vegetation cover is to ensure that intensive grazing is restricted to the period after the seeds have been scattered. Unfortunately, during drought the animals had nowhere else to go and so the plants were consumed during the 'rainy' season before seeding took place. Under such conditions the vegetation cover in the following season was seriously depleted, with some species dying out. Erosion potential was intensified as a result of depleted vegetation cover and trampling of the soil by the animals. Around the wells trampling also seriously damaged the roots of perennial species. Given these conditions it was not surprising that the forage potential fell markedly (Figure 5.5).

5.1.5 Conclusion

To ensure the survival of the nomadic system in the Sahel it is crucial that pasture rotation takes place, especially during the rainy season, so that the grass seeds can be self-sown before the period of intensive grazing begins. With the advent of boreholes this has become increasingly difficult as many nomad groups now prefer to remain close to assured water supplies rather than to follow the seasonally available pastures on the margin of the desert during the rainy season. It is also important to ensure that a balance is maintained among the animal species. During drought it is the cattle numbers which seem to suffer most, and yet cattle do have a positive role to play in seed generation. Therefore, if their numbers decline too much the productivity of the pastures is reduced. What has happened is that the stable relationship between animals and pastures, which had evolved on a trial and error basis over the years, has begun to break down for a variety of reasons, of which perhaps the most important is population pressure. With the loss of the traditions which maintained the efficient use of the pastures, the government of Niger has been forced increasingly to intervene with a management strategy which has tried to maintain as many beneficial aspects of the old system as possible. Unfortunately, with the increasing population numbers such a task has proved to be extremely difficult to implement.

Besides the problems of management within the pastoral zone itself, the government of Niger has also had to face the problem that as a result of population pressure from the south, the northward movement of cultivation is confining a growing number of pastoralists and dispossessed pastoralists into an ever-decreasing area. The government has been aware of these trends for more than 30 years, and in the early 1960s it did legislate to limit the northward movement of cultivation. However, in a time of food shortage there is no way in which the government can forcibly remove the farmers and destroy their holdings. Indeed, what now seems to have happened is that the government has chosen to protect these new cultivated lands at the expense of the pastoralists, who are often fined for destroying crops in areas which a few years earlier were definitely a part of the pastoral zone.

A satisfactory balance between the cultivators and the pastoralists in the Sahel is going to be difficult to achieve. It is impossible to deny that cultivation is capable of supplying many times more food per unit area than pastoral activities. As cultivation advances northwards, deeper into the pastoral zone, the chances of crop failure in any year increase. However, given the fact that the decision-makers in the arable colonisation process are individual farmers, the risks they are willing to accept may be considerably greater than a government would consider acceptable. As a result both pastoral and arable farmers may well be producing far below their optimum levels.

On the pastoralists' side it is true that the meagre vegetation resources of the Sahel can only be efficiently harvested by animals which move over a range of pastures during the annual cycle. To achieve this it is essential that nomadism, or at least transhumance, is permitted to continue. Such movements are, of course, only possible if there are adequate grazing grounds in the southern part of the Sahel pastoral zone which are available during the dry season. Unfortunately these are the very areas which since the Second World War have been increasingly colonised by cultivators moving northwards.

In effect, what is happening is that the nomads of the Sahel are being squeezed from the south on to their summer or rainy season grazing areas on the margin of the Sahara. Although these lands have a high carrying capacity during the late summer and autumn, their ability to sustain livestock during the dry season is low. The issue has been further complicated since the drought of 1968–73 by the growing numbers of dispossessed pastoralists and their dependents who have been forced to take up a sedentary existence within the pastoral zone without any adequate means of livelihood. Although not grazing any animals, these people can have a considerable environmental effect by collecting shrubs and woody plants for firewood.

5.2 CENTRAL AUSTRALIA

5.2.1 Historical development

The UN classification of the arid zone delimits all of central Australia as arid, with a broad belt of semi-arid land to the north and east, and a much more restricted one to the south and west (Australia, 1962–73, CSIRO, 1960). This aridity has meant that, since the advent of the first European settlers in the early nineteenth century, the major land use of central Australia has been commercial livestock farming, carried out on a very extensive scale, with the products destined largely for overseas markets.

The arid heartland was first occupied by European settlers moving inland through the Blue Mountains of New South Wales. At first the British government discouraged the occupation of the interior drylands in an effort to promote the development of the coastal regions, but by the early decades of the nineteenth century the controls which they had imposed were breaking down. From the 1830s onwards it was recognised that pastoral farming provided a pioneering mode of settlement which would open up the interior and so stockmen were granted temporary six-month permits for grazing (Alexander and Williams, 1973).

Although a few settlers had established farmsteads in the 1820s and 1830s, the real exploitation of the interior began in 1847 when a system of leases was established for the use of the grazing resources. With this scheme, sufficient land to provide feed for up to 4,000 sheep, or a lesser number of cattle, was leased for up to 14-year periods at low annual rents, for it was recognised that financial returns were too low to generate sufficient capital to permit the farmers to buy the land outright. A form of land occupation without ownership was therefore established, which still prevails today throughout the arid interior. With the system

Figure 5.6 Major land use areas in Australia

Land use
Pastoral – Intensive
Pastoral – Extensive
Agricultural
Relict landscapes/Reserves

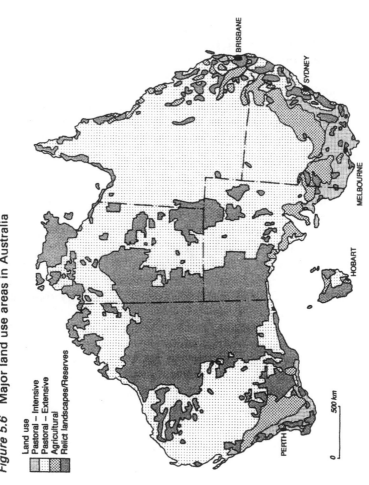

BRISBANE
SYDNEY
MELBOURNE
HOBART
PERTH

0   500 km

the occupier was granted a lease for only a restricted range of the available environmental resources. These usually covered the grazing and watering of the occupier's stock, and the use of timber for personal purposes or to ensure the survival of the animals. If he or she proposed changing to arable farming, the pastoral lease had to be relinquished and all the land purchased.

With increasing population growth there was, from the 1860s onwards, increasing competition for land in the south-eastern part of the interior. This led to a retreat of extensive pastoralism towards the more arid regions and its replacement by more intensive animal husbandry or even arable farming in the wetter regions. As a result of these increasing pressures, most of the southern grazing lands, where winter rainfall predominates, had been occupied by about 1870, and 20 years later the same was true of most of the summer rainfall lands of the north.

After 1880 many of the original pastoral leases were subdivided, with half the land remaining with the lessee. On the other half, several smaller holdings were established to promote new settlement by farmers with limited capital resources. The net result was to produce a greater density of population and a greater degree of capital investment per unit area. Thus, by the closing years of the nineteenth century, the basic pattern of grazing in the continental interior had been established and since that time very few changes have taken place. By the 1980s extensive pastoral stations, supporting either cattle or sheep, are found fringing almost all of the desert interior. In terms of area, they cover as much as 57 per cent of the continent, yet they only account for approximately 5 per cent of the population (Heathcote, 1975). Almost 99 per cent of these stations are still occupied by leaseholders (Figure 5.6).

5.2.2 Pastoral practices

The farmers who brought animals into central Australia used imported livestock from Britain and had little experience of the local environmental conditions. Sheep were first introduced for wool production, but later cattle for hides and local meat supply became more important. The harsh and unforgiving nature of the environment quickly meant that the farmers had to adapt their methods to the demands and limitations set by the land if they were to survive. From the 1850s onwards the demand for labour

217

elsewhere in Australia caused manpower on the extensive pastoral stations to decline. As a result ranges had to be fenced and the animals were left to roam free within these large 'paddocks' for many months: they were only rounded up for activities such as shearing. Originally hand shearing was carried out, but from the 1880s the introduction of mechanical shears allowed large numbers of animals to be processed rapidly by bands of shearers moving from one station to another. Experience showed that the animals benefited from keeping their fleeces during the hot summers, as the wool acted as an effective insulating layer. Lambing time was also moved from spring to autumn, as it was found that the falling temperatures at this time permitted the young animals to grow faster than during the scorching summer months (Barnard, 1958, 1962; Roberts 1968).

Over the years the original stock breeds adapted to local conditions, mostly by natural selection, but also as a result of breeding policies to improve the quality of the fleece and the meat. In the cooler areas the initial British breeds proved well able to tolerate the moderately hot and dry conditions, and prospered well. Under more extreme conditions, the merino (of Spanish origin) became the dominant animal (Carter, 1964). Much less attention was paid to the rearing of cattle. The Hereford and the Shorthorn were the two major breeds introduced from the UK, and once on the ranges they were largely left to fend for themselves. It was really only from the 1950s onwards that a policy to improve the cattle stock on the extensive stations was embarked upon with any commitment. Humped Brahmin cattle from India and similar strains from the hotter parts of the USA were introduced to the tropical grasslands to lessen the environmentally related problems of heat stress, parasites and ticks, and these animals have proved well suited to the Australian conditions.

The greatest difficulty to pastoral development was the lack of surface water supplies (Perry, 1966). In the wetter parts of the interior small reservoirs were constructed along stream courses to collect flood flows. However, their water volumes were small and they soon dried out under drought conditions when they were most needed. Shallow wells were also dug in dry alluvial gravels to provide watering holes, but once again their utility was limited owing to the meagre nature of groundwater supplies. Later geological exploration revealed abundant groundwater resources under much of the arid interior. However, it was not until the 1880s that rotary drilling technology had evolved to the level that deep wells could be sunk at a cost which could be afforded by the pastoral farmers. Such boreholes discovered water at depths of 600 to 1,000 m in confined aquifers. This water was under considerable hydrostatic pressure, so it flowed out of the wells without the need for high pumping costs. These wells were instrumental in opening up the northern shrublands and grasslands, where the lack of surface waters had previously precluded pastoral farming.

The provision of borehole water did not in itself solve all the problems of the sheep and cattle farmers. Indeed, there have been many instances, right up to the present day, of animals having access to water in a drought and yet still dying due to the fact that all edible plant material within walking distance of the well had been consumed. Cattle fare somewhat better than sheep under drought conditions (Lovett, 1973; Newman and Condon, 1969). They are able to forage over distances of up to about 24 km from a watering hole, compared with only 8 km for sheep, and by their greater size they are able to reach higher into trees and shrubs to crop the available vegetation.

The nature of the vegetation also presented problems. Grasses and shrubs were entirely different from those in Europe (Slayter and Perry, 1969; Williams, 1977). Some species were toxic to sheep, others to cattle and some to both. Certain seeds which caught in the fleeces of sheep seriously reduced their value, while spines and thorns penetrated the skins of animals to cause irritation, loss of condition and even death in extreme cases. Perhaps the most important factor, though, was the low density of vegetation per unit area which necessitated stocking levels many times lower than those in Britain. In the more arid parts of the continent more than 64 ha were and still are needed to sustain each beef animal and more than 8 ha for each sheep. Over the years range management has been practised in the wetter parts of the semi-arid zone to improve productivity. Exotic species, such as rye grass in the south and kikuyu grass in the north, have been sown, and for the most part prospered well. In the more arid regions financial returns have not been great enough to warrant the investment needed to improve pastures and here, animals today remain dependent on the native species.

The limited nature of the natural vegetation has meant that the pastoral farmer has always been in competition with the larger native herbivores such as the kangaroo and the wallaby (Low and Low, 1975; Newsome, 1975, 1977). However, these animals have only posed a serious problem under drought conditions when

219

competition for vegetation becomes particularly intense. Nevertheless, it has given rise at certain times to policies to eliminate these animals. The biggest threat to the farmer has been caused by the rabbit. This animal was brought into Australia in the 1860s and by the turn of the century had spread round most of the desert margins of the south and east. Their numbers increased rapidly and they were soon competing with the cattle and sheep for the limited grasses available. Over the years the pastoral farmers have waged a war on the rabbit, but it was not until the 1950s with the introduction of the myxomatosis virus and its periodic recurrence that the rabbit problem was eventually brought under control.

In many parts of Australia the dingo, the only large predator on the continent, has caused difficulties for farmers by attacking sheep and calves. The animal has proved difficult to control, and the only effective measure has been the construction of high fences which the dingos cannot cross. These fences are extremely expensive to erect, but do at least provide protection for the smaller animals. A major dingo-proof fence runs from South Australia northwards to Queensland, and this marks a division between sheep farming in the non-dingo areas of the south and east and the cattle stations to the west. Cattle are, of course, also found to the east of the dingo-proof fence.

A special feature of Australian pastoral farming has been the scientific interest which has been taken in the activity by government research institutions such as the CSIRO (Commonwealth Scientific and Industrial Research Organisation). Research into introduced grass species has permitted higher carrying capacities, while detailed work on soils has identified critical trace element deficiencies, such as cobalt, copper, molybdenum and zinc, which have seriously hindered animal development and growth. Government policy has also encouraged pastoral stations to improve their carrying capacities by financial incentives. Lower rents or longer leases have been used, together with financial assistance for the sinking of new wells. In times of drought, government policy has permitted leases to be extended, rents reduced and provided cheaper freight rates for the movement of stock or foodstuffs — all in an attempt to ensure that the occupancy of the land continues.

The picture of extensive pastoralism which emerges in Australia is one of national ownership of resources, and in particular of land. This is the result of the low and variable financial returns obtained from such activities, which has meant that even poor quality semi-arid land is regarded as having more intrinsic 'value' than can be afforded by any farming method utilising the available biotic resources. This has produced a system in which the resources of the arid interior are held in perpetuity by the government. This government then leases the land to the pastoralist, often on favourable economic terms, in an attempt to ensure that the meagre resources of this region are utilised and that human occupancy continues. So great is the commitment of the government that it has made generous financial assistance for stress conditions such as those caused by drought, that it has also been willing to invest heavily on research into arid conditions. In effect, much of this work is another form of subsidy for extensive pastoral farming. What is interesting is that the system, as it has evolved, does permit the government to have a certain measure of management control over the environment, at least in terms of the location, size and duration of the leases which are available. Where it does not have any management input is with regard to the stocking rates which individual occupiers are free to choose. This is, indeed, a crucial factor which can lead through overstocking to a collapse in the productivity of a particular ecosystem (Condon, *et al.*, 1969a, b, c, d).

5.2.3 Case study — Gascoyne basin

In the next section two areas of Australia are selected as detailed case studies of the pastoral industry. In the west is the Gascoyne basin, which is a sheep-rearing region. The other is the Alice Springs district, situated in the heart of Australia. This is cattle land and is certainly the most isolated part of the country.

5.2.3.1 The Environment

The Gascoyne basin is situated in the north-western part of Western Australia. Topographically it consists of a subdued landscape with a gentle slope upwards from west to east (Figure 5.7). Small amounts of groundwater are often present in shallow aquifers at depths of less than 30 m. Climatically the area is arid, with an annual rainfall of about 200 mm almost equally distributed between the summer and winter seasons. The basin experiences high temperatures in summer, and during the whole period from November to March temperatures make living and working difficult without artificial cooling. The dominant vegetation of the

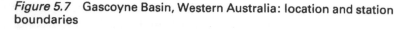

Figure 5.7 Gascoyne Basin, Western Australia: location and station boundaries

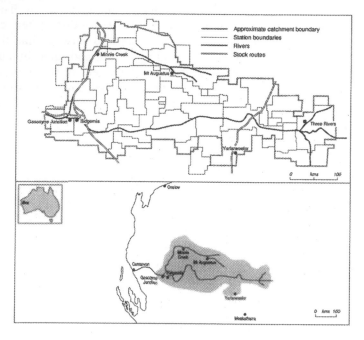

Source: Redrawn from Williams O.B. *et al.*, 1980, 'The Gascoyne Basin', in Biswas M.R. & Biswas A.K., *Desertification*, Pergamon Press, Oxford. Reproduced with permission from Pergamon Books Ltd.

basin consists of acacia shrubland, mainly *Acacia aneura*, but with other acacia species as well. These shrubs have average heights in excess of 2 m, and have lower shrubs beneath them. A ground layer of sporadic perennial grasses is also found.

5.2.3.2 *Pastoral development*

The European settlement of the Gascoyne basin took place from 1876, at a time when speculation in pastoral activity in Australia was considerable. The regulations which governed the early settlement were introduced in the 1860s in the colony of Western Australia. With these rules settlers were permitted to choose up to 40,469 ha (100,000 acres) which could be used rent-free for grazing for up to three years. Thereafter the lease could be continued for a further eight years at an annual rent of £1 sterling/

405 ha (1,000 acres). Over the years the length of the lease was increased to 14 years in 1872 and to 21 years in 1887, and other changes were also made (Williams *et al.*, 1980). The most important of these was a rental charge which varied according to land quality (previously, a constant rent had been levied irrespective of land conditions).

Merino sheep were introduced into the basin from the south-east and were shepherded by aboriginal labour. Although the early leases were large, the lack of surface water and the distribution of vegetation meant that the sheep were farmed on a semi-nomadic basis, travelling hundreds of kilometres to crop the best pastures. This practice, which was only possible because of cheap labour, continued for many years. It meant that small parts of the lease were subjected to very severe grazing pressures, while huge areas were left virtually untouched.

In 1917 a Pastoral Appraisement Board was established to monitor farming activity and to provide an organisation to protect the interests of the state with regard to the crown land which was being leased. An early survey by the Board in 1920 revealed a substantial vegetation cover in the Gascoyne basin, despite a very marked increase in sheep numbers since their introduction in the 1880s. In the 1920s and 1930s economic recession throughout the Western world made sheep farming a marginal activity. Local farmers reacted to this by further increasing animal numbers in an attempt to maintain their incomes. During this period there can be little doubt that the short nature of the tenancies and the general economic conditions resulted in severe over-stocking and consequent degradation of the ranges. By the mid-1930s, at the height of the depression and at the beginning of a major drought, animal numbers in the Gascoyne basin were estimated at 650,000 sheep equivalents (ibid.). This figure was well beyond the available pasture resources and the inevitable crash in numbers occurred. By 1936 only 200,000 animals remained. Since then animal numbers have never attained the high values of the mid-1930s, though peaks of around 400,000 were reached in the early 1950s and almost 500,000 in 1971 (Figure 5.8).

A survey of rangeland types and erosion susceptibility in 1969–70 concluded that the widespread deterioration of the rangeland which was observed was the result of the introduction of sheep from the latter part of the nineteenth century onwards (Wilcox and McKinnon, 1972). These animals, by their heavy stocking levels and continuous grazing of selected areas, had caused a major

Figure 5.8 Relationships between sheep numbers and annual precipitation in the Gascoyne Basin, Australia

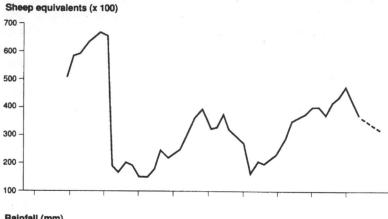

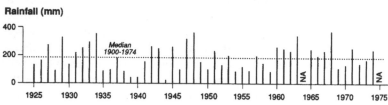

Source: Redrawn from Williams O.B. *et al.*, 1980, 'The Gascoyne Basin', in Biswas M.R. & Biswas A.K., *Desertification*, Pergamon Press, Oxford. Reproduced with permission from Pergamon Books Ltd.

reduction in the quantity and quality of vegetation. The grazing resources of the basin were effectively 'mined', particularly in the period from the late nineteenth century to the time of the great drought in 1936. The report stated that 15 per cent of the area would be irreversibly degraded if grazing was not stopped, and a further 52 per cent was damaged and in need of remedial treatment. It was recommended that stocking rates would have to be lowered and animal numbers reduced from 417,000 to 237,000. It was also stressed that certain rangelands should be retired from continuous grazing activity to permit regeneration (Figure 5.9).

The government of Western Australia accepted the major recommendations of the report and subsequently informed the stations of the number of livestock they were permitted to graze, and the areas which were to be retired from grazing. Not surprisingly, the lessees resented the notion of lower carrying capacities,

Figure 5.9 Erosion vulnerability in the Gascoyne Basin, Australia

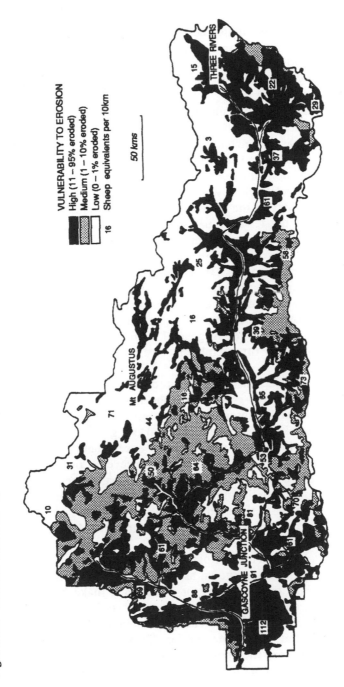

Source: Redrawn from Williams O.B. *et al.*, 1980, 'The Gascoyne Basin', in Biswas M.R. & Biswas A.K., *Desertification*, Pergamon Press, Oxford. Reproduced with permission from Pergamon Books Ltd.

as it greatly reduced their incomes. They felt that they had not been involved in the major destruction of the range resources and yet they were having to pay the financial penalties. Despite these difficulties individual agreements were reached to reduce livestock numbers over a ten-year period and to exclude animals from certain parts of the lease.

This action provides an example of how the public agencies began to supervise land use. Although technically the stations are granted on leases, which in the case of the Gascoyne basin all have an expiry date in the year 2015, the situation is regarded by many as a form of *de facto* freehold. The conditions of the lease only demanded a relatively low rental charge based on the estimated carrying capacity of the land. One of the few requirements of the lease is that the lessee must spend two-and-one-half times the annual rental on improvements to the station.

Within the Gascoyne basin there are 31 major sheep and cattle stations, with just over 300 people resident on them. Individual station communities vary from three to ten people and the stations are usually separated from each other by about 35 to 70 km. With increasing costs, labour has had to be cut and now family help is used wherever possible. In general, families are small, with few having more than three children, and attitudes are middle-class and urban.

Livestock in the basin consist predominantly of sheep, though cattle numbers have grown since the late 1950s. The sheep are sheared in early winter and then left to roam free in large paddocks until the next year. This marks a great change from the 1880s when the sheep were herded every night by aboriginal labour. Dingoes are still a problem, however. Competition with wild species for vegetation in the Gascoyne basin does not seem as great a problem as in other parts of Australia. Rabbits are not numerous owing to the high temperatures and the lack of surface water, but kangaroos are present in large numbers and up to 30,000 have been shot in a single year.

The link between the pastoral farmer and the market is provided by finance companies known as 'pastoral houses'. These collect, store and auction the wool, for which they make handling and commission charges. Many of the smaller stations are in debt to these finance companies. The profitability of sheep farming depends on the prices obtained for the wool, the prices paid for necessities to keep the operation going, and the supply of feed for the animals. In general, either economic depression or drought causes profits to decline or even disappear, and when the two coincide the effects can be devastating.

Over the years economic conditions within the basin have shown major fluctuations. From about the turn of the century up to 1929, pastoral farming prospered, though major droughts did occur in 1911, 1924 and 1928. From 1939 to the end of the Second World War, conditions were much less favourable, especially during the massive drought from 1935 to 1941, when animal numbers declined sharply. The post-war period up to the late 1950s was once again a time of favourable trading. However, droughts beginning in the mid-1950s initiated a period of decline which has continued to the present day (Williams *et al.*, 1980). This decline is revealed in the ever larger number of animals which are needed on a station for it to become economically viable. In the mid 1970s it was estimated to be about 5,000 sheep, but by the late 1980s it will probably be double that figure.

One of the particularly worrying points about the Gascoyne basin — and indeed, about Australia as a whole — is that the domesticated animals are dependent on long-lived vegetation species for sustenance during drought periods. Work in eastern Australia has shown that many of these species which are so valuable for grazing purposes have a half-life of over 100 years (Crisp and Lange, 1976). Since the introduction of sheep and cattle these vegetation species have been greatly reduced in numbers. Over large areas they have been completely removed and it seems that their ability to reproduce has also been adversely affected. A situation has arisen, therefore, in which animals are becoming increasingly dependent under stress conditions on ageing plant species which will not go on living indefinitely. When such shrubs begin to die out in large numbers, the pastoral industry of the basin will suffer greatly and experience tremendous difficulty in gaining access to suitable replacement species.

5.2.4 Case study — Alice Springs

5.2.4.1 Historical development

The centre of Australia, consisting of semi-arid land surrounded by desert and epitomised by the Alice Springs district, exhibits a pattern of development different from that of both east and west (Figure 5.10) (Parkes *et al.*, 1985). The idea of opening up the centre was developed by the graziers of South Australia in the

227

Figure 5.10 Alice Springs district, Australia

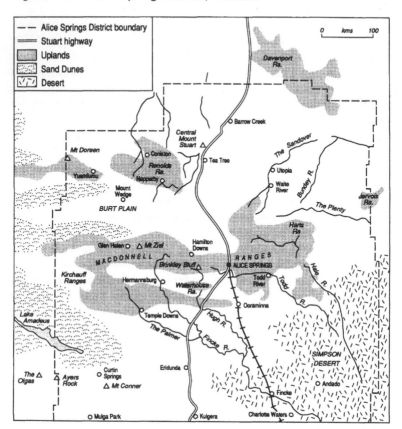

early decades of the nineteenth century, in an attempt to gain access to new pasture lands. The first explorer into the arid heartland, John McDouall Stuart, described it in 1859–60 as '. . . as fine a pastoral hill country as a man could wish to possess' (Stuart, 1861). It was the building of the overland telegraph line between the coasts of north and south Australia, however, which prompted a rapid development of the centre in the 1870s. Work began on the line in 1871, and the following year saw the first application for a holding near the Alice Springs telegraph station.

In the mid-1870s, interest in the land was for cattle, and it was of a speculative nature by the large landholders from South Australia. By the end of the 1870s most of the best land was allocated. The earliest regulations, which had been framed to reduce speculation, stated that the land had to be occupied and stocked before a lease could be applied for. These terms were soon relaxed, though, to permit stocking over a three-year period and even extensions on this were granted. One of the reasons why land applications during the 1870s were rather slow was that rents were high — sometimes four times the cost of similar land in South Australia. In 1881 this was changed with a lowering of the rents to values comparable to those elsewhere. The result was a speculative boom in central Australia, which coincided with similar activity in the east. Almost 72,000 mi^2 of land were applied for in 1881 alone, and by 1885 all available land had been claimed, though not necessarily occupied (Table 5.3). In the last year of the decade the town of Alice Springs was founded.

Table 5.3: Amounts of land applied for in the Alice Springs district, 1872–85

Year	Land applications (square miles)
1872	1,475
1873	800
1874	–
1875	5,280
1876	3,280
1877	3,790
1878	675
1879	355
1880	2,050
1881	71,980
1882	–
1883	310
1884	5,210
1885	590

Source: Bauer, F.H. (1983).

At this time the only feasible route to the centre was from the south and cattle movement over this long trail was a hazardous business. However, in the 1870s a run of favourable seasons meant that many animals survived the journey inland. By 1880, 3,000 mi^2 were stocked with 13,200 cattle, 1,250 horses and 6,850 sheep (Bauer, 1983). Poorer seasons in the early 1880s meant that few properties remained profitable, despite the lower rents. Then, as happens with all semi-arid regions, conditions changed for the

229

better and the late 1880s led to high expectations being generated. By the end of the decade there were over 50,000 cattle in the Alice Springs district. At this time stock were limited to the better-watered areas as no artificial water sources had yet been created. Not surprisingly, the overgrazing of pastures soon began to show itself.

The last decade of the nineteenth century saw a marked decline in the fortunes of the pastoral industry of the centre. Financial depression occurred, followed by drought, and livestock prices fell. The farmers had difficulty in moving their animals south over long and poorly watered stock routes and at the same time there was growing competition from Queensland for the South Australia market. Not surprisingly, the speculators withdrew their capital from the centre, properties were sold and stock numbers declined. By the beginning of the twentieth century all the major pastoral operators had moved out and were replaced by owner-occupiers on moderate-sized holdings, but with little money for improvements and expansion (Davison, 1966; Duncan, 1967).

Little changed in the centre until after the First World War, when a slow expansion began. The key to this expansion lay in the provision of water wells which were drilled from the mid-1920s onwards and which allowed new areas of pasture to be exploited. Even more important, however, was the opening of the railway to Alice Springs in 1929. For the first time this permitted the relatively easy entry into the region of materials for fencing and machinery, and at the same time facilitated the export of cattle to South Australia. The long period of isolation of the centre was over and the region became much more closely integrated into the Australian, and later the world, economy (Hooper, 1983).

The 1930s were also a time of modest development and technological change. In 1932 the government drilled water wells along important stock routes and in the same year the first road train for the movement of animals began. From the communications point of view a landmark was reached in 1939 with the introduction of the pedal radio, which for the first time permitted the cattle stations of the centre to communicate with each other and with services such as the flying doctor. It was, however, the Second World War which caused the major spur to development in the region. With the cessation of hostilities, surplus equipment from the Pacific war was both abundant and cheap and many of the farmers took the opportunity to purchase lorries, jeeps and other machinery. High cattle prices prevailed throughout the 1940s and

1950s, and a widespread optimism was generated as the economy of the centre grew. New management methods were introduced, with fencing becoming a much more important tool in the control of the animals.

As with all semi-arid regions, a 'bust' always follows a 'boom'. This came in 1959 with the advent of a major drought which was to last into the mid-1960s (Foley, 1957). Pastures showed a progressive deterioration over a seven-year period and cattle numbers declined from a peak of 350,000 in 1958 to only 136,000 when the drought broke in 1965 (Bureau of Agricultural Economics, 1970; Chisholm, 1983; Newsome, 1983). During the expansion phase of the 1950s the farmers had borrowed money to finance expansion and improvements, but they were now unable to service their debts. They were only saved by a government programme of well-drilling, freight subsidies and rental remissions.

5.2.4.2 The current situation

The end of the drought in 1965 was followed by a series of favourable seasons. Higher prices for cattle permitted farmers to pay off their debts and herd numbers were built up rapidly. The post-1965 period was also a time of change from the management point of view. New and cheaper drilling systems to sink water wells were introduced and the all-steel suspension fencing systems began to be widely adopted. Restrictions of animal movements to control diseases such as tuberculosis and brucellosis meant that from now on almost all cattle movement had to be by road transport. In the paddocks, cattle monitoring and movement became increasingly undertaken by helicopter and light planes. The use of helicopters revolutionised the mustering system for, with the aid of portable yards, it enabled a small labour force to round up 80 to 90 per cent of the cattle in an area in a single day and to transport them directly by vehicle from the mustering point (Chisholm, 1983). It had previously taken ten men on horseback almost a week to muster only 50 per cent of the animals in any region. Problems returned once more in 1972 when there was suddenly an oversupply of cattle, and prices on the world market collapsed. Herd numbers built up, as it proved impossible to sell cattle, reaching 430,000 by 1975 and up to 550,000 in 1978 (Bastin et al., 1983). Many property owners were forced to borrow money and some only survived with loans from the Pastures Protection Board and Development Bank.

The difficulties of the centre were further compounded in the mid-1970s by escalating inflation and the rapid rise in energy prices as a result of the Arab–Israeli War. Inflation meant that labour and operating costs rose sharply, while increasing fuel costs served to accentuate the isolation of the centre in terms of the distance of 1,500 km over which the animals and materials had to be hauled. The net result was that the long-term economic competitiveness of the Alice Springs district was reduced, largely due to the increasing 'friction' of distance. This did not mean that the area could no longer compete, but rather that higher levels of production had to be attained before the threshold of profitability was reached. In fact, this level was reached in 1978 when cattle prices improved and more than 150,000 cattle (a record) were shipped from the centre. However, the period up to 1978 did result in considerable overgrazing and a cessation of improvements on many properties.

The key resource for any pastoral industry is, of course, the vegetation cover. Before European penetration to the centre, the native herbivores, the kangaroo and related species, appear to have had their numbers controlled during droughts by lack of water rather than lack of vegetation. This meant that the animals would die and that the vegetation would therefore be spared (Fleming, 1983). With the domesticated species this was not the case, especially after the 1920s when a major drilling programme created watering places. With little water shortage the pressure on the vegetation built up considerably. This was made even worse because the pastoralists were not able to adjust their stock numbers or grazing patterns to any major degree. Cattle had to remain within the station boundaries, be dispatched for sale, or sent elsewhere at a high cost. When the profit motive is the aim of the operation, it is not surprising that vegetation deterioration was viewed as the least of a number of evils. This almost total lack of flexibility in being able to adjust cattle numbers to ecological conditions, on both a seasonal and longer-term basis, naturally meant that the vegetation on most stations suffered severe over-grazing during drought.

It is obviously possible in theory to devise a management strategy, by control of animal numbers, to permit the vegetation to provide sustenance under differing environmental conditions without long-term degradation (Gentilli, 1969; Heathcote, 1966). One scheme suggested after the major drought of the late 1950s and early 1960s, for example, known as the Condon scheme, recommended a safe grazing capacity for the centre of only

290,000 animals (Condon *et al.*, 1969a, b, c, d). However, in reality it is much more difficult to achieve this goal and it would almost certainly require government aid to permit more rapid destocking and restocking.

On any property the key to the grazing strategy is the number of wells and their positioning. If there are numerous wells all the pastures can be utilised, but if there are only a few then large areas will not be subjected to grazing pressures. Even around individual wells there are usually variations in vegetation types, which result in selective grazing by cattle. This means that around each well there can be certain parts which are overgrazed, whilst close by relatively little grazing has taken place. In general, though, it is possible to recognise a series of concentric rings of lessening vegetation degradation away from the well.

Ideally, what is needed is a plan to vary cattle numbers in response to the seasonally changing carrying capacities of the paddocks. However, this would involve adjusting stock numbers, either upwards or downwards, in a manner which would not be financially viable. Inevitably, therefore, mistakes in decision-making will occur which will lead to some element of the pastoral operation suffering. In general, what seems to occur is that the vegetation is sacrificed, in preference to changes in animal numbers, as the station manager seems to possess a belief in the vegetation's ability to regenerate following drought. Whether the vegetation does have the resilience which is assumed is a matter of some conjecture, as vegetation changes or trends are extremely difficult to monitor (Newsome, 1983).

In the early 1980s there were 93 leases in the Alice Springs district covering an area of about 305,000 square km, though not all of this was actively grazed (Hooper, 1983). Individual properties ranged from 393 to 10,885 square km, with an average size of 3,200. Stocking rates depended on the quality of the land, but 43 per cent of the properties carried between 5,000 and 8,000 cattle each. However, cattle numbers do fluctuate greatly in response to climatic conditions and have varied from as low as 136,000 in the late 1950s drought to over 500,000 in the wet seasons of the late 1970s (Griffin and Freidel, 1985) (Figure 5.11).

The forage for these animals is mainly grasses and forbs, though up to 20 per cent of the diet can be obtained from shrubs and trees during drought. Most of the fodder plants are warm-season species, resulting in maximum growth after the summer rains. To cope with this period of 'plenty followed by scarcity' with regard to

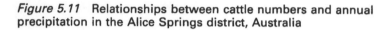

Figure 5.11 Relationships between cattle numbers and annual precipitation in the Alice Springs district, Australia

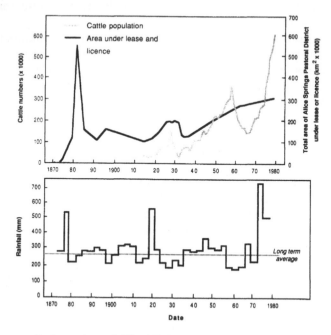

Source: Redrawn from Griffin J.F. & Freidel M.H., 1985, 'Discontinuous change in Central Australia: some implications of major ecological events for land management', *Journal of Arid Environments*, v. 9, pp. 63–80. Reproduced by permission of Academic Press Inc. (London) Ltd.

vegetation, the cattle have to be grazed over large areas with densities of 4 to 20 animals per square kilometre. Of the different forage types, Mulga has the lowest carrying capacity with only four animals per square kilometre.

Newsome (1983) suggested that human-induced vegetation changes had affected the abundance of native animal species, thereby providing a warning of possible long-term degradation of the resource base. To illustrate this theory, he examined relationships between kangaroos and cattle in central Australia. He found that cattle ate almost twice as many species of plants as kangaroos, with the proportion increasing from grasses to trees. Kangaroos were much more dependent on grasses, with eight species accounting for 75 to 80 per cent of their diet. During damper conditions there seems to be little competition between the native herbivores and the domesticated exotics (Low and Low, 1975). In time of drought, kangaroos need access to green herbage and therefore concentrate on open plains, whereas in damper conditions they are usually found in dense *mulga* woodlands. In general they move out on to the plains as drought increases and move back into the woodlands when it ends. In contrast, the cattle seem to prefer open plains in damp conditions and *mulga* scrub in drought, suggesting that vegetation use by the two species is complementary.

What seems beyond dispute is that at the present day, red kangaroos are much more abundant than they were formerly. The reasons are not altogether clear, but one important factor seems to be the well-sinking programme of the 1920s and 1930s, which opened up permanent water holes where none had previously existed. These wells also allowed cattle numbers to expand from less than 60,000 before 1935 to as many as 353,000 in 1958. One of the impacts of increased cattle numbers was to crop closely the perennial grasslands, and in turn these grazed grasses proved more attractive to kangaroos and undoubtedly encouraged their numbers to expand.

However, other changes to the vegetation pattern were noted which did not augur well for the future: for example, around each water point in drought it was noted that beyond about 12 km the vegetation was virtually uncropped, whilst up to 2 km almost no vegetation remained. What was particularly worrying was that during drought the number and diversity of plant species fell by almost two-thirds. Once rains came again the vegetation recovered, but it never attained its original state. What seems to happen is that the nutritious grasses were slowly replaced by *spinnifex*, which has a low value as forage for both cattle and kangaroos. Effectively, this means a continual decrease in the carrying capacity of the pastures and, therefore, the profitability of the pastoral stations.

5.2.5 Conclusion

The utilisation of the arid and semi-arid lands of Australia by Europeans has only taken place for at most 150 years and in many parts for much less. Settlement has been primarily for pastoral farming, with the product (wool, hides or meat) being marketed

235

outside the area of production, and in many cases overseas. This activity has always been carried out on a commercial basis and subsistence agriculture, at least by Europeans, has never existed.

An unusual feature of this farming has been a dependence entirely upon introduced species, in this case domesticated sheep and cattle of British origin. The pastoral activity has been limited in spatial extent by the boundaries of the individual stations, though these are huge by British standards. Apart from the first few years of European settlement there has been no concept of the movement of herds and flocks in search of the best pastures available, such as is practised by nomadic pastoralists.

Given the driving force of the profit motive, the farmers have always used high technology solutions to overcome their difficulties. This is illustrated by the use of water well drilling equipment in the 1920s; the advent of personal radios and the flying doctor service in the 1930s; the widespread movement of cattle by road in the 1940s; and currently, the use of the helicopter and the light plane as a replacement for the horse. The lifestyle of the people on the stations, despite their isolation, is very much 'Western urban', with electricity, air conditioning, and all the electrical gadgets which one associates with the affluent middle class being commonplace.

The main pressures on the pastoral system have been threefold: the incidence of drought; the fluctuations in prices for cattle and wool on the world markets; and variations in input costs, particularly of energy. Each of these factors varies in an unpredictable manner, rendering management decision-making little more than guesswork. When two or more of these factors work together (such as drought and low world market prices), the effects on the Australian arid zone can be so severe that the industry is only able to survive with government subsidies. This suggests that true commercial farming in these arid lands is not viable if the full extent of market prices were permitted to apply. In effect, what exists is a management system which is capable of exploiting the vegetation resources of moist times in a profitable manner, but which in drought has to resort to substantial injections of capital from outside in order to maintain the infrastructure until the next boom phase.

The effect that the pastoral industry has had on the environment, and in particular on the vegetation, is difficult to assess because little detailed monitoring of change has taken place until recently (Griffin and Freidel, 1985). What is obvious, though, is that the earliest pastoralists had an over-optimistic view of the carrying capacity of the arid and semi-arid lands, which led to animal numbers in excess of the available vegetation resources. Over the years, and especially since the 1920s when animal numbers began to rise rapidly, the vegetation of the interior has been degraded from the original plant associations in existence a hundred or so years ago. The actual process of degradation or desertification has been slow, with spurts of activity being registered in times of drought. In moister conditions, vegetation has recovered to some extent, but never to the state prior to a drought event. In effect, what has been occurring is a 'mining' of the vegetation cover over a period of between 100 and 150 years. It has permitted large number of animals to be raised in the short term, but the consequence has been that the carrying capacities of the pastures/paddocks has continued to decline, and will continue to do so unless new management policies are introduced.

6

River Basin Development: The Nile and the Colorado

Even in the most arid parts of the world extensive human settlement is occasionally possible along the banks of major rivers which import water surpluses from better-watered regions. The Nile and the Colorado are two such rivers whose waters have permitted the desert to bloom. Yet their histories are very different. Along the Nile there is a history of traditional irrigation development going back at least 6,000 years. This irrigation, which has been continuous since that time, has given rise to some of the greatest civilisations of the ancient world. Even today irrigated agriculture in Egypt alone supports almost 50 million people. The story is very different along the Colorado, for here irrigation by white Americans is only about 100 years old, and all major developments have taken place in the twentieth century. The Colorado can claim to be the river on which modern dryland irrigation systems were first developed, and in the Hoover Dam it has one of the USA's most potent symbols of human conquest of the environment.

6.1 THE NILE BASIN

6.1.1 Introduction

The Nile basin has a long history of water resource management, but until the nineteenth century the scale of operation of the irrigation networks was small and centred around one or more village communities. Each of these units functioned separately and so the need for integrated irrigation management along the Nile was much less important than in areas like the Tigris–Euphrates lowlands, where large canal systems had to be developed. This meant that the Nile system suffered only minor disruption from natural events such as floods: for even if the infrastructures were damaged, their small sizes meant that they could easily be repaired once the flood-wave had passed. The irrigation system itself proved to be highly productive and was well able to support large populations, with sufficient extra crops to generate considerable wealth.

During the nineteenth century the scale of irrigation operations began to change as first the Egyptian government and later the controlling colonial power (Britain) became increasingly involved in irrigation management issues. However, it was not until the first decade of the twentieth century that water storage facilities, in the form of the Aswan Dam, were constructed on the Nile. Since then the scale of projects has increased, culminating with the Aswan High Dam of the 1960s. Surprisingly, though, even today there are relatively few large water resource management structures on the Nile, despite the huge watershed and the fact that 100 million people live within its boundaries. The chief reason for this is that all the countries with land in the basin are poor and lack the necessary capital resources to invest in major projects.

6.1.2 The basin

The Nile basin covers an area of 2,978,000 square km, which is almost one-tenth of the land surface of Africa (Figure 6.1). On leaving Lake Victoria via the Owen Falls Dam, the White Nile flows through a series of rapids before entering the swampy Lake Kioga. From here the river descends in a series of cataracts and waterfalls into Lake Albert, which it almost immediately leaves to traverse swamps until the Sudanese border is reached. Further cataracts follow before the river enters the huge region of the Sudd swamps. Downstream from the Sudd region the Nile flows across the flat plains of the Sudan until it reaches Khartoum. Here it is joined by its main tributary, the Blue Nile, which rises in Lake Tana in Ethiopia at a height of 1,829 m. From Khartoum to the sea — a distance of 3,000 km — the Nile receives no other perennial tributaries. The Atbara, which enters the Nile about 300 km north of Khartoum, only flows during the flood season although it contributes a sizeable portion of the total discharge of the river.

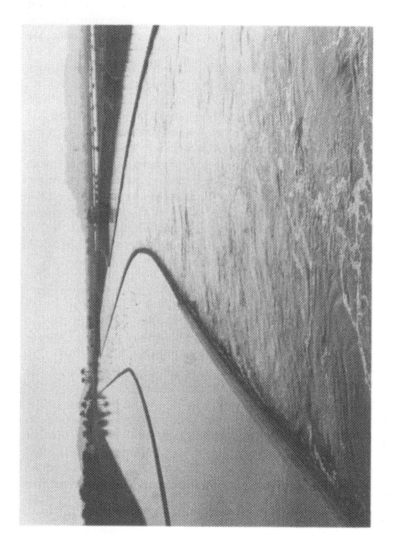

Plate 6.1 Canals taking water from the Colorado River at the Imperial Dam.

Plate 6.2 Gillespie Dam on the Salt River near Phoenix, Arizona. This illustrates sedimentation in the reservoir behind the dam. The top of the dam is seen as a straight line just below the bridge abutment. The dam was completed in 1920 as part of the Salt River Project.

Figure 6.1 The Nile basin and its major hydraulic works

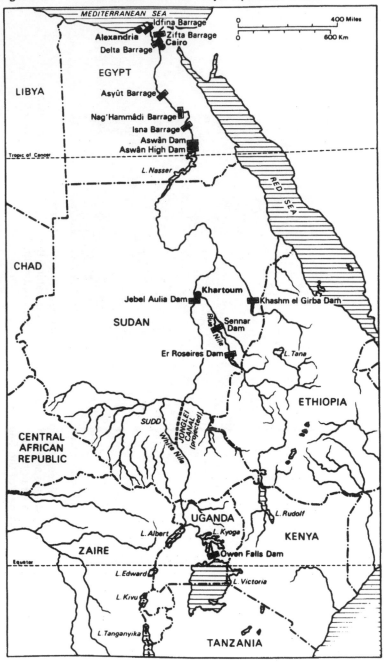

The southernmost part of the basin in Kenya, Ethiopia, Uganda and south Sudan includes areas receiving more than 1,000 mm of precipitation each year, with a marked concentration in the period from March to September. In contrast, the north of the Sudan and almost all of Egypt, with the exception of the coastal fringe, receive less than 100 mm of precipitation each year. At Aswan the mean annual value is 1 mm, and it is not uncommon here for no rain to fall for a number of years at a time. Even at Cairo the average is only 22 mm. Interestingly, about half the population of the basin lives in the arid zone of the Sudan and Egypt, and all of these people depend totally on the waters of the Nile for their livelihood.

6.1.3 Hydrology

The amount of water entering Lake Victoria is estimated to be about 114,000 million cubic m per year, of which about 98,000 million cubic m (86 per cent) is in the form of precipitation falling directly on the lake (Hammerton, 1972, p. 174.). Over 80 per cent of this total volume is lost through evaporation, leaving only 21,000 million cubic m per year (18.4 per cent) available as discharge into the White Nile. On leaving Lake Albert, the flow of the river, known in this region as the Bahr el Jebel, has increased to almost 30,000 million cubic m per annum. The annual volume entering the Sudd swamp area is of the order of 28,000 million cubic m, but only about 14,000 million cubic m per annum are discharged into the Sudanese part of the White Nile owing to enormous evaporation losses (Sutcliffe, 1974, pp. 237–8). The Sudd also acts as a reservoir for the flood waters flowing from the East African plateau. As a result of both these factors the regime of the White Nile varies little throughout the year. At Malakal a minimum flow of 525 cubic m per second occurs during April, rising to a maximum of 1,200 in October and November (Table 6.1). A considerable proportion of this increase in discharge during autumn is the result of water from the Sobat River, which derives most of its flow from the Ethopian Highlands.

The Blue Nile rises in Lake Tana, which is a shallow water body with an area of 3,150 square km. The average annual discharge from the lake is 3,800 million cubic m. Daily discharge varies from 0·9 million cubic m in May and June to approximately 33 million cubic m in September (Talling and Rzoska, 1967, p. 640). From

Table 6.1: Monthly average discharge for the major rivers in the Nile basin

	White Nile	Blue Nile (m³/sec.)	Atbara	Nile–Aswan
October	1,200	3,040	340	5,200
November	1,200	1,030	79	2,270
December	1,100	499	25	1,400
January	829	282	8	1,100
February	634	188	2	1,020
March	553	156	0	834
April	525	138	0	819
May	574	182	1	698
June	742	461	35	1,340
July	897	2,080	640	1,910
August	1,030	5,950	2,100	6,570
September	1,130	5,650	1,420	8,180
Annual average	872	1,640	389	2,650

Source: UNESCO (1969) *Discharge of selected rivers of the world. Vol. 1 — General and regime characteristics of stations selected* (UNESCO, Paris, pp. 43–4).

Lake Tana the river passes through a deep gorge section 800 km long before emerging on to the plains of the Sudan near Roseires. It is in this section where most of the river water originates as run-off from the high Ethiopian plateau. Two important tributaries, the Dinder and the Rahud, join the mainstream between Roseires and Khartoum. These rivers only carry water during the flood season from July to October, and they contribute about 8 per cent of the annual discharge of the Blue Nile.

Apart from Lake Tana there is little water storage capacity in the valley of the Blue Nile, but at the same time the altitude of the Ethiopian Highlands and consequent lower temperatures tend to limit evapotranspiration losses. With the occurrence of heavy rainfall associated with the summer monsoon, torrents of water and eroded soil pour down the valley of the Blue Nile to meet the White Nile at Khartoum. As a result the regime of the Blue Nile reveals a maximum flow of 5,950 cubic m per second in August, which is about 40 times the minimum flow in April. The regime of the Atbara closely resembles that of the Blue Nile.

The superimposition of the relatively steady flow of the White Nile and the very flashy discharges of the Blue Nile and the Atbara produce a characteristic flow pattern at Aswan. Here the maximum

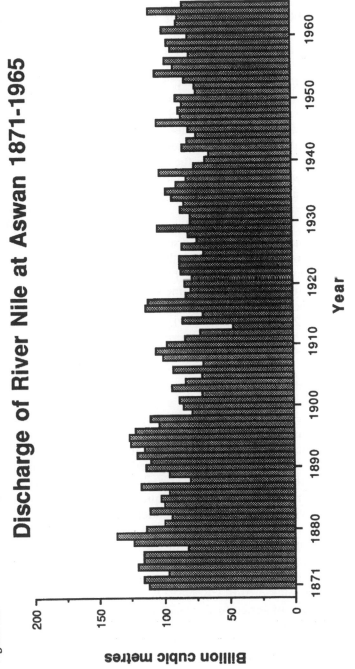

Figure 6.2 Annual variability of flow of the River Nile at Aswan, 1371–1965

Discharge of River Nile at Aswan 1871-1965

water volume is attained in September, with an average flow of 8,180 cubic m per second; this is eleven and a half times the discharge of the month of lowest flow (April). The average annual discharge at Aswan is 84,000 million cubic m.

Besides the monthly variations in discharge, what is perhaps an even more important aspect is the fluctuation in total water amounts from year to year, which in the cases of the Blue Nile and the Atbara can be considerable (Figure 6.2). This annual variability in discharge spelled good fortune or ruin for the farmers of the floodplain. Severe floods devastated the fields and settlements, whilst low flows meant insufficient water for the crops to mature. During recent times the annual discharge of the Nile has varied from an annual low of 42,000 million cubic m in 1913–14 to a maximum of 151,000 million cubic m in 1878–9 (United Arab Republic, Ministry of the High Dam, 1968, p. 2). A number of workers have attempted to discover if trends in the annual flood levels were present. Owing to the lack of detailed evidence, much of this work is inconclusive, but a growing body of evidence is pointing to a progressive decline in the annual flood level, at least during the period of the First and Second Dynasties from about 3,100 to 2,800 BC (Bell, 1970, pp. 569–73; Butzer and Hansen, 1968).

6.1.4 Basin irrigation

Agriculture seems to have begun in the valley of the Nile about 7,000 years ago, following a long period in which humans were predominantly engaged in hunting and food gathering (Butzer, 1959, p. 47). At this time and, indeed, until the latter part of the nineteenth century, large-scale farming was only possible by use of a simple river engineering technique known as basin irrigation, a system which was entirely dependent upon the vagaries of the annual inundation. It consisted of the division of the lower parts of the floodplain into a series of basins separated by natural or human-made levees, into which the floodwater was channelled and held for as long as possible to permit the replenishment of the depleted soil moisture storage.

This type of agriculture seems to have developed rapidly, being greatly assisted by a strong central government with power concentrated in the person of the ruling Pharaoh. As the power of the head of state was dependent to a large degree on the agricultural productivity of his subjects, it is not surprising that the early Pharaohs took an active role in the construction of large-scale control works to enable more successful irrigation to be practised (Willcocks and Craig, 1913, p. 299).

The basins in which agriculture was practised varied in size from 1,000 to 4,000 feddans (420–1,680 ha) and were divided into smaller units of four to five feddans (1·68–2·10 ha) paralleling the irrigation canals (Hurst, 1952, p. 38). Settlements were concentrated on the levees adjacent to the basins for flood protection. The floodwave along the Nile begins to affect upper Egypt by late July. Water levels rise rapidly during August and reach a peak in September. As soon as the river stage begins to rise the irrigation canals would be opened and the basins inundated to a depth of about 1·5 m. This water would be retained within the basins for a period of 40 to 60 days and then drained back to the river, or, in a time of water shortage to the next basin downstream (ibid., p. 40 *et seq*.). Once the water has drained from the fields, no more water was added until the next flood season, and so the growth of crops was dependent on the accumulated soil moisture reserves.

Owing to the long dry season which broke up the soil, little ploughing was necessary, and the crops were sown on the thin film of newly deposited silt and clay once the waters had receded. Plant growth took place during the winter months and the crop was harvested in late spring. The ground then remained fallow until the next crop was sown following the succeeding annual flood. The basin irrigation system as practised in Egypt allowed cultivation during the winter season. As a result the range of available crops was limited, and dominated by the production of barley until about 500 AD, and by wheat after this time (Hamdan, 1961, p. 123). Crop rotation was practised, with the cereal crop alternating with beans and lucerne.

Within the framework of the basin irrigation system, small areas were perennially irrigated, usually from wells, but also directly from the Nile. These areas were devoted to the cultivation of valuable summer crops such as sugar cane, rice, cotton, onions and tobacco. The major advantage provided by perennial irrigation was the greater range of crops that could be cultivated during the summer months. It did, however, have its drawbacks. As this form of irrigation was usually undertaken on the higher levees above the level of the average flood, it meant that there was no annual deposition of silt to maintain soil fertility and so a complex rotation system, together with the heavy application of animal

247

manure, had to be adopted. Another problem was the expenditure in time and energy involved in lifting the water which was to be used on the fields. Over the years numerous ingenious methods have been employed for this task including the shadouf (a counter-balanced pole and bucket); the tamboor (the Archimedean screw), and the sakieh (water wheel) (Hurst, 1952, p. 43 *et seq.*). Even as late as 1965 there were more than 200 sakiehs in operation on the Nile between Kareima and the Third Cataract (Adams and Holt, 1985, p. 75).

One of the truly amazing features of the basin irrigation system was that it survived almost unchanged in the Nile valley for some 7,000 years until the twentieth century. The reasons for its success are set down by Hamdan (1961) as follows. Firstly, the system had evolved naturally, and was well adapted to the environment. It needed a relatively small labour force to operate and yet at the same time was capable of supporting a dense population. Secondly, the flood season and the growing season were well timed, unlike the situation in Iraq where the maximum discharge occurs in spring when the crops are already growing in the fields (Beaumont, 1978a). Thirdly, it possessed certain physical attributes which allowed annual cropping of the soil. These included the yearly deposition of silt providing a nutrient supply for crop growth; good natural drainage through the sands of the floodplain, ensuring that salinity was rarely a problem; and the long-enforced fallow period, permitting aeration of the soil and the destruction of crop pests.

The system did, however, have its problems. Being technically primitive, it meant that control of the river was limited and so most of the floodwater was discharged into the Mediterranean unused. The timing of the flood peak limited farming activity to the winter portion of the year, while the relative levels of land and water meant that extension of the cultivated area was extremely difficult. The greatest problem of all, though, was that it was not possible to control or even to predict the size of the annual flood. In effect this meant that the human occupation of the floodplain was at best a hazardous venture, with disasters a frequent occurrence.

6.1.5 Perennial irrigation

The change over from basin irrigation to perennial irrigation began in the nineteenth century, and it was eventually completed with the damming of the Nile waters by the Aswan High Dam in 1964. The importance of perennial irrigation is that it allows the production of commercial crops, grown during the summer months, on a large scale. The economic effects on the Egyptian economy have been very great.

The development of perennial irrigation in Egypt began from 1820 onwards, with the deepening of the intakes of the flood canals in the delta region in order to permit water to flow into them when river levels were low during the summer. The high cost of lifting water from the canals to the fields, coupled with the problem of rapid siltation, meant that the scheme quickly proved unsatisfactory (Hamdan, 1961, p. 126). This was followed by projects which attempted to raise the river level by the construction of barrages, so that the canals could carry water all the year round. The Delta Barrages were commenced in 1843 with the aim of supplying three major canals and a number of smaller ones. Unfortunately, they never functioned effectively until they were modified and repaired by British engineers in the 1890s. In the Nile valley itself, the first summer canal — the Ibrahimiya canal — supplying water to the sugar cane plantations of Ismail was not constructed until 1873.

Throughout the nineteenth century, although the area of perennial irrigation gradually increased, the methods of water supply consisted entirely of raising summer water levels. No attempt was made at this time to store the floodwaters of the Nile. With the construction of the Aswan Dam in 1902 a new phase of irrigation was initiated in Egypt (Addison, 1959). The dam, which was built in the extreme south of the country, meant that at least a portion of the huge annual flood volume could be stored and used for agricultural activity in the north. The capacity of the dam was 1,000 million cubic m which provided sufficient water to allow an extension of the area of perennial irrigation, and at the same time assured the availability of water during the summer months. To ensure the rational utilisation of the water downstream, a series of barrages and canals were constructed at Assyut (1902), Zifta (1905), Isna (1909) and Nag Hammadi (1930). The Aswan Dam soon proved a success, and as a result it was further heightened on two occasions to increase the storage capacity to 2,500 million cubic m in 1912 and to 5,700 in 1933. Even with these increases in size the reservoir was still only able to hold about one-fifteenth of the annual flood: the rest had to be passed through the sluices of the dam into the Mediterranean Sea. The danger of flooding was reduced but certainly not eliminated. Additional storage capacity was created on the Nile by the construction of the Jebel Aulia Dam above Khartoum in 1937. The reservoir behind this dam has a capacity of 2,500 million cubic m.

The major irrigation works which were constructed in the first 60 years of the twentieth century increased the crop area of Egypt from 7,624,620 feddans (3,202,340 ha) in 1907 to 10,367,730 feddans (4,354,447 ha) in 1960 (Frood, 1967, p. 368). When perennial irrigation was first introduced on a large scale, crop yields increased as the fertility reserves of the soil were mined, and then declined (Crouchley, 1938, p. 154). The most obvious cause of this decline was the decrease of the annual silt deposition, which in the basin irrigation system had helped maintain fertility over thousands of years. With less silt, nutrients were quickly depleted from the soil by the growing crops, and for the first time in Egypt the large-scale introduction of chemical fertilisers, especially nitrates and superphosphates, was essential to maintain agricultural productivity. This demand for fertilisers was made even greater by the fact that in many areas the fallow period was suppressed and continuous cultivation occurred.

With the absence of a long fallow period, insect pests began to multiply in the soil. Crop losses grew as pests such as the boll-weevil spread through the cultivated areas of cotton (ibid., pp. 155–6), and pesticides were now also a necessity. In nearly all regions the methods of application of water to the cultivated areas were primitive and little attempt was made to see that the water supply matched demand. All too often excess water was applied resulting in rising water tables, waterlogging of the soil, and consequent crop failure. In other regions the rising water tables led to upwards capillary movement of water, increasing soil salinity. Cotton was one of the crops most susceptible to this salt build-up and yields often dropped radically. Following the First World War, both waterlogging and salinity problems were remedied by the construction of drainage channels.

Irrigation opportunities along the Blue Nile in the Sudan have been greatly enhanced by the construction of the Sennar and Roseires Dams. The Sennar Dam, built in 1925 with a capacity of 800 million cubic m provides water for 2·2 million feddans (924,000 ha) in the Gezira scheme near Khartoum (Talling and Rzoka, 1967, p. 641). The Roseires Dam holds back irrigation water for 1·3 million feddans (546,000 ha) and — if heightened a further 10 m as has been suggested — could supply an extra 2·6 million feddans (1,092,000 ha) (Field, 1973, p. 13). One of the most important functions of the Roseires scheme is that of acting as a regulating system for the Blue Nile runoff, so enabling it to provide releases of water to refill the Sennar reservoir during periods of water need. As a result of this extra water being available, the Sudan has been able to proceed with the Managil irrigation project of 838,000 feddans (351,967 ha), which, together with the Gezira scheme, permits the cultivation of about 1·8 million feddans (756,000 ha).

6.1.6 The Aswan High Dam

The current stage in the development of water resources of the Nile is represented by the Aswan High Dam, seven km upstream from the Aswan Dam, which is the first river engineering project in Egypt to store more than a single season's floodwaters (Little, 1965). The idea of a large dam in southern Egypt to control water distribution in the Nile valley to the north is not new, and dates back to the latter years of the nineteenth century when the Aswan Dam was being planned. At this time it was thought more prudent to embark upon a smaller project, leaving the construction of a larger dam for the future.

In October 1952 an international committee was set up by the Egyptian government to make a detailed feasibility study of the construction of a large dam on the upper Nile (Holz, 1968, p. 230). This committee reported favourably in 1954, and it was immediately decided that the construction of the Aswan High Dam should commence. Before work on the project could begin, it was essential that a new agreement on the distribution of the waters of the Nile should be made between Egypt and Sudan in accordance with the Anglo-Egyptian agreement of 1929. This was achieved in 1959 with the signing of a new Nile Waters Agreement.

The estimated costs of the Aswan High Dam project amounted to £415 million (sterling). The International Bank for Reconstruction and Development, following a survey of the economic and technical aspects of the dam, originally offered to finance the scheme but, with the deterioration of the political situation in the Middle East, withdrew its approval in July 1956. A week later President Nasser nationalised the Suez Canal and a few months later, British, French and Israeli forces invaded the canal zone. Following the cessation of the fighting, the problem of financing the High Dam project still remained. Eventually, however, in 1958 an agreement was reached between the UAR and the USSR according to which the latter offered a long-term loan of £348 million (sterling) to finance the first stage of the project. In 1960 a second agreement was reached whereby the USSR supplied an additional loan of £78 million so that the project could be completed (United Arab Republic, Ministry of the High Dam, Aswan High Dam Authority, 1968, p. 6).

The storage capacity of Lake Nasser, held up by the Aswan High Dam, is 164,000 million cubic m: this is approximately 30 times larger than the storage of the heightened Aswan Dam (Arab Republic of Egypt, Ministry of Culture and Information, State Information Office, 1972, p. 11). It is also about twice the annual flow of the Nile. During the planning stage it was decided that the storage volume would be divided into 30 million cubic m for sediment accumulation and 37 million cubic m for flood protection, leaving the rest for water for beneficial uses (Al-Barawy, 1972, p. 90). Sediment deposition is estimated at about 60 million cubic m per annum, so that it should take 500 years to fill the dead storage capacity of the reservoir (Field, 1973, p. 13). The hydroelectric generating equipment of the dam provides a total capacity of 2,100 megawatts (MW). With the full operation of the generating capacity of the dam, the electricity output of the country was doubled.

The chief benefits gained from the construction of the dam have been an expansion of the cultivated area of Egypt by 1.3 million feddans (546,012 ha) and the conversion of 700,000 feddans (294,006 ha) from basin to perennial irrigation. Flood protection along the lower Nile is now guaranteed and the productivity of the land has been increased in many areas by a lowering of the water table. To all intents and purposes the Nile below Aswan has now been reduced to the status of an irrigation canal (Figure 1.11). With the closure of the Aswan High Dam the flood of 1964 was the last Nile flood to reach the Mediterranean Sea. It was also an unusually large one with 52,000 million cubic m flowing out into the sea between August and December (Halim et al., 1967, p. 401).

The construction of the High Dam has given rise to a number of problems, some of which were only vaguely appreciated before construction began. One of the more obvious ones is that of sediment movement and deposition. The Aswan Dam, built in 1902, was so designed that the sediment load could be passed through the structure and so continue to benefit the downstream farmer. In contrast the High Dam has no provision for passing sediment which will, therefore, accumulate in Lake Nasser. Downstream from the dam, the water, as the result of a much-reduced sediment load, has more energy available for erosion and the foundation of many structures, such as bridges and barrages, soon began to be undermined (Hollingworth, 1971).

Besides erosion in the Nile Valley itself, increased erosion has also been observed along the Mediterranean coastline, where rates of up to 5 m a year have been noted. The absence of silt in the lower Nile also appears to have damaged the Nile fishing industry, by reducing the nutrient supply to the minute plants and animals on which the fish feed. As a result fish catches in Egypt began to decline steadily once the Aswan High Dam became operational in 1964 (Arab Report and Record, 1970c, p. 103). The effect has even been noted in the south-eastern part of the Mediterranean, where fish landings, particularly of sardines, were much lower (Arab Report and Record, 1970a and b, pp. 677 and 346 respectively). To balance these declines, a new fishing industry has been established on Lake Nasser (Middle East Economic Digest, 1970, p. 238).

The spread of human disease, especially bilharzia (schistosomiasis), has also been blamed on the High Dam. Bilharzia is a debilitating disease carried by parasites living on freshwater snails in irrigation canals. With the continued spread of perennial irrigation and the construction of new canal systems, some workers felt that the incidence of the disease would rise rapidly (McCaull, 1969, p. 6). However, it has been claimed that since 1952 the number of sufferers from bilharzia has been substantially reduced as a result of the work of medical teams and the destruction of snails by chemicals (Little, 1971, p. 6). In fairness, it should be recognised that both the reduction of silt amounts carried by the Nile waters, and the spread of bilharzia, are problems which date back to the initiation of widespread perennial irrigation in the late nineteenth and early twentieth centuries. The High Dam may well have increased the severity of certain problems, but it was certainly not the initial cause of them.

To make an objective assessment of the environmental, social and economic effects of the Aswan High Dam is an extremely difficult task. A voluminous literature already exists on the subject, but with many of these works it is hard to separate fact from prejudice. It is perhaps worth remembering that at the time of the completion of the dam, it was in the political interests of the

253

USA and Israel to foster the idea that the Aswan High Dam was a failure, just as much as it was for Egypt and the USSR to maintain that it was a success. The West was, of course, subjected to a greater information flow from the former than the latter source, and this undoubtedly had an impact on people's perception of the value of the High Dam scheme. This is not to deny that the dam has caused widespread environmental effects. However, given that the population of Egypt was increasing at 2·6 per cent per annum, there was little doubt that something had to be done to try to alleviate the falling standards of living of the people. The government chose to build the High Dam. By so doing it did not solve the basic Egyptian problem of ever-increasing human pressure on resources, but it did at least buy a period of time in which more fundamental changes in the outlook of the people could at least be attempted. One fact does seem certain — namely, that the Aswan High Dam was seen by the Egyptian people as a symbol of progress and as such provided a psychological spur which helped in the initial stages of the modernisation of the Egyptian economy.

Since the completion of the High Dam, Egypt has been using the waters of the Nile within her boundaries to the maximum possible use. Any further development of water resources will need basin-wide agreements involving all of the countries within the watershed. Over the years various major projects have been proposed to permit the more efficient use of the available water resources and the relative merits of many of these schemes have been evaluated (Morrice and Allan, 1959, pp. 101–56; Simaika, 1967, pp. 214–22).

One of the main schemes for the further development of the Nile basin is to reduce evaporation from the Bahr el Jebel in the Sudd swamp and from the Sorbat in the Mashar swamps further north. It is estimated that about 31,000 million cubic m per annum are lost through evaporation in the upper part of the basin (Field, 1973, p. 14). The main project to conserve this water is the Jonglei Canal, which leaves the Bahr el Jebel below Bor and rejoins the mainstream near Malakal (Mageed, 1984; Moghraby and Sammani, 1985). At the intake of the canal is a regulator which will direct half of the discharge into the canal, leaving the rest to flow along the main channel. The feasibility of the project was evaluated between 1946 and 1954 by the Jonglei Investigation Team brought in by the Sudanese government (Jonglei Investigation Team, 1954). The team concluded that the project was feasible, but that the way of life of the local peoples would be severely disrupted (Howell, 1953, pp. 33–48; Sutcliffe, 1974, pp. 237–55). Work on the canal began in the early 1980s, but no work is currently being carried out because of political unrest. Three other canals with similar aims of reducing evaporation losses have been suggested. In the Sorbat basin a canal would divert flow from the Baro around the Mashar swamps and into the White Nile. A collecting canal would gather waters from the upper parts of the Bahr el Gazal system and divert them north of the Sudd into the Bahr el Jebel. Finally, to the south of the Sudd swamp a canal would divert the flow of a number of streams which empty into the Sudd and carry it eastwards to the Bahr el Jebel near Bor. If all the above schemes were to be implemented, it is thought that the increase in the discharge at Aswan would amount to 9,000 million cubic m (Field, 1973, p. 14).

There has also been considerable discussion about constructing a series of regulation dams throughout the Nile basin so that water flow can be more easily controlled. One such scheme — the Owen Falls Dam at the outlet of Lake Victoria — was built in the mid-1950s with the major aim of supplying 150 MW of power to Uganda, but also to control the flow of water into the Nile. Similar projects for Lake Albert and Lake Kioga on the White Nile and Lake Tana on the Blue Nile have been suggested, as well as a hydro-electric project on the White Nile at the Sixth Cataract in the Sudan. None of these schemes seem likely to implemented in the near future, however. A much more complex study has been proposed for the Lake Victoria catchment by the United Nations, to see if there is any way in which the amount of water leaving the lake and entering the White Nile can be increased. At the present time only about 7 per cent of the precipitation falling on the catchment actually discharges into the river.

6.1.7 Current position

Egypt and the Sudan are currently experiencing rapid population growth. In 1983 the populations of Egypt and the Sudan were 45·9 million and 20·6 million respectively. By the year 2000 it is estimated that Egypt's population will have risen to 67·3 million and that of the Sudan to 33·2 million (Population Reference Bureau, 1985). Since 1974 Egypt has been a net importer of agricultural products, and since the early 1980s imports have accounted for almost half of all basic foodstuffs consumed. This position seems likely to deteriorate still further in the future.

Egypt's crop area is currently about 2·47 million ha (FAO, 1986). Since the completion of the High Dam, perennial irrigation is practised almost everywhere, with two crops being produced each year on average. Crop yields are high for certain crops, despite a low level of agricultural technology. For many crops, growing conditions are near optimum, with high temperatures, fertile soils and adequate water resources. Water for irrigation is delivered from the Nile by the Ministry of Irrigation, normally on four- or seven-day rotations. Water volumes are not monitored and no charges are made for the water. Delivery of water into the irrigation canals is in sufficient quantity so that even farmers at the lower ends do not suffer from water shortages. Since the advent of perennial irrigation one of the major problems facing farmers has been that of obtaining adequate soil drainage. This has become so severe that, in the mid-1970s, a USDA (United States Department of Agriculture) survey claimed that 80 per cent of the agricultural area of Egypt was suffering from waterlogging and salinity problems (USDA, 1976).

To help combat the growing food shortages the Egyptian government has committed itself over the last decade and a half to a programme of irrigation expansion, which, it is claimed, will increase the cultivated area by 1 million ha by 2000. Much of this land is situated to the east and west of the Nile delta and its development often necessitates high water lifts. This irrigation development has received 40 per cent of investment given to the agricultural sector in recent years, but results have been disappointing as few schemes have lived up to their planned outputs. Despite increases in the area of arable land, the cropped area *per capita* has shown an almost continuous decline since the early part of the nineteenth century as a result of the rapid population growth (Figure 6.3).

One of the first land reclamation schemes in lower Egypt was the West Nubarya Project, dating from 1952. Calculations made at this time revealed that as much as 91,056 ha (225,000 acres) could be reclaimed by the more efficient use of already-existing water resources. In 1954, 12,140 ha of this area, known as the North Tahrir sector, were selected to act as a pilot project (Schulze and De Ridder, 1974, p. 12). A large-scale land classification programme was undertaken with the assistance of FAO/UNDP (Food and Agriculture Organisation/United Nations Development Pro-

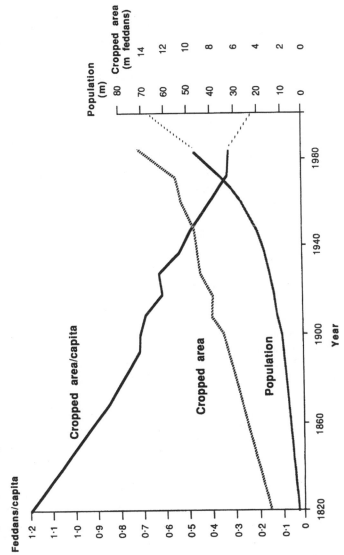

Figure 6.3 Relationships between cropped area and population growth in Egypt from the early nineteenth century to the present day

gramme), which revealed that 486,000 ha would be suitable for reclamation and the introduction of irrigated agriculture. Almost 40 per cent of this land was located within the area of the West Nubarya project.

The soils of the project area were highly calcareous, but also permeable. They presented two major problems with regard to irrigation. Firstly, there was a tendency for thin surface crusts to form when water is applied, which hindered seed germination and restricted aeration of the soil. Secondly, they only retained relatively small amounts of water between field capacity and wilting point. This meant that for efficient crop growth water had to be applied in small doses at frequent intervals. Irrigation water for the site was obtained from the Nubarya Canal which runs at between 4 and 5 m above sea level. As parts of the project area rise to 50 m above sea level, irrigation of these areas was only feasible by the use of pumping stations connected by gravity flow canals.

Before the project was initiated, ground water levels over the site varied in depth from about 20 to 60 m. However, since widespread irrigation commenced, the water table rose rapidly in places at a rate of about 4 m per annum, so that by the early 1970s it was at depths of only 3 m below ground level (ibid., p. 13). Two large groundwater mounds, the result of irrigation water recharge, were identified in the Maryat area. Groundwater flow from these mounds has caused a reversal of the flow direction, with water being carried towards the Nubarya Canal, instead of away from it as was the case before irrigation commenced. The chief problem here is that the groundwater is saline in places with electrical conductivity values of between 8,000 and 80,000 microhms/cm (ibid.). If this water enters the Nubarya Canal in any quantity — as it seems likely to do if the high water table levels are maintained — the irrigation water will become contaminated and perhaps even rendered unsuitable for crop growth.

In the Sudan almost 80 per cent of the labour force still earns its living from agriculture and more than one-third of the GDP is generated by the agricultural sector. Out of a total of 12·44 million ha of arable and permanently cropped land, as much as 10·75 million ha (86 per cent) are cultivated by dry-farming techniques (FAO, 1986). Of this land, approximately three-quarters is cultivated by traditional methods, often involving shifting cultivation. With increasing population numbers the pressure on land resources has increased significantly and desertification and degradation of the land have become severe problems. Recently crop failures have led to mass starvation.

Within the Sudan there are large areas of potentially arable land, claimed by some workers to attain 35 million ha (Whittington and Haynes, 1985). Most of the best land is found on the clay plains adjacent to the Blue Nile, and a large proportion of it could be irrigated without too much difficulty. A feature of Sudanese agriculture has been the vary rapid spread of irrigation in the post-Second World War period. In the 1930s there were 125,000 ha of irrigated land in the Sudan. By the mid-1960s this had risen to 1·18 million ha and to 1·85 million in 1981. If development plans for irrigation schemes progress as scheduled — which seems unlikely given the economic conditions of the Sudan — the irrigated area should reach 2·4 million ha by the early 1990s. Mechanised agriculture in rain-fed areas is also a feature of the post-war period. Much of this is concentrated on the clay plains to the south-east of Khartoum where rainfall averages between 400 and 800 mm per annum. Sorghum is by far the most important crop.

The critical factor affecting future development in the Nile basin concerns water availability. The first formal treaty allocating the waters of the Nile was the Nile Waters Agreement of 1929 between Egypt and the Sudan. This was negotiated as a result of increasing concern in Egypt about obtaining an adequate supply of water for the summer cotton crop between January and July. It should be remembered that, at this time, the storage behind the Aswan Dam was insufficient to provide enough water for irrigation throughout the year. The agreement stated that 48 billion cubic m of water were to be allocated to Egypt and 4 billion to the Sudan leaving 32 billion unallocated (United Arab Republic, Ministry of the High Dam, Aswan High Dam Authority, 1968). Importantly, the flow of the Nile during the crucial low discharge period (20 January to 15 July) was granted entirely to Egypt. The water rights of other states within the basin were not considered at this time.

With the decision to build the Aswan High Dam and, therefore, the prospect of much greater storage capacity available, it was felt politic to update the agreement between the two countries. Since 1929 the Sudan had become an important user of Nile water for irrigation and the 1959 Nile Waters Agreement recognised this fact by producing a more equitable distribution of available resources. With the construction of the High Dam it was predicted that the annual surplus at Aswan would be reduced to 22 billion cubic m owing to evaporation and seepage losses from the reservoir Some

14·5 billion cubic m (66 per cent) of this surplus was allocated to the Sudan and the remaining 7·5 billion (34 per cent) to Egypt. Bearing in mind the original divisions made under the 1929 Agreement, this increased the total water rights of the Sudan to 18.5 billion cubic m per year and those of Egypt to 55.5 billion. The actual share for the Sudan was 20.35 billion cubic m to be withdrawn at Sennar, leaving 65.5 billion to flow into Lake Nasser. With this agreement any above-average flows entering Lake Nasser would become the property of Egypt. Evaporation and seepage losses from the lake were estimated as 10 billion cubic m per annum. Following the construction of the dam, it appears that the annual evaporation and seepage losses are closer to 15 billion cubic m per annum than the 10 billion used in the treaty (Little, 1971, p. 5). The assumed mean flow at Aswan, on which the treaty was based, was 84 billion cubic m per annum, which represents the mean flow of the Nile over a 60-year period (United Arab Republic, Information Department, 1963, p. 6). It was also agreed that Egypt would pay the Sudan the equivalent of £15 million (sterling) as compensation for the flooding of parts of the Nile valley within the Sudan as a result of the construction of the High Dam.

The 1959 Agreement established the Permanent Joint Technical Committee which had the responsibility of implementing the decisions reached. It was also agreed that the Sudan and Egypt would share investment costs in building any new projects on the upper Nile, as well as the extra water resources which would be provided by them. The rights of the other riparian states within the basin were considered and it was decided that any new allocation of water rights would have to be met by lowered quotas for Egypt and the Sudan.

Both the 1929 and 1959 Agreements have taken the view that Egypt's claim to the waters of the Nile using the doctrine of 'prior appropriation' is the correct approach to take from the standpoint of international law. However, it must be recognised that the irrigation demands of the other states, and in particular Ethiopia, have been minimal but that this position may well change in the future.

6.1.8 Future

By the 1990s the water needs of Egypt and the Sudan will be

Table 6.2: Estimates of water supply and demand in Egypt, 1990

	Egyptian Water Master Plan	Waterbury
DEMAND		
Old Lands	29·4	33·0
New Lands	8·5	11·2
Munic. net loss	2·2	4·0
End net loss	0·8	2·0
Navigation	1·6	1·6
Unaccountable and evap.	2·2	7·0
Drainage	14·2	14·2
Total demand	58·9	73·0
SUPPLY		
Aswan	61·7	58·9
Drainage re-use	5·4	6·0
Drainage return flow	–	4·0
Total supply	67·1	68·9
Balance	+8·2	−4·1

Source: Waterbury, 1982; Whittington and Haynes, 1985.

placing extreme pressure on the available resources of the Nile. Estimates of Egypt's water use suggest that by 1990 demand will be between 58·9 and 73 billion cubic m per annum, while water availability values vary from 67·1 to 68·9 billion (Whittington and Haynes, 1985, p. 38) (Table 6.2). The figures of the Egyptian Water Master Plan indicate that there is ample water available for Egypt to go ahead with its new and ambitious irrigation projects. In contrast, Waterbury's estimates suggest that a deficit situation, with a shortfall of 4 billion cubic m per annum, will arise (Waterbury, 1982). With figures like these it is always difficult to reach an objective assessment of which scenario is closest to reality. However, experience has shown that government statistics do tend to put forward the most optimistic interpretation possible. Thus, for example, the official estimate of water use in the Old Lands of Egypt is 12,000 cubic m per annum for surface irrigation, while Waterbury (1979) claims that actual usage is in fact 17,000 cubic m per annum (Adams and Holt, 1985).

261

There is no doubt that water use in Egypt is high compared with many other parts of the world. If more sophisticated application systems (such as sprinkler and drip) were to be used, then crop water requirements could be significantly reduced. In Israel the average use of irrigation water by sprinkler and drip systems is 6,000 cubic m per ha, but here there are higher precipitation amounts and summer temperatures are cooler. The great problem, though, is that these high technology irrigation systems are expensive both to install and to manage. In the Sudan the World Bank (1982) has calculated that the costs of new surface irrigation are US$4,800 per ha, while in Egypt the equivalent costs for auto-mated sprinkler and drip systems are at least double that figure at between $9,600 to 14,400 per ha (Adams and Holt, 1985, p. 82). Further savings could also be made from a more efficient management of existing systems and the introduction of quantitative estimates of actual crop requirements. Such changes would not be expensive to introduce, and savings could be considerable.

In the early 1980s the Sudan was using between 15 and 16 billion cubic m of water per annum from the Nile. With its plans for new irrigation schemes water demand could rise by as much as 10 billion cubic m, to way above its quota of 18·5 billion. However, given the present situation of the Sudanese economy, it does seem unlikely that adequate funds will be available for such development in the near future.

Currently the basic problem in the Nile basin is that Egypt controls the majority of the water resources and yet the Sudan possesses the best land for future irrigation development. Ideally what is needed is a plan which would permit an integrated development of the lower Nile basin. This would permit Egyptian waters to be used in the Sudan on the clay plains of the Blue Nile to the south-east of Khartoum. A possibility for development may be joint ventures between the two countries, whereby the Sudan supplies the land on a lease basis and Egypt provides the water (Whittington and Haynes, 1985). However, the political difficult-ies in both countries probably preclude such schemes for the near future. Much more likely is that each country will continue with its own projects, despite the fact that in economic terms the Egyptian ones have not been showing high returns on the capital and water utilised.

Such development has to be seen in the context of rapid population growth which will see an increase of 65 million people within the basin between 1983 and 2000. Urban and industrial demands for water will grow sharply and, given the higher costs of water that these demands can sustain, increasing pressure will be put on supplies of irrigation water at a time when food demand will be escalating. Just how the Egyptian and Sudanese governments will cope with this dilemma must remain to be seen.

6.2 THE COLORADO BASIN

6.2.1 Introduction

The Colorado river is unusual, if not unique, in that all of its waters are now committed to uses beneficial to humanity (Brede-hoeft, 1984; Fradkin, 1981). As a result, very little of the total discharge of the river flows into the sea at its mouth (Figure 1.10). The waters of the river have only been developed on a large scale since the end of the nineteenth century with the opening up of the American West. Before this time various Indian tribes had utilised irrigation in parts of the basin, but only on a small scale. However, once large water management schemes commenced in the 1920s the pace of development quickened rapidly. Within a 70-year period all of the water resources of the basin had been committed and, from now on, any new water demands will have to be met by the reallocation of existing resources (National Academy of Sciences, 1968; Skogerboe, 1982a).

6.2.2 The Basin

The Colorado River has a drainage area of 94,200 square km and covers parts of seven of the western states. For river planning purposes it is divided into an upper and lower basin, upstream and downstream from the town of Lee Ferry, near the Glen Canyon Dam (Figure 6.4). The upper basin in which most of the water supplies are generated has an area of 42,085 square km, and almost two-thirds of this area remains in public ownership (Iorns et al., 1965). The topography is complex in detail, but is outlined by north to south ranges marking the eastern and western flanks of a high plateau. These ranges attain heights in excess of 4,250 m in the front range of Colorado, and over 3,950 m in the Wind River Range and the Uinta Mountains. This varied topography has a

263

Figure 6.4 The Colorado River Basin showing the major hydraulic facilities

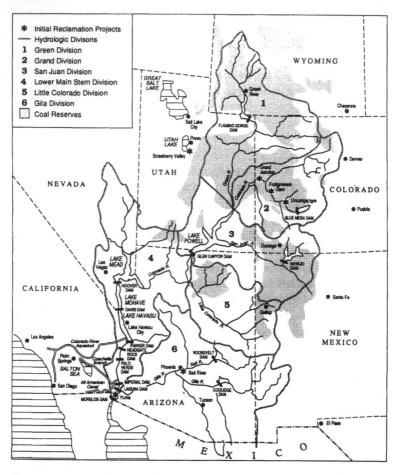

marked influence on climatic conditions. The dominant air masses affecting the region are brought by winds from the Pacific Ocean, though they do have to cross more than 950 km of land before they reach the Colorado basin. These air masses move over the northern parts of the basin and produce precipitation totals in excess of 1,250 mm on the high mountains. A large proportion of this precipitation falls as snow in the period from late October to mid-April. Temperature ranges are considerable, due largely to altitudinal variations. In the lower valleys temperatures can approach 40°C in summer, while in the mountains winter temperatures may remain below freezing point for weeks at a time.

6.2.3 Water resources

It is, of course, the water resources of the basin which are most important for human activities. It is estimated that about 117,800 million cubic m (95 million acre feet (maf)) of water fall on the basin of the Colorado each year, but of this almost 98,678 million cubic m (80 maf) are lost by evapotranspiration. This leaves approximately 18,502 million cubic m (15 maf) (15·8 per cent of total rainfall) for stream discharge and groundwater recharge (Skogerboe, 1982c). Extreme discharges for the upper Colorado basin have varied from a minimum of 6,907 million cubic m (5·6 maf) in 1934 to a maximum of 29,603 million (24·0) in 1917. For planning purposes the key value is the average volume of water which is available in the Colorado River. Like so many other statistics, it all depends on which historical period one is examining. The longest available record, from 1896–1978 (a span of 83 years) suggests an average flow of 18,131 million cubic m (14·7 maf). However, the 57-year measured flow is only 17,022 million cubic m (13·8 maf) (Holburt, 1982a) (Table 6.3). In terms of the driest and wettest decades, the range is from 14,555 to 23,189 million cubic m (11·8 to 18·8 maf). These figures illustrate

Table 6.3: Virgin flow of the Colorado River at Lee Ferry, Arizona

Period	Average annual virgin flow (million m³)	Million acre feet	Comments
1896–1978	18,131	14·7	83-year period of record and estimates
1896–1929	20,722	16·8	34-year wet period
1922–78	17,022	13·8	57-year period of measured flow
1930–68	16,159	13·1	39-year dry period
1969–78	17,145	13·9	10-year recent period
1914–23	23,189	18·8	10-year wettest period
1931–40	14,555	11·8	10-year driest period
1917	29,604	24·0	Maximum single year
1977	6,784	5·5	Minimum single year

Source: Holburt, 1982.

the importance of historical sequences in determining how much water is likely to be available for beneficial uses. It is interesting to note that, at the time of the 1922 Colorado River Compact, the river was experiencing its wettest decade. As a consequence, the negotiators believed that the flow of the river at Lee Ferry was as high as 20,229 million cubic m (16·4 maf) a year. Although there is still some disagreement about average flow values, a figure of 17,022 million cubic m (13·8 maf) per annum is widely accepted for planning purposes. However, from now on it is not average flows, but rather, actual flows which will have the controlling influence on water resource development.

6.2.4 Historical development

The European development of the Colorado basin dates from 1732 when Spanish Jesuits began to open missions in what is now Arizona. This led to irrigation farming along the Santa Cruz River, near Tucson, from the mid-eighteenth until the early decades of the nineteenth century. It was, however, the early exploration of the West by Americans which really transformed the region. In 1869 John Wesley Powell explored the south-west and made his now famous trip through the Grand Canyon. Powell's report had a great impact on the shaping of policy for the development of the whole of the arid south-west throughout the rest of the nineteenth century. From the 1860s onwards the earliest settlers in the Colorado basin were miners, searching for gold and other precious metals. Numerous settlements sprang up, which in turn provided a spur to railroad development. Mining activity was sporadic and soon many of the miners settled down to become farmers or foresters. Many of these farmers made diversion works and canals to promote irrigation, but no major storage facilities were completed.

The real development of the south-west of the USA stems from the passing of the Reclamation Act of 1902, with its objective of opening up the arid and semi-arid lands. A Federal agency, the Reclamation Service, was created to achieve this aim. In 1923 this agency became known as the Bureau of Reclamation, a title it still carries today, though for the short period between 1979 and 1981 it was renamed the Water and Power Resources Service. Since its formation the Bureau of Reclamation has played the dominant role in the management of the water resources of the Colorado

River. Its first task was the establishment of five major irrigation projects on which work began soon after the passing of the Reclamation Act. These were the Salt River Project, Arizona; the Yuma Project, Arizona and California; the Strawberry Valley Project, Utah; the Uncompahge Project, Colorado; and the Grand Valley Project, Colorado. The major change signalled by the work of the Bureau compared with earlier schemes was the fact that large water storage facilities were now able to be constructed. Also new was the involvement of a large Federal agency with the resources, influence and incentive to get projects built.

6.2.5 Major projects

From 1903 onwards a series of major construction projects began on the Colorado, which are still continuing at the present day (Skogerboe, 1982d). It is easy to forget that major projects take many years to complete: for example, the Salt River Project, which was authorised in 1903, and which began delivery of water in 1907, was not finally completed until 1946. It involved the construction of six dams with a combined storage capacity of 2,467 million cubic m (2 maf) and permitted the irrigation of 101,474 ha (250,000 acres). It also included five hydro-electricity plants with a generating capacity of 232 MW. This project was typical of the large multi-purpose projects which were to become a symbol of the work of the Bureau of Reclamation.

From the early 1900s a series of agreements have been reached by the users of the waters of the Colorado River, which together have subsequently become known as the 'Law of the River'. One of the earliest of these was the Colorado River Compact, signed in 1922 by the states of the basin. This divided the waters of the river between the upper and lower parts of the basin, but it did not make any attempt to allocate waters amongst the seven states of the basin.

All the early projects were, however, dwarfed by the Boulder Canyon Project which was authorised in 1928, following planning going back to 1918 (Worster, 1984). To facilitate its building the Boulder Canyon Project Act was passed in 1928, which authorised the construction of the Hoover Dam and the All-American Canal system, as well as approving the Compact of 1922. The Project was to be multi-purpose, with prime objectives of flood control, water

267

storage and hydro-electric power generation. Situated in the lower part of the basin, it was ideally placed to provide control of most of the water resources of the Colorado River. The original capacity of Lake Mead, the reservoir behind the dam, was 39,965 million cubic m (32·4 maf) (about twice the annual flow of the river). In total it provided a dependable water supply for irrigating 263,051 ha (650,000 acres) in southern California and south-west Arizona and a further 161,878 ha (400,000 acres) in Mexico. In power terms it yielded 4 billion kW-hours per annum of firm output. Associated with the Hoover Dam was the All-American Canal system. This obtained water from the Imperial Dam, 480 km south of the Hoover Dam and conveyed it in a concrete lined canal to irrigate 214,488 ha (530,000 acres) in the Imperial valley and a further 31,780 ha (78,530 acres) in the Coachella valley.

The construction of the Hoover Dam had an important impact on Mexico. Before its construction in 1935, the vast variations in flow of the river had meant that Mexico had been unable to use more than 925 million cubic m (750,000 acre feet) each year. With the assured summer flows once the dam was in operation, Mexico's use grew rapidly to 2,220 million cubic m (1·8 maf) in 1943. To recognise this increased use the Mexican Treaty was signed in 1944, by which the USA agreed to deliver to Mexico 1,849 million cubic m (1·5 maf) each year.

Besides the Boulder Canyon scheme, a whole series of other projects were implemented by the Bureau of Reclamation between the 1920s and the 1950s. These were the Moon Lake Project, Utah; the Gila Project, Arizona; the Fruitgrowers' Dam Project, Colorado, the Mancos Project, Colorado; the Schofield Project, Utah; the Collgran Project, Colorado; the Parker-Davis Project, Arizona, California and Nevada; and the Palo Verde Project, California and Arizona. In common with earlier schemes, they consisted of water storage facilities and distribution networks, with the emphasis on the provision of irrigation water.

Just as the Boulder Canyon Project commanded the lower Colorado in the 1920s and 1930s, so the upper Colorado was dominated by the Colorado–Big Thompson Project in the 1930s and 1940s. Unlike the Boulder Canyon Project, which consisted of one enormous structure, the Colorado–Big Thompson Project was a complex scheme made up of more than 100 structures, operated in an integrated manner. One of the major objectives of the project was to divert 320 million cubic m (260,000 acre feet) of water from the Colorado basin to demand-areas to the east of the

Rockies. It was also to provide supplemental irrigation water for 291,380 ha (720,000 acres), water for industrial and domestic uses, to generate hydro-electric power and to facilitate water-based recreation. Construction of the project began in 1938 following the enactment of the Colorado Water Conservancy Law by the Colorado legislature in the previous year. An unusual feature of this Law was that people were to contribute to the costs and

Figure 6.5 Colorado River Storage Project (CRSP). The location of storage capacity and participating projects

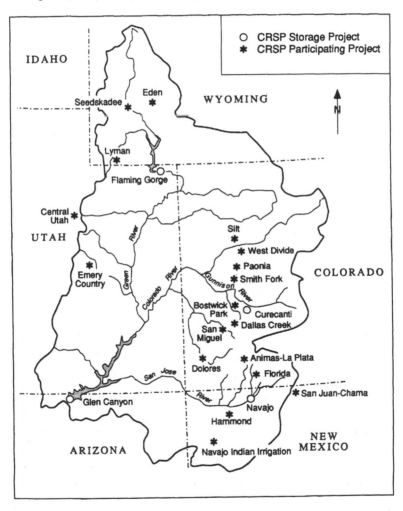

operation of the project in proportion to the benefits they received from it.

By the late 1940s it was becoming obvious that the piecemeal development of the water resources of the Colorado basin were not permitting the most efficient use of the available resources. Partly to overcome this, the Upper Colorado Compact was established in 1948, which divided the waters of the upper part of the river amongst the upper basin states. A particular problem was that in dry years there was insufficient water available in the upper basin to permit this area to increase its water use as permitted under the 1922 Colorado River Compact, and at the same time make the necessary deliveries of water to the lower basin. What seemed to be needed was yet more storage capacity in the upper basin. Accordingly four major reservoirs were authorised for construction in 1956 as part of the Colorado River Storage Project. These were the Glen Canyon Project, Colorado; the Flaming Gorge Project, Utah; the Navajo Project, New Mexico; and the Curecanti Project, Colorado (Figure 6.5). The aim of the scheme was to operate these reservoirs in such a way that, in dry years, the water flow into the lower basin at Lee Ferry would meet the Compact requirements. The total water storage provided by these four reservoirs was 41,938 million cubic m (34 maf), a figure slightly larger than that originally available in Lake Mead. Of the reservoirs Lake Powell, behind the Glen Canyon Dam, is by far the largest with a storage capacity of 30,303 million cubic m (27 maf). Hydro-electricity plants were constructed at three of the dams, the exception being the Navajo Dam.

Associated in part with the Colorado River Storage Project Act was the Central Utah Project, which abstracts water from the Colorado, and other sources, for use in the Uintah and Bonneville basins. Like many other projects it is multi-use in concept. There are six units in this scheme — Vernal, Bonneville, Jensen, Upalco, Uintah and Ute Indian. Four of them were authorised by the 1956 Colorado River Storage Project Act and the remaining two by the later 1968 Colorado River Basin Project.

The need for yet another overview of water resource development in the basin was recognised by the authorisation, in 1968, of the Colorado River Basin Project. Its comprehensive nature is seen in a list of its aims: regulating flows on the Colorado River; controlling floods, improving navigation; providing for storage and delivery of water of the Colorado River for reclamation of lands, including supplementary water supplies and municipal, industrial

Figure 6.6 Central Arizona Project

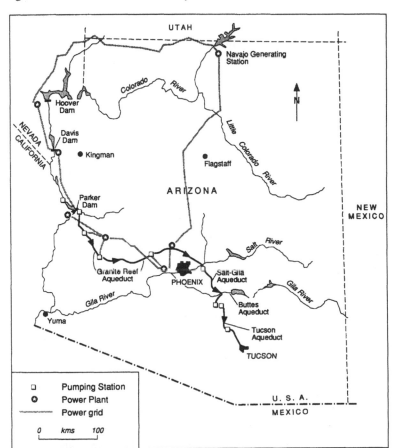

and other beneficial purposes; improving water quality; providing for outdoor recreation facilities; improving fish and wildlife conditions; and generation and sale of electric power. It will also provide the framework for a regional water plan; for the satisfaction of the Mexican Water Treaty; and long-range augmentation studies of the Colorado River.

This Act, the result of widespread discussions between all interested parties, also authorised the Central Arizona Project (Figure 6.6). This is a multi-purpose project to provide supplemental irrigation for 404,694 ha (1 million acres) of land and 617 million cubic m (500,000 acre feet) per annum for urban and industrial use in Phoenix and Tucson. It will also generate power, provide flood and sediment control and facilitate outdoor recreation. The main feature of the scheme, on which construction began in 1973, is the Granite Reef aqueduct, which carries water 300 km to Phoenix and has a maximum capacity of 85 cubic m (3,000 cubic feet) per second. It is expected that the project will be in full operation by the late 1980s.

6.2.6 Future water supply and demand

Detailed estimates of water use in the Colorado basin in 1980 have been provided by the US Geological Survey, and these indicate that total surface water withdrawals are 47·69 million cubic m per day (12,600 million US gallons/day (mgd)) and groundwater 17·56 million cubic m per day (4,640 mgd) (Solley et al., 1983). The surface water figure, it will be noted, is slightly higher than the mean average flow of the Colorado River at Lee Ferry. In terms of the two divisions of the river, the upper and lower Colorado, water withdrawals in each are surprisingly similar. In the upper basin the average figure is 32·17 million cubic m per day (8,500 mgd), against a value of 32·93 million (8,700) in the lower basin. However, in the lower basin slightly over half of the total (17·03 million cubic m per day) (4,500 mgd) is from groundwater sources, whereas in the upper basin the equivalent figure is only 0·53 million cubic m per day (140 mgd).

By far the largest user of water in both parts of the basin is irrigation, accounting for 88·2 per cent in the upper basin and 87·4 per cent in the lower basin (Frederick and Hanson, 1982). Given the higher temperatures in the lower basin it is not surprising that consumptive use at 16·28 million cubic m per day (4,300 mgd) is more than double the figure of 7·57 million (2,000) observed in the upper basin. Throughout the lower basin, the contributions from surface water and groundwater to irrigation are almost equal, with groundwater supplying the slightly larger amount. In contrast, 98·6 per cent of the irrigation water of the upper basin is derived from surface water sources.

A feature of irrigated agriculture in the Colorado basin is the high technology and investment levels employed. Sprinkler systems, particularly the central pivot systems, were widely introduced during the 1960s and early 1970s to reduce labour inputs and make more efficient use of available water resources. In the late 1970s laser-levelling techniques were used on certain projects, such as the Welton–Mohawk, to produce large basins of about 4 ha in extent with almost zero surface slopes. This has enabled rapid flood irrigation to be practised with minimum soil erosion and relatively uniform water application rates. On many of these systems the headgates to the basins are controlled electronically from a central control point, so permitting field irrigation to take place with minimum labour inputs. This allows a highly centralised system where all decision-making can be evaluated scientifically before being put into operation. Such systems involve heavy investment to establish, but operating costs are low.

Figure 6.7 Welton-Mohawk irrigation project

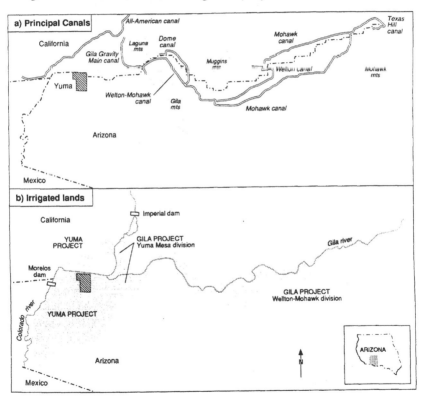

Source: Redrawn from Schilfgaarde J.V., 1982, 'The Welton-Mohawk dilemma', in 'Water and development in an arid environment: the Colorado River Basin', *Water Supply and Management*, v. 6, pp. 3–10. Reprinted with permission from Pergamon Journals Ltd.

The total reservoir capacity in the Colorado basin is now about 78,943 million cubic m (65 maf) or about 4·7 times the mean annual flow. Of this total, 85 per cent is accounted for by the two lakes behind the Hoover and Glen Canyon Dams. These are Lake Mead with 35,154 million cubic m (28·5 maf) storage and Lake Powell with 33,303 million (27·0). The total active storage is about 61,674 million cubic m (50 maf).

One of the major problems in the lower Colorado basin in the post-Second World War period has been the marked increase in water salinity associated, in particular, with the establishment of the Wellton–Mohawk irrigation project, to the east of Yuma, Arizona (Evans et al., 1982; Law and Hornsby, 1982; Schilfgaarde, 1982) (Figure 6.7). This was authorised in 1947 and completed in 1952 by the Bureau of Reclamation. Unfortunately the irrigated lands soon became waterlogged, necessitating the construction of a drainage project which began operation in 1961. The waste waters from the scheme were discharged into the Colorado River below the last US diversion point, but above the Morelos Dam, which diverted water for Mexico. The groundwater pumped into the Colorado was highly saline, with an initial total dissolved solids value of over 6,000 parts per million (ppm). When this discharge began, river flows were low and so the dilution effect of the Colorado River water was minimal. As a result the water delivered to Mexico increased in salinity from 800 ppm in 1960 to more than 1,500 ppm in 1962. Not unexpectedly, the Mexican government protested strongly to the USA. Prolonged discussions between the two nations eventually led, in 1973, to an agreement referred to as Minute 242 of the International Boundary and Water Commission. By this the USA was to ensure that the quality of Colorado water above the Morelos Dam would have an average salinity of only 85–145 ppm above the average salinity value at the Imperial Dam. The necessary legislation to implement this agreement was passed by Congress in the following year (1974), and it included the necessary finance to construct a number of desalination units to treat the irrigation return waters near Yuma.

Strong differences of opinion exist between the states of the upper and lower basin as to the interpretation of the Colorado River Compact, particularly with respect to the Mexican Water Treaty (Holburt, 1982a). Article IIIc states that, if there is a water surplus after the Compact requirements have been fulfilled, and if this surplus is insufficient to supply Mexico's demands, this deficiency is to be met equally by the upper and lower basin. What is at issue is whether or not there is a surplus. The lower basin states maintain that, in future, no surplus will exist and, therefore, the upper basin states will have to provide half of the 1,850 million cubic m (1·5 maf) per annum (together with losses) which has to be delivered to Mexico. In contrast, the upper basin states claim that there will still be a surplus and, as a result, they are not required to provide any of the water for the Mexican Treaty. The 1968 Colorado River Basin Project Act recognised that the Mexican Treaty was, in fact, a national obligation; it went on to state that the Treaty would be able to make first claim on any project built to increase the Colorado River flow by more than 3,084 million cubic m (2·5 maf) per annum.

A potential problem, the seriousness of which is very difficult to assess, concerns Indian water rights (Smith, 1982). In a famous case, often referred to as the Winters Doctrine or the Reservation Doctrine, Indian water rights were recognised as paramount over all others, because the Indian reservations were created before any river diversions were established by Americans of European origin. The crucial judgement states: 'The Indians had command of the lands and the waters, command of all their beneficial use, whether kept for hunting, and grazing roving herds of stock, or turned to agriculture and the arts of civilisation.' As a result of this judgement, the Indian tribes could theoretically claim rights to utilise much more of the water of the basin than they at present make use of. How modern courts would deal with such claims is as yet unknown.

The increasing demand for water is related to the growth of economic activity within the basin of the Colorado and in adjacent areas (Mann, 1975; Petersen and Crawford, 1978; Skogerboe and Radosevich, 1982). Population projections suggest that major growth will occur in the lower part of the basin, where numbers are expected to increase from the present value of around 3 million to about 7 million by 2020 AD. In contrast, the predicted growth in the upper Colorado is much less, with an increase from the current value of 400,000 to 700,000. In the Mexicali valley of Mexico, detailed population statistics are not available, but it is thought that the present population of around 500,000 will double by 2020 AD.

Associated with these problems is the fact that the Colorado basin contains a huge potential for the development of energy resources including coal, oil shale, oil, gas, uranium, and hydropower (Abbey, 1979; Blackman et al., 1973; Flug et al., 1979,

1982). Which, and where each will be developed, depends almost entirely on market conditions within the US economy. What is beyond doubt, though, is that the development of these resources is bound to place even greater strains on the available water resources of the basin. Indeed, certain estimates suggest that by the year 2000 the demands from the energy industry may be up to 1,000 million cubic m per annum.

It is now obvious that the development of the water resources of the Colorado River has reached the limit in terms of those currently available, and that any further demands will have to be supplied by the reallocation of existing supplies (Engelbert and Scheuring, 1984; Westcoat, 1984). In effect this means the reallocation of irrigation water as agriculture is by far the largest water consumer. The critical demand areas look as if they will be those around Phoenix, Tucson and the Mexicali Valley.

From the late 1980s onwards, when the Central Arizona Project reaches full capacity, water use in the basin will exceed the annual flow and from then on water will have to be taken out of storage, either from groundwater reserves or surface water reservoirs. The best available estimates suggest that by 1990 water use will be approximately 17,022 million cubic m (13·8 maf) per annum, compared with a figure of 15,172 million (12·3) in the late 1970s (Holburt, 1982a). Everyone agrees that by the year 2000, the competition for water resources will be extremely severe. What is still disputed, however, is how the deficiencies are to be coped with. Providing that there are not long periods of low flow, it does seem that the available water resources in storage can be mined to provide adequate water to last well into the twenty-first century. This is not, however, a long term solution to the basic problem of excessive water use.

7

Oases: Isfahan and
Salt Lake City

7.1 INTRODUCTION

Throughout the arid zone, oases have developed wherever reliable
water supplies have permitted permanent settlement. The size of
the settlement which has come into being depends on many
factors, such as communications, a fertile hinterland and security,
though perhaps the most important is the volume of water available. In the West the most common image of an oasis consists of a
small pool of groundwater, set amongst date palms and surrounded
by small white flat-roofed houses. A camel or two and people in
flowing robes would complete the picture. Such oases do, of
course, exist throughout north Africa and the Arabian Peninsula.
Their importance normally lies as staging posts along lines of
communication and trade. Much more important, though, are the
oases which have given rise to major urban centres. These are
often found in inland locations and are watered by a major river or
series of smaller streams. For them to be prosperous large areas of
fertile land have to be located nearby. In the Old World,
Damascus, Isfahan, Samarkand and Kairouan stand out, while in
the New World, Salt Lake City, Phoenix and Tucson provide
recently created examples.

In this section the setting and evolution of two oases will be
examined: from the Old World, the oasis of Isfahan, and from the
New World, Salt Lake City have been chosen. It is important to
realise that the success of an oasis depends upon its agriculture and
that the growth and prosperity of the major urban centre rests
initially on wealth being created from agricultural production.
Once above a certain population size, self-sustaining growth
within the city might be generated through craft, industrial and

commercial activity. The basic resources for producing agricultural prosperity are abundant sunshine owing to the arid climate, ample water resources and adequate flat land which can be irrigated. The utilisation of these resources requires an efficient management system which becomes increasingly complex and integrated as the oasis grows. The two oases studied are very different. Isfahan has a history spanning at least hundreds of years and has been characterised by a traditional agricultural system based almost exclusively on solar energy imputs up to about 1960. In contrast Salt Lake City is only 150 years old and its original settlement and growth is closely associated with the Mormon religion. Although originally using traditional agricultural methods, it quickly adopted modern high technology systems with high fossil fuel inputs. Today it has one of the most sophisticated irrigation systems in the world.

7.2 THE ISFAHAN OASIS

7.2.1 Introduction

Isfahan is the second largest city in Iran. It is situated in the Isfahan basin which is one of the world's great oases, and is now a major industrial centre as well. The basin is located on the western margin of the Iranian plateau adjacent to the Zagros Mountains. The water for the oasis comes from the Zayandeh River which rises high in the Zagros Mountains (Beaumont, 1973) (Figure 7.1). For most of its upper course the river flows in a gorge-like channel, with its valley only broadening into a floodplain about 20 km upstream from Isfahan. The main agricultural area stretches from this point to about 40 km downstream from the city. Given the very low annual rainfall (about 100 mm in Isfahan) agriculture can only be carried out using irrigation. Most of the water comes from the river itself, but on the edge of the basin many villages utilise groundwater from the alluvial fans on which they are situated. There are more than a thousand villages in the basin.

The site of Isfahan appears very old, but knowledge of its early beginnings are sparse. During the last one thousand years the city has enjoyed two periods of considerable prosperity, separated by intervals of political instability and economic decline. The first such period occurred under the Seljuks in the eleven and twelfth centuries when Isfahan became capital of the empire. The site of

Figure 7.1 The Isfahan Oasis and the Zayandeh River basin, Iran

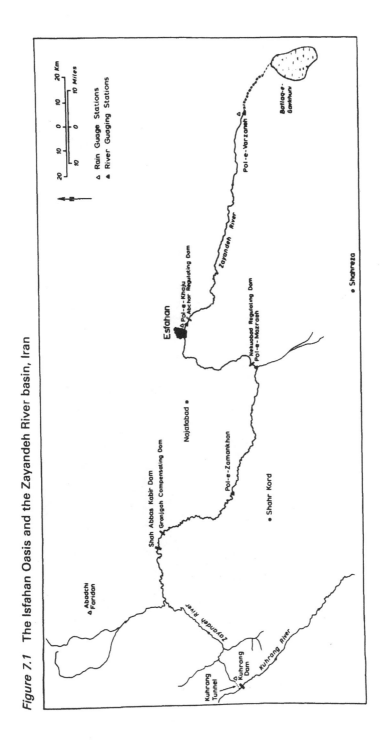

Plate 7.1 The Shah Mosque in the centre of the city of Isfahan.

Plate 7.2 The Isfahan Oasis from the air.
The photographs covers the area to the east of the city.

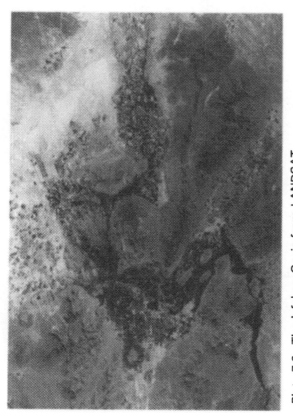

Plate 7.3 The Isfahan Oasis from LANDSAT.
The Zayandeh River flows from left to right across the photograph. The dark areas are irrigated lands. These are mostly small irrigation fields with many tree crops on the left hand of the photograph. On the right hand side as water quality declines the pattern of vegetation cover becomes less dense, with many fewer trees present. The city of Isfahan is a light coloured area in the centre top. The major roads can be clearly distinguished. New irrigation development with large rectangular fields can be seen in the bottom centre.

the town was then north and east of the present centre. The Mongul invasions led to a long period of instability and economic depression in Iran, which continued until the mid-part of the sixteenth century when Safavid rule began. The Safavid period marked Isfahan's greatest prosperity prior to the twentieth century and it was at this time that many of Isfahan's great public buildings were constructed. The real key to the improvement of Isfahan's fortunes occurred in the last decade of the sixteenth century, when Shah Abbas I moved his capital from Ghazvin to Isfahan. The prosperity of Safavid rule finally collapsed in 1722 when Afghan invaders overran the city. From then until the early part of the twentieth century, the region experienced a period of stagnation and decline.

There seems little doubt that even during times of economic depression large parts of the oasis were still cultivated. Under these conditions, irrigation is most likely to have occurred close to the river itself, where water could easily be obtained. In effect what probably happened is that the cultivated area expanded and contracted throughout history in response to economic and social conditions as well as environmental factors. During periods of political stability the area of cultivated land is likely to have expanded, whilst during times of unrest the more marginal lands and villages were abandoned. In times of severe drought, such as occurred between 1869 and 1871, large areas of irrigated land had to be abandoned owing to lack of water (Okazaki, 1986). During the twentieth century there has been the extra factor of a rapidly rising population within both the basin itself and the country as a whole.

7.2.2 Environment

Because of its high elevation (mostly above 1,400 m) and its inland location, the Isfahan basin experiences large temperature fluctuations between winter and summer. In January the mean temperature is 3·3°C, while in July it rises to 28·2°. Frosts have been recorded in six of the twelve months, but in most years the growing season is of eight months duration. Within the watershed of the Zayandeh River there are large variations in precipitation totals. In the headwater region annual falls are in excess of 1,400 mm, while close to the inland lake into which the river drains they are less than 50 mm. (Table 7.1). Throughout much of

283

the cultivated area of the basin, annual precipitation varies from about 150 mm to 50 mm in a west to east direction. Almost all of the precipitation falls in winter from November to May inclusive. Isfahan itself receives about 60 rain-days each year, with precipitation coming in bursts of 1 to 4 days associated with the passage of eastwards-moving depressions.

Table 7.1: Annual precipitation values at four stations in the Isfahan oasis, Iran (mm)

Month	Kuhrang (1966–72)	Zamankhan (1966–72)	Isfahan (1966–72)	Varzaneh (1966–72)
October	33	8	2	1
November	128	26	10	5
December	210	44	12	5
January	145	68	16	11
February	291	40	20	9
March	272	39	14	8
April	253	52	17	8
May	100	29	11	4
June	3	3	6	3
July	0	1	0	0
August	0	0	0·5	0
September	3	0·5	0	0

Source: Ministry of Roads, Iran, *Meteorological yearbooks*, 1964–72.

It is, of course, the Zayandeh River which has permitted the growth of the Isfahan oasis. The river rises high in the Zagros Mountains and then flows in a south-easterly direction before losing itself in the Ghav Khuni salt flats. It is characterised by a marked snowmelt regime with maximum discharge occurring in April and May. Minimum flow takes place in September, October and November. What is remarkable about the Zayandeh River is that it exhibits a marked reduction of flow in a downstream direction (Figure 1.8 and 7.2). At Varzaneh, on the margin of the Ghav Khuni salt flats, the annual flow is only one-tenth of that at Zamankhan in the Zagros foothills. The reasons for this decline are believed to be diversion of water for irrigation in the Isfahan basin and percolation of water through the bed of the stream in the floodplain areas to recharge the alluvial aquifer system. Little is known about the water quality along the Zayandeh, but the available evidence suggests that quality decreases through the oasis area as a result of return drainage from irrigated fields.

Certainly on the margin of the Ghav Khuni water in the river has a high salt content except under high discharge conditions.

Figure 7.2 Flow of the Zayandeh River at Pol-e-Zamankhan and Pol-e-Varzaneh (1964–65)

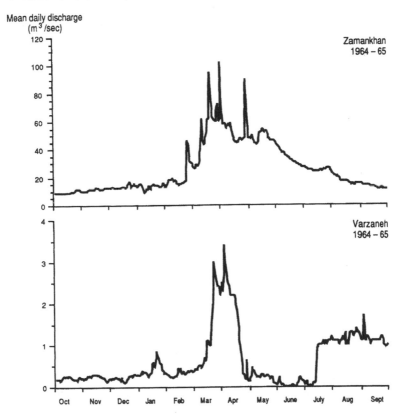

7.2.3 Land use

Although most of the villages close to the Zayandeh River use surface water for irrigation, many villages on the periphery of the basin rely on groundwater. These villages are often situated on alluvial fans and the groundwater beneath them was traditionally tapped by the use of qanats. Since the 1950s many pumped wells have been installed throughout the Isfahan basin and these have permitted the irrigation of areas which formerly were used only for grazing purposes. A detrimental effect of some of these new wells has been to lower the water-table in a downslope direction, causing many qanats to decrease in discharge or to dry up completely (Beaumont, 1981).

Within the basin itself the allocation of irrigation water rights has always been of vital importance to the traditional agricultural system. Over the years a code of water distribution known as the 'Tumar' had been built up which has survived into the twentieth century (Lambton, 1953). The most important aspects of the water distribution system are as follows. Between late November and the end of May anyone throughout the basin can abstract as much water from the Zayandeh River as they wish. During the rest of the year, when water is scarce, strict controls are enforced. Basically the water was divided into 33 primary shares, and then these were allocated amongst the districts of the oasis. The upstream districts of the oasis, Lenjan, Alenjan, Marbine and Jey, were the most privileged in terms of water availability (Figure 7.3). Those downstream from Isfahan, Kararadj, Baraan and Rudashtin only received water at the beginning and end of the growing season. In the 'Tumar' the 33 main shares were further subdivided into 276 secondary shares associated with the major irrigation canals and into 3,105 tertiary shares at the village level. With such complex water rights the successful operation of the system depended on the honesty and efficiency of the officials administering them. Maintenance of the canals was a communal responsibility and was carried out by the farmers of the villages which received water from a particular canal.

One of the main problems with the traditional system was the very favourable treatment of the upstream districts compared with their downstream neighbours. Water allocation has always been imprecise owing to the lack of accurate measuring instruments and it was always possible for unscrupulous officials to divert water away from the rightful owner at the village level. With water being allocated on a time basis, for example the flow of a canal for a half-day period, it meant that under low-flow conditions, less water was actually delivered to the fields than when water levels were high. It also meant that during low flows insufficient water was left for downstream users.

The total area which could be irrigated within the oasis by the use of traditional irrigation canals is believed to be about 80,000 ha. However, water shortages meant that the actual cropped area in any year was usually below 45,000 ha. Cereals,

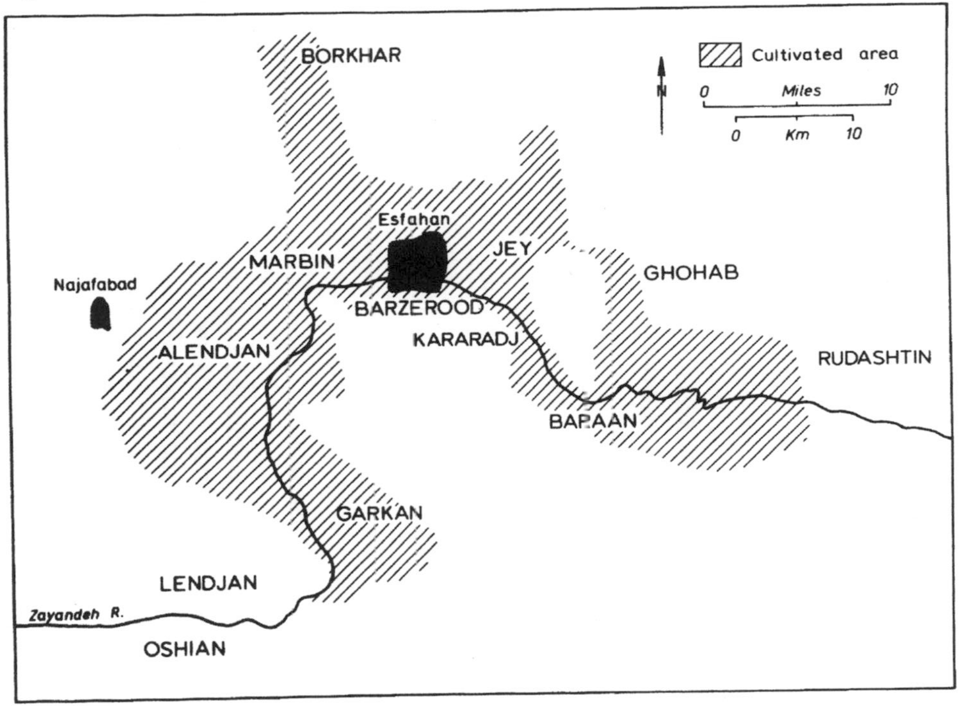

Figure 7.3 Cultivated area and districts of the Isfahan Oasis

mainly wheat and barley, were the most important crops and accounted for at least half the cropped area in many districts. Planting of the cereals took place in October and November and harvesting in late May and early June. In the mid-1960s the highest yields of between 2·4 and 3·4 tonnes per ha were found in the better-watered areas of the oasis upstream from Isfahan. In the downstream districts much lower yields of less than 2·0 tonnes per ha were recorded, which reflects the lower quality of both soils and water in a downstream direction. Around the city of Isfahan itself and in many of the larger villages in the western part of the oasis, orchards and gardens were and still are very important, and they make an important contribution to the local economy.

An idea of the traditional pattern of life in the oasis of Isfahan can be gleaned from a survey undertaken by the Institute of Economic Research, University of Tehran, in 1965 and 1966. This revealed local conditions prior to the construction of the Shah Abbas Dam and the main phase of industrialisation in the region. For the 14 districts of the oasis the survey recorded a rural population of 615,000 in 814 villages. The population of the urban area of Isfahan at this time was about 424,000, which gives a total population of the basin of around one million people. The area of arable land, including fallow, was about 80,000 ha, while the land actually being cultivated was 45,000 ha. The years of the survey had approximately average river flows.

7.2.4 River management

The chief problem within the Isfahan basin has always been the variability of river flow. In some years devastating floods have occurred, while in others river volumes have been very low. Low river levels meant that the cultivated area had to be drastically reduced and this sometimes led to acute food shortages, though rarely famines. It is not surprising, therefore, that attempts have been made to increase the flow of the Zayandeh River. One of the oldest ideas, dating from the sixteenth century, was to divert the Kuhrang River, a tributary of the Karun, into the headwaters of the Zayandeh. Nothing came of this until 1946 when work on a three-kilometre tunnel under the Zagros Mountains began. It was completed in 1953 and diverted a maximum flow of up to 35 cubic m per second. In most years this diversion augments the flow of the Zayandeh by about 200 million cubic m.

It soon became obvious that the unregulated flow of the Zayandeh River would not be able to supply the growing water needs of the Isfahan basin. In the 1950s, it was therefore decided to build a dam on the river to store the spring and early summer flood waters, which had previously run to waste into the Ghav Khuni salt flats. This water could then be used to supply the peak summer demands of the oasis, as well as HEP for the Isfahan region. Work began in 1967 and the dam was completed in 1970. The reservoir behind the dam provides a total usable storage of 1,260 million cubic m. Average flow into this reservoir was estimated to be 1,240 million cubic m each year. However, this figure may well be optimistic as it was based on what now seems to have been a wetter than usual period during the 1950s and early 1960s.

With the extra water available it was calculated that the total area of land which could be irrigated was 95,000 ha, though the actual land receiving water was thought likely to be 86,000 ha. This was approximately 90 per cent more than the average irrigated area observed in 1965 and 1966. With the new project the basic idea was that all land capable of being irrigated would receive water each year. The practice of fallowing would be done away with and replaced by cash crops, such as sugar beet and fodder.

Figure 7.4 Regulators and main canals of the modern irrigation network, Isfahan Oasis

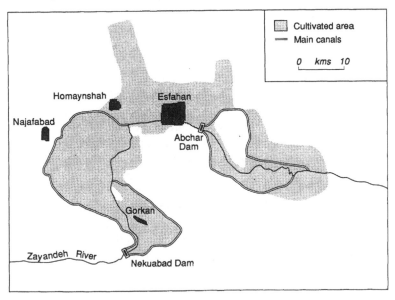

Associated with the new Shah Abbas Dam was the construction of a major irrigation network for the oasis. The key features of this were two river regulators at Nekouabad and at Ab-char. These raised river water level and so permitted diversion of water into concrete canals on either bank of the stream (Figure 7.4). From these canals water is distributed to the fields by means of a series of secondary and tertiary canals. A new drainage system has also been constructed.

The average water needs for the irrigation project were calculated at 14,730 cubic m per ha per annum, giving a total requirement for the 86,000 ha of 1,265 million cubic m (Beaumont, 1981). It was believed, however, that about 10 per cent of this total could be met from groundwater sources. When the project was completed the original idea had been to use the new concrete canals to supply water to the traditional canals or mahdis. However, technical difficulties arose which meant that the traditional canals had to be used for much longer than expected for abstracting water directly from the river. The most difficult problem was the subsidence of sections of the main canals when they crossed certain strata.

One of the most remarkable features of the Isfahan oasis has been the growth of the urban area of Isfahan itself. No accurate population statistics are available for the early years of this century, though estimates suggest that following a series of droughts in the late nineteenth century the number of inhabitants in the city may have fallen to about 100,000. A detailed map of Isfahan, dating from 1919, reveals a small and compact urban form. Since this time the built-up area has expanded noticeably, devouring some of the most fertile gardens of the oasis, while new roads have smashed through the older parts of the town to cater for the motor car. A master plan of Isfahan, drawn up in 1968, reveals a planned urban area about eight times larger than the 1919 city (Figure 7.5). This potential loss of fertile agricultural land has so worried the planners that they have made an attempt to encourage future urban growth along the Tehran road to the north-west of the city where land quality is low.

Another major problem is providing water for the new industrial and urban demands (Figure 7.6). The largest single water user to date is the Russian steel mill to the south-west of the city. This has a design water need for 300 million cubic m of water a year, though it is hoped that a large proportion of this can be met by recycling. Other large prestige works commissioned during the

Figure 7.5 Isfahan Master Plan (1968)

Figure 7.6 Major new projects in the Isfahan basin

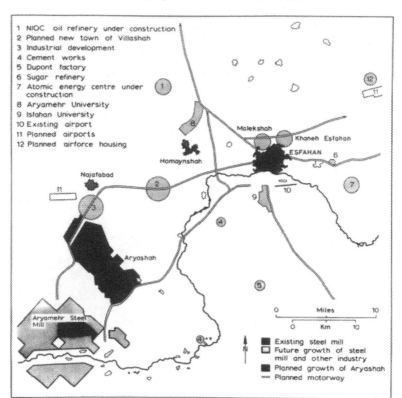

reign of the Shah were a second steel plant built by the British and a Bell helicopter plant. Other major projects included an oil refinery, a cement works, a sugar refinery and an atomic energy plant. Under the new regime of the Islamic Republic it is obvious that the economic development planning is not as ambitious as was the case under the Shah. If further industrial development goes ahead water demand will increase dramatically.

Yet further growing water demand is being generated within the city of Isfahan itself, which now has a population of three-quarters of a million people. With rising standards of living, *per capita* demand is also growing, thus imposing an extra burden on water supplies. Besides Isfahan there are other growing centres: for example, the new town associated with the steel mill complex is predicted to have a population of 200,000 by the end of the century and other new towns are planned to the north of Isfahan. If these developments go ahead the population of the Isfahan basin could be around three million by the year 2000.

Given these extra demands it is obvious that there is insufficient water in the river to meet all the needs, as well as to supply the proposed increase in irrigated area. This is despite the fact that yet more waters (probably about 250 million cubic m each year) are to be diverted into the Zayandeh River. The inevitable result will be that growing urban and industrial needs can only be met by less use of water for irrigation. At first this must mean a decrease in irrigated area, or — perhaps more correctly — that the irrigated area will not be able to expand to the target figure. Some modest expansion might still be possible if the farmers were to change to a more efficient method of water distribution than the inundation and furrow techniques which are widely used. Sprinkler irrigation would certainly reduce water demands, but it is difficult to see how the peasant farmers will be able to afford such infrastructures.

7.3 SALT LAKE CITY AND THE WASATCH OASIS

7.3.1 Introduction

The term Wasatch oasis is given to the belt of irrigated land stretching from Logan in the north, where the waters of the Bear River were used, to Provo in the south, which the Mormons developed in the nineteenth century (Figure 7.7) (Meinig, 1965). Today it includes about 178,065 ha (440,000 acres) of irrigated land in the counties of Box Elder, Cache, Rich, Weber, Davis, Morgan, Summit, Salt Lake and Utah (US Department of Commerce, Bureau of the Census, 1984). Almost all of this land is irrigated by water generated within the watershed of the Great Salt Lake as water flowing from the Wasatch Mountains or as groundwater. The range of crops is large but is dominated by wheat, fruit, vegetables, sugar beet and potatoes. Irrigated fodder crops are widely grown for beef and dairy cattle.

The story of Salt Lake City and the Wasatch oasis is the story of the Mormons. It can be said to have begun on 4th February 1846 when the Mormons crossed the Mississippi River on their trek westwards under the leadership of Brigham Young. Eventually, after a long and exhausting journey, Brigham Young reached the Great Salt Lake valley on 24 July 1847 where he pronounced his famous statement, 'This is the place.' Settlement began

Figure 7.7 Wasatch Oasis and Salt Lake City, Utah

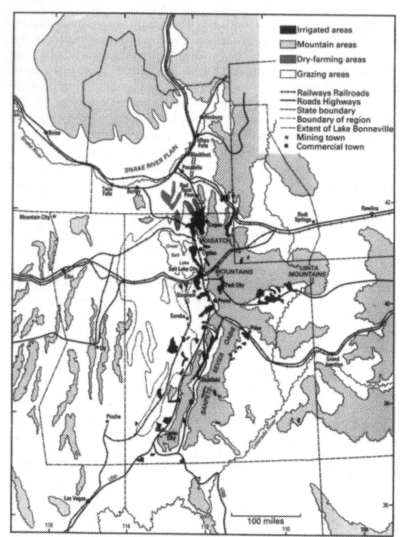

immediately: by early August plans for the new settlement were being drawn up and parties were sent out to explore the terrain. Brigham Young returned to spend the winter in Nebraska but returned in September 1848 with a new group of 2,400 settlers. On his return Salt Lake City had 1,500 inhabitants and ten other settlements had already been established in the valley.

There has always been controversy as to whether the settlement site was the result of divine inspiration given to Brigham Young when he viewed the Great Salt Lake valley for the first time, or whether it was the result of a planned choice made much earlier. Most of the evidence suggests that it was the latter. During the 1830s and early 1840s the Mormon leaders collected information from fur trappers and explorers about conditions in the valleys of the Rockies. The reason for their interest seems to have been that this appeared an isolated region which would permit the development of a distinct society based on their beliefs and values without interference from other peoples. As a result of the writings of Fremont and Hastings the Mormon leaders seemed to have formed a favourable view of conditions on the eastern margin of the Great Basin (Jackson, 1978). In particular the Bear valley and the Utah valley were described as fertile areas with adequate water, grassland and timber, which were well suited to agriculture. The only constraint appeared to be a statement by Hastings that north of 42°N the winters were very severe and conditions too cold for maize cultivation. The Mormon leaders also seem to have appreciated that the low precipitation would necessitate the use of irrigation and discussion of the techniques which might be necessary.

A letter written by Brigham Young to President Polk in August 1846 clearly indicates the planned destination of the Mormons:

> The cause of our exile we need not repeat, it is already with you, suffice it to say that a combination of fortuitous, illegal and unconstitutional circumstances have placed us in our present situation, on a journey which we design shall end in a location west of the Rocky Mountains, and within the basin of the Great Salt Lake, or Bear River Valley, as soon as circumstances shall permit, believing that to be a point where a good living will require hard labour, and consequently will be coveted by no other people, while it is surrounded by so unpopulous but fertile country. (Quoted in ibid.).

Even on their journey the Mormons collected all the information on the Great Basin they were able to glean. In June they made contact with an experienced explorer, Jim Bridger, who reinforced their views that the best land was to be found in the Bear and Utah valleys. Bridger also stated that good land stretched south of Utah Lake for a distance of 320 km.

When the Mormon party arrived at the Great Salt Lake valley in July 1847 their view of it seems to have been a very favourable one. One of the travellers, William Woodruff, wrote in his diary:

295

. . . we came in full view of the valley of the Great Salt Lake, or the Great Basin — the land of promise, held in reserve by the hand of god as a resting place for the Saints.

We gazed with wonder and admiration upon the most fertile valley spread out before us for about 25 miles in length and 16 miles in width, clothed with a heavy garment of vegetation, and in the midst of which glistened the waters of the Great Salt Lake, with mountains all around towering to the skies, and streams, rivulets and creeks of pure water running through the beautiful valley. (Quoted in ibid.)

There can be no doubt that the area the Mormons came to — namely, the Great Salt Lake valley — was one of tremendous agricultural potential. It was certainly not, as many Mormon writers were later to make out, a desert area almost totally lacking in natural resources. The key to its potential was the juxtaposition of highlands and lowlands at the eastern side of the Great Basin. These mountains were high enough to produce extra precipitation compared with the arid and semi-arid lands to the west and south. Although precipitation totals throughout the highlands are not high, reaching figures of around 500 mm in the Wasatch Mountains, they were sufficient to ensure a steady water flow along the rivers and streams draining into the Great Basin. The height of the mountains also meant that a large proportion of the precipitation, which comes mainly in the winter months, was stored as snow. When this melted in spring and early summer it provided extra water at an ideal time for irrigation in the lowlands.

7.3.2 The new settlements

In the first ten years up to 1857, 95 Mormon settlements had been established, mostly around Salt Lake City itself (Figure 7.8). Unlike in many other parts of the American frontier, the pattern of settlement in the valley was largely directed by the Mormon leadership, with people being sent to settle in specific locations. The aim of this seems to have been to gain Mormon control of key points such as access to water resources. It soon became obvious that Brigham Young could not single-handedly direct the colonisation process and so he divided the area into large territorial units with senior members of the Church in charge of settlement within each of them. A feature of the settlement pattern directed by

Figure 7.8 Major sites of Mormon settlement 1847–57

MORMON EXPANSION 1847–1857

NEVADA

Carson Valley 1851

CALIFORNIA

Great
Salt
Lake

Tooele 1849 ●

Utah Lake

● Cache Valley 1856
● Brigham City 1851
● Ogden 1848
● Bountiful 1847
● Salt Lake City 1847
● Lehi 1850
● Provo 1849
● Payson 1850

● Nephi 1851

Ephraim 1852 ●
● Manti 1849
● Fillmore 1851

UTAH

● Beaver 1856

● Parawan 1851
● Cedar City 1851

Santa Clara 1854 ● ●

St George 1861

Elk
Mountain
1855

Virgin R.

Little Colorado R.

Las Vegas
1855 ●

San Bernardino 1851 ●

ARIZONA

Colorado R.

Pacific
Ocean

M E X I C O

Source: Redrawn from Jackson R.H., 1978, 'Mormon perception and settlement', *Annals of the Association of American Geographers*, v. 68, pp. 317–34. Reproduced by permission of the Association of American Geographers.

Brigham Young was that originally, most of the emphasis was placed on the area to the south of Salt Lake City. Many of the settlements were established in arid regions with few natural resources and many of the settlers were unhappy to go there. However, they felt that it was God's will as interpreted by their leaders and so were willing to overcome their own doubts.

The reason why Brigham Young concentrated settlement in the

region to the south of the Great Salt Valley is not known with any certainty (ibid.). Two factors do seem to be important. The first was the belief that to the north of 42°N climatic conditions were so severe that settlement was not worthwhile. This idea was taken up by the Mormon leaders as a result of the writings of Hastings. The second seems to have been a desire to make the settlement in the Great Basin as self-sufficient as possible. To achieve this required agricultural activity in areas with higher temperatures where crops such as cotton and tobacco could be grown. This naturally meant settling in areas further south. Unfortunately these areas did not possess the rich resource base of the Salt Lake City region, and considerable resistance developed to the idea of moving to the new southern centres. Jackson (ibid.) states that the Mormon leaders appear to have reacted to this by making out that the original settlement experience in the Great Salt Lake valley was much more difficult than in fact had been the case. Emphasis was laid on the desert nature of the climate and the severe obstacles which had to be overcome to make the region agriculturally prosperous. In a sermon given in 1877 Wilford Woodruff, who had spoken so eloquently in praise of the area in 1847, was to say:

> But when we came to this country, what did we find here? A barren desert, as barren as the desert of Sahara; and the only signs of life were a few black crickets, some coyote wolves, and a few poor wandering Indians. (Quoted in ibid.)

Perhaps the intention was to convey the impression that the Salt Lake City region was similar to the more arid regions to the south and that these too could be made like the original settlement by faith and hard work. Although the settlers moved to the new areas they were never able to reproduce the success which the rich resource base of the Salt Lake City region had made possible. It probably also represented a desire to make the achievements of the original settlers seem even more remarkable than they actually were.

The second decade of settlement from 1857 to 1867 resulted in the establishment of 135 new settlements (O'Dea, 1957). In contrast with the earlier settlement much of this occurred to the north of Salt Lake City. During the 1860s more than 20,000 settlers from Europe arrived in the region. This phase can be said to have ended with the coming of the railway to the basin in 1869, because from now on a regular link to the east was established. In the third decade of settlement up to the death of Brigham Young in 1877 a further 127 settlements came into existence. Salt Lake

City was the capital of the Territory of Utah and the regional centre for a Mormon population of about 140,000 people.

7.3.3 Irrigation

It is interesting to note the attitudes towards the management and use of resources which were held by the Mormons. On only his second day in the Great Salt Lake valley in 1847 Brigham Young announced that land was a gift from God and that the land, water and timber would be the property of the people (ibid.). The Church also believed that agriculture was the foundation of community life and the source of its wealth and well-being (Widtsoe, 1947a and b). In the first period of settlement areas of about 4 ha (10 acres) were allocated to each person. The unit of settlement was the small village with the farms nearby. All activities were supervised by organisations of the Church.

The success of the Mormons as agriculturalists was based on their ability to use the abundant water resources for irrigation. Successful large-scale irrigation has always required close supervision and organisation from a strong central authority. With the Mormons it was the Church which provided this leadership. The irrigation system in the Wasatch oasis developed rapidly, with the first waters diverted from City Creek in what became Salt Lake City as early as 23 July 1847, the day before Brigham Young arrived in the valley. The key to the success of the new irrigation networks was the co-operation between the people guided by the leaders of the Church. It was the Church which gave Mormon society its cohesive nature and in the early days of the settlement the Church played an important role in water distribution.

In each Mormon settlement an informal water association developed which was supervised by the local Church leader. This association established a water distribution network by either building a reservoir in which water could be stored, or, if a perennial stream was close by, by constructing a diversion structure for directing water into the fields. The irrigation network was built as a co-operative venture by the labour of the local farmers. The amount of water the farmers received was in proportion to the amount of labour and draft animals they supplied to the project. With this system the volume of water allocated to each farmer was relatively equal as labour inputs were broadly similar. It also possessed the advantage of preventing any one individual or group from gaining control of the water resources.

The distribution of water into the fields was made the responsibility of the water master. The first water master was appointed on 22 August 1847, only one month after the Mormon party arrived in the valley (ibid.). The Church continued to organise the irrigation system until 1853, when a city government was established in Salt Lake City, and thereafter, a civil administration took over the water network. A general city water master was appointed who had control over the water masters in each of the city wards. By 1865, less than 20 years since the original settlement, 1679 km (1043 miles) of canals had been constructed and the irrigated area was close to 62,322 ha (154,000 acres) (ibid.).

The need for irrigation had meant that the water laws and regulations of the eastern United States had to be modified to meet the needs of the new conditions. As a result the Riparian Rights rules were abandoned and a new system of laws adopted based on the Spanish practices which have come to be known as the doctrine of 'Prior Appropriation' (Meyer, 1984). In 1852 the territorial legislature passed a law which made many natural resources come under the direct control of the courts:

> The county court shall have control of all timber, water privileges, or any water course or creek, to grant mill sites, and exercise such powers as their judgement shall best preserve the timber and subserve the interest of the settlement in the distribution of water for irrigation or other purposes. (Quoted by O'Dea, 1957.)

In 1865 a law passed by the Territorial Legislature permitted the establishment of irrigation companies (Haws, 1973; Israelsen et al., 1946). In effect this was a formalisation of the rules and regulations which had evolved in connection with the original water associations. However, the key legislation with regard to water and its use was the Water Code of 1903, which even today forms the basis of water regulations. The Code stated that water in streams and rivers could not be privately owned, though it did permit individuals to own rights of diversion and usage. The Code formalised new abstraction procedures and at the same time recognised existing rights which had developed over the years.

Even as late as the 1930s the Wasatch oasis was still largely dependent upon agriculture for its livelihood, with about 700 irrigation companies in existence. Salt Lake City had a population of 150,000 and was the centre of the state of Utah with 500,000 people (Harris, 1940). During the Second World War and in the immediate post-war period, major changes took place, particularly with the introduction of industry, which have changed the area to one more like other American cities. Population has grown rapidly and by 1980 the state of Utah had almost 1·5 million people (Table 7.2).

Table 7.2: Population growth of the state of Utah, USA (thousands)

1850	11
1860	40
1870	87
1880	144
1890	211
1900	277
1910	373
1920	449
1930	508
1940	550
1950	689
1960	891
1970	1,059
1980	1,461

Source: US Department of Commerce, Bureau of the Census, 1975, 1981.

7.3.4 Escalante — a pioneering Mormon settlement

The Mormon occupation of what became the state of Utah was a colonising settlement pattern in an arid and semi-arid environment. It was unusual in a number of ways compared with settlements elsewhere in the USA. Firstly, the actual location of the settlement was often dictated by the Mormon Church as part of a strategy to colonise a given region. Secondly, the leadership of the settlement was often vested in a Mormon bishop guided by instructions from Salt Lake City. Even the type of settlement was unusual as nearly everywhere it took the form of a nucleated village, as opposed to the dispersed homesteads found elsewhere in the western USA. In the early days of these settlements survival was the most important concern. Survival against hostile Indians, as well as against an environment which through water scarcity or insect pests severely limited food production.

We are fortunate in possessing studies of how these villages have changed in the 140 years since their establishment. One of the

301

most interesting of these is a work by Nelson (1952) which surveyed three settlements in Utah in the mid-1920s and in 1950. One of these was Escalante. Escalante is an isolated village in the highlands of southern Utah which was settled in 1875 (Figure 7.9). The site, at 1,740 m, has high mountains to the west and broken desert terrain to the north, east and south. Average precipitation is about 320 mm and the growing season is approximately 130 days

Figure 7.9   The site of Escalante, Utah

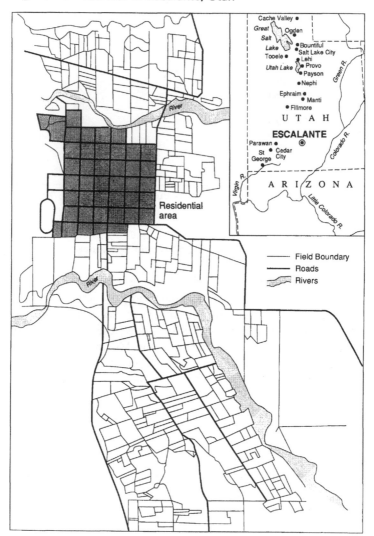

long. The original settlers divided the land into 64.75 ha (160 acre) tracts and then subdivided these into 9.1 ha (22.5 acre) holdings. The aim behind this was to establish a community quickly and to encourage others to come and settle there. By 1876 the town site had been laid out and work begun on the first irrigation canals. The canals were dug by the labours of all the settlers and a mutual irrigation company established with one share of water for each acre (0·4 ha) of land. Since the land holdings were equal, so too were the water volumes. The farmers built their houses on lots within the village while their fields were often considerable distances away.

The basis of agricultural activity was livestock production, with arable farming confined to subsistence food and forage production. The mountains provided summer pastures for the livestock, whilst in winter, grazing of the desert ranges was possible. At the time of the first survey in the early 1920s, there were 9,400 cattle and 33,000 sheep in the village. After 1905, when the National Forests were established, many of the summer grazing areas were located in the new Powell National Forest. In these forests stringent limitations on grazing were introduced with each individual only being allowed up to 800 sheep and 60 cattle.

The total area of land in farms in 1923 was surprisingly small and only totalled 4,604 ha (11,376 acres).Of this figure, 2,226 ha (5,500 acres, 48 per cent) were pasture and a further 1,659 ha (4,100 acres, 15 per cent) alfafa and hay. Cereal and other food crops only amounted to about 405 ha (1,000 acres, 9 per cent). Farm size varied now from 2 to 275 ha (5 to 680 acres), though about 50 per cent of the holdings were less than 20 ha (50 acres). A feature of all the Mormon settlements, of which Escalante was typical, was the large distances between the farmers' house in the villages and their fields, with almost two-thirds of all farmers having to travel between 1·5 and 8 km. This was obviously inefficient in time, but living in the village did convey significant social benefits. Food supply was ample. Each family kept a dairy cow and grew vegetables and fruit in their large gardens. Pigs and poultry were also universal.

The population of the village grew rapidly from its creation in 1875 to 623 people in 1880. Thereafter, there was little change, with the population growing to just 723 by 1900. In the early years of the twentieth century a more rapid increase was noted, reaching 1,030 in 1920 and 1,161 in 1940. From the 1920s onwards most of the natural increase in terms of numbers seems to have left the village. A feature of Escalante in 1923 was the almost total dependence on agriculture, with 135 out of 180 heads of households classified as farmers and stock rearers. As far as amenities were concerned electricity had been introduced to the village in 1923, but at that time few people had obtained electrical appliances. Drinking water was still largely obtained from wells.

By 1950 the isolation of Escalante had been broken down with the advent of the automobile, the radio and a road delivery service which changed the way of life of the villagers. Goods such as bread and milk were now shipped in over a distance of 150 km where formerly self-sufficiency prevailed. The village became dependent on the rest of the state and the USA for its goods and livelihood. The most important events which precipitated such massive changes over so short a period were the depression and droughts of the 1930s and the Second World War. The depression and droughts combined caused many of the farmers to incur large debts. As a result the banks sold off the animals to recoup some of the money they were owed. At the same time Federal relief programmes were introduced, many of which benefited the inhabitants of Escalante, including the construction of a water supply system.

Following the droughts of the 1930s the Federal government began to control the use of the public domain for grazing purposes. This meant that the land to the north, east and south of Escalante which had been used freely as grazing land was after 1936 given livestock quotas by the Bureau of Land Management. This was not before time, as these lands had been severely overgrazed and the productivity of the rangelands reduced. Despite these conservation efforts, even by the 1950s range quality had not recovered to its former levels. For many farmers these restrictions meant that they could not survive, as they often already lacked summer grazing rights in the National Forests. Such farmers had to sell their stock and sometimes their farms as well. Animal numbers dropped dramatically. In 1923 there had been 33,000 sheep and 9,500 cattle in the village. By 1950 there were only 7,000 sheep and 4,200 cattle. Perhaps even more revealing was that at the earlier date 54 families owned sheep, whereas by 1950 the figure had dropped to only five. By 1950 only 25 per cent of the families received some income from sheep and cattle, whilst formerly almost everyone had this income source. Similar declines were also recorded with dairy cows, pigs and poultry, as people gave up their self-sufficiency and joined the market economy in which foodstuffs were brought in from elsewhere. The irrigated area remained remarkably similar between the two dates, though the number of owners declined by about one-third. Ownership was being concentrated in fewer hands and in larger farms. Mechanisation had taken place with six tractors and 53 trucks being recorded.

Occupation had also changed. The number of farmers was reduced by a half, and the number of labourers increased five-fold. People now worked in a variety of different trades as well as for the Forest Service and the State Highways Department. For the first time the service industries were represented with café, motel and petrol service station operators. A change had clearly taken place from a farming village to a village with urban trappings.

Perhaps equally important was a change in attitudes noted by the researcher. In 1923 the feeling of belonging together and the willingness to help each other was a tangible part of the village community. By 1950, self-interested free market competition was the order of the day, and this attitude had the effect of increasing the gulf between the more entrepreneurial members of the community and those with lower ambitions. In 1950 almost 20 per cent of the people of Escalante were dependent on public welfare for at least part of their income.

Escalante, therefore, provides an illuminating example of how a farming village had been used to colonise an alien and harsh environment. Initially it depended on self-help and community-help and was guided and controlled by the Mormon Church. By the early years of the twentieth century Escalante could be classed as a success as a settlement, though this success had depended on the 'mining' and subsequent deterioration in carrying capacity of the rangelands. The Federal government, through the establishment of the National Forests in 1905 and the control of the public domain in 1936, had introduced conservation measures in the interests of the environment. However, in turn these measures had destroyed the livelihood of the farmers a few years earlier than would have occurred with the final destruction of the rangelands. Interestingly and coincidentally, these events were taking place at a time of almost revolutionary change in the American economy, with the introduction of the motor car and the commercialisation of the rural economy. In a mere 25-year period Escalante had changed from a traditional self-sufficient society using techniques little different from those developed in the Neolithic to a modern market economy based on cash, food imports and a degree of mobility which was unthinkable 25 years earlier.

Similar changes were also taking place elsewhere in Utah at the same time, with population statistics suggesting that the agricultural village and the small market town reached its heyday in the early years of the twentieth century (Nelson, 1952). Thereafter, as the influence of the US economy became more dominant, subsistence and self-sufficiency in rural communities declined. Smaller communities became less important and their population numbers fell as well. In contrast, centres with specific locational or resource advantages have grown rapidly to become important nodes in the economy of the state.

7.3.5 The Future

In the 1980s the Wasatch oasis has a population of 1·1 million, representing three quarters of the total population of the state of Utah and including major cities such as Salt Lake City, Ogden and Provo. Since the Second World War the area has registered a remarkable growth from a population of only 380,000. Most of this population increase has been the result of urbanisation and has produced what some describe as an 'urban corridor' along the foot of the Wasatch Mountains. As a centre Salt Lake City dominates the region with a population of around half a million people.

The massive population growth in the post-war period has changed the ethos and economy of much of the state of Utah. Until the 1930s Utah was a state whose economy was dominated by agriculture and mining, but today a much broader economic base exists, centred on Salt Lake City itself and including banking, commerce, industry and tourism. As a result the agricultural production based on the Wasatch oasis is now of much less importance than formerly and seems likely to decline further in the future. Urban growth is now the controlling factor in economic activity. As growth continues it seems likely that increasing quantities of water will be transferred from relatively low value uses, such as irrigation, to supply high-value urban and industrial needs (Evensen, 1985). As a result irrigated lands, especially near the urban areas, will cease agricultural production and revert once again to semi-arid scrubland (Price, 1985). This will mean that the phase of agriculture marked by the first settlement of the Mormons in the 1840s will be over, and replaced by a period when it will be urban growth which will dominate land management decisions.

8

The Great Plains of the USA:
Changing Patterns of Exploitation

8.1 INTRODUCTION

The Great Plains of North America stretch in a north–south belt
from the Mexican border to the Arctic Ocean (Figure 8.1). They
comprise level or gently sloping land which rises about 1,000 m in
a westerly direction to the base of the Rocky Mountains. Situated
as they are in the continental interior, they represent a transition
zone between the humid east and the arid west. The Great Plains
are unusual, if not unique, in that they permit an examination of
rapidly changing land use over a period of 120 years. Documentary
records of events are good and photographic evidence is also
available. It is possible to follow the change from a traditional
hunting and gathering society carried on by the Plains Indians in
the mid-nineteenth century, through pioneering ranching and
arable farming, to today some of the most capital-intensive
irrigated agriculture found in the USA. The detailed evidence
available permits us to study the decision-making processes with
regard to dryland management in a manner which is certainly not
possible throughout most of the Old World.

Human occupancy of the Great Plains, and particularly the
western part or High Plains, has been beset by environmental
problems, of which the most important has been drought. In
Paterson's words:

> On the east and north the boundary between the problem area
> and the rest of the Great Plains province fluctuates year by year;
> it is a boundary not of relief but of risk; it differentiates between
> the relatively secure farming of the Agricultural Interior and the
> relatively hazardous business of occupying the dry plains.
> (Paterson, 1971.)

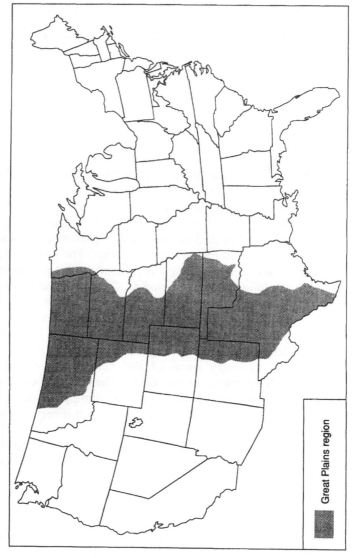

Figure 8.1 Location and extent of the Great Plains

Great Plains region

8.2 THE ENVIRONMENT

The climate of the Great Plains is best described as sub-humid in the north and east, grading to semi-arid southwards and westwards. Precipitation is chiefly brought by moist air masses moving northwards from the Gulf of Mexico. As a result a decline in precipitation is witnessed towards the west and north from about 750 mm in central Texas to less than 380 mm in the north and west. Over most of the Great Plains at least two-thirds of the precipitation falls during the summer months. Unfortunately the air masses which bring these summer rains can be fickle, and sequences of dry and wet years are a feature of the region. Indeed, average precipitation values have little meaning for agricultural planning purposes on the Great Plains as they suggest a stability of conditions which in reality does not exist. This is especially the case in the drier, western High Plains.

Temperature variations over the Plains reveal marked differences. During summer the continental interior heats up rapidly and temperatures over the whole region can exceed 38°C. In winter a much greater contrast is noted. Even within the USA the southern part of the Great Plains in Texas will be experiencing mean January values close to 10°C while in Montana and North Dakota the equivalent figure will be below −10°C. As a result of these low temperatures the growing season in the north of the USA is only about 110 days, whereas in Texas it reaches 240 days.

Of all the climatic elements it is probably the incessant wind which has the greatest impact on both the inhabitants and visitors (Blouet and Luebke, 1979). Given the lack of trees and the smooth nature of the plains the wind blows unchecked across the region with a constancy rarely experienced in the undulating and wooded lands of the eastern USA. Its effect on unprotected non-cohesive soils can be devastating, producing giant dust storms which achieved world-wide publicity during the Dust Bowl years of the 1930s. In winter these same winds can produce ferocious blizzards blowing out of the north which can reach as far south as Texas.

The soils and vegetation of the Great Plains broadly reflect the changing moisture conditions of the area. In the wetter east, deep black soils predominate, grading into dark brown and then brown soils in the drier western parts. On the black soils, tall grass is the characteristic vegetation, while in drier conditions short grasses predominate. A critical factor in this change seems to be the depth to the calcium accumulation layer. When it is more than between 635 and 760 mm in depth, black soils and tall grass prairie are found; when shallower than this, short grass vegetation predominates. The line between the two types of grassland also delimits the drier High Plains region to the west, where droughts impose enormous risks on farming activity, from the humid east where farming is relatively secure.

8.3 THE HISTORY OF SETTLEMENT

Prior to their settlement by white people the grasslands of the Great Plains provided sustenance for vast bison herds. In turn these animals were sources of food, clothing and shelter for the Plains Indian tribes, such as the Blackfoot, Sioux and Cheyenne. These tribes proved aggressive to white settlers and were only subdued following numerous bloody encounters by the development of quick firing weapons, such as the Colt revolver.

Widespread settlement of the Great Plains did not begin until the 1870s (McKee, 1974; Webb, 1931). Large-scale cattle ranching developed first in Texas, with the animals being driven northwards along trails to railheads in Kansas. From here they were moved to the great stockyards in Omaha and Chicago. The opening up of the cattle empire of Texas began with the Missouri–Pacific railroad reaching Sedalia, Missouri, in 1866, thus providing the first outlet for the cattle drives from the south. Later the railroads moved further west to places such as Abilene and Dodge City, which were destined to ship the largest numbers of Texas cattle to the north.

Cattle ranching moved northwards from Texas and as it did so the bison herds were systematically exterminated. In 1870 it was believed that there were probably 15 million bison on the Plains. By 1885 only a handful remained. With the removal of the bison the Indian communities could no longer survive, and so the spread of cattle ranching increased in pace. Sheep farming also grew in importance in the south. The success of animal husbandry on the High Plains of Texas was greatly assisted by two inventions made at this time. The first was the windpump, which permitted groundwater sources to be developed cheaply for animal watering. The second was barbed wire, which enabled the ranchers to control their animals and to protect their water sources.

A new wave of pressure on the High Plains began with their opening up to homesteading from the mid-1870s (Wessel, 1977).

The 1862 Homestead Act gave a farmer the right to buy a 32·37 to 64.75 ha (80–160 acre) lot at very low prices, provided that the farmer had lived on the land and cultivated it for five years. This led to wheat cultivation moving into the eastern margins of the Great Plains as a result of promotional activities by land companies and railways. The initial spread of wheat farming took place in the late 1870s, at a time when the Great Plains were experiencing above-average rainfall conditions. The farmers believed their futures were assured. However, this was not to be, for during the 1880s a series of droughts caused huge crop losses and drove many farmers back to the east. This was the first time that there was any realisation that the Great Plains experienced a very variable climatic environment, and that whilst good crops could be raised in wet years, crop failure was inevitable in dry ones.

Alongside cultivation the ranching enterprises continued to prosper as the railway network edged further south and west. Better animal breeds were developed by crossing Texas Longhorns and Herefords, whilst refrigerated rail wagons, introduced in the 1870s, meant that meat could be shipped to wider markets. The 1880s also proved costly to the ranchers as a result of another form of natural disaster. This time, though, it was a huge blizzard in 1886–7 which wiped out half the cattle population of the Plains. After it the great cattle empires never recovered to their former sizes.

In the 30-year period up to the early 1890s the traditional land use of the Great Plains (Indian tribes living off the bison herds) was swept away. It was replaced first by ranching and then from the east by dryland farming. However, both these activities suffered greatly from natural disasters in the late nineteenth century, causing them to retreat from their maximum extent. The 'boom and bust' character of the Great Plains at last began to be appreciated by the settlers.

8.4 DROUGHT AND RISK

It is the variability of precipitation which has produced the 'boom and bust' economy of the Great Plains (Saarinen, 1966). Rain is mainly brought by the warm air masses from the Gulf of Mexico during summer and it is the erratic nature of these air masses which creates a severe drought problem, especially on the western margins of the Plains. Over the last century dry and wet years have

311

been grouped together, so compounding the difficulties of arable farming in a semi-arid environment. During the wetter-than-normal phases, farming has expanded, only to be curtailed when a long dry phase follows. While farmers are usually able to survive a single year of drought, they lack the resources and the reserves to cope with longer periods of stress.

One of the images held by the early settlers crossing America was that the Great Plains, with its grassland vegetation, was an area of low fertility. Indeed, so strong was this image that for the early part of the nineteenth century, following the expedition of Long (1819–20) to Yellowstone, the region was known as the 'Great American Desert'. Long stated:

> In regard to the extensive area between the Missouri and the Rocky Mountains, we do not hesitate in giving the opinion that it is almost wholly unfit for cultivation and, of course, uninhabitable by a people depending on agriculture for their subsistence (Watson, 1976, p. 25).

One should also not underestimate the intimidating nature of the immense open space which the Great Plains must have created for the early settlers (Blouet and Lawson, 1975). Certainly the people who were used to the forested environments of the east regarded the Great Plains as an area to be avoided and crossed as quickly as possible on their way to the woodlands of Oregon. The reality of the situation was that their agricultural techniques had been developed for woodlands and that a change to a grassland environment would have required unnecessary risks and time to acquire new skills. The Great Plains were avoided while ample land still existed in Oregon. The myth of the 'Great American Desert' prevailed through the mid-part of the nineteenth century, and attained its most widespread acceptance, according to Webb (1931), during the 1850s. Thereafter, though, views changed rapidly.

When the first farmers moved on to the Great Plains in the 1880s and early 1890s, precipitation totals were above the long-term average values. The farmers, however, believed that such conditions were the norm and so they were severely hit by drought in the 1890s. The early years of the twentieth century, yet another wetter than normal period, saw the continued expansion of settlement, but the 'boom' nature which had characterised the latter part of the nineteenth century was now gone. New lands were opened up, but in a much more cautious manner than previously and with a greater appreciation of the inherent environmental limitations. Severe drought struck once again between 1910 and 1915, bankrupting thousands of farmers, many of whom emigrated back to the east coast of the USA.

For two decades in the 1920s and 1930s the Great Plains experienced below-average precipitation, with very severe drought conditions prevailing between 1931 and 1937. Agriculture on the Plains was everywhere devastated. The 1920s were a time of economic depression throughout the USA, and nowhere was this more pronounced than in rural areas. Many ranchers were forced to sell their farms and land speculation was rife. Others tried to increase their income by ploughing extra lands, but instead they only saw their top soil blown away. New farmers still came into the region, but with low rainfall few were able to profit and debts for land and machinery built up.

When the real drought began in 1931, hard on the heels of the Wall Street Crash of 1929, the result was devastating. Thousands of farmers were ruined as crops failed year after year and top soil was blown away in what has become known as the 'Dust Bowl' years (Coffey, 1978, Worster, 1979). The mood of utter despair of this period has been well captured in the novel *Grapes of Wrath* by John Steinbeck. Many farmers either left to seek a new life in California or returned back east; others managed to hold on and hope for better things to come.

Better conditions did arrive in the 1940s. Rainfall was once again above average and the demand for food from war-torn Europe meant that agricultural prices picked up. 'Boom' conditions prevailed once more and the wheat acreage expanded (Kraenzel, 1955). It was not to last, however. In many parts of the Great Plains, the early and mid-1950s produced drought conditions which were more severe than those of the 1930s (Hewes, 1965). Although bankruptcies did occur, many lessons had been learnt from the 1930s, particularly with respect to soil conservation practices. General economic conditions were also much more favourable than in the 1930s as the post-war economic boom got into full swing.

From the end of the 1950s till the late 1960s, national economic prosperity was accompanied by more humid conditions on the Great Plains. As a result farmers enjoyed a level of prosperity which has not been surpassed either before or since (Borchert, 1971). Drought conditions returned between 1968 and 1972, and 1971 and 1980 were severe drought years, but the impact on the farmer has been much reduced by the introduction of drought-resistant grains, moisture conservation cultivation techniques, and carefully designed cropping patterns to minimise soil erosion.

What the farmer has not been able to gain protection against has been the much less favourable economic conditions of the late 1970s and early 1980s. On the Great Plains farmer indebtedness is once again a serious problem. Already many farmers have had to sell their land and move to other occupations. If a severe drought phase were now to be superimposed on these depressed economic conditions, the plight of farmers on the Great Plains could become very serious indeed.

For the farmers, what has effectively been a 40-year period of prosperity (1940–80) now seems to be over as harsher economic conditions prevail. During this time the 'bust' portion of the 'boom and bust' cycle in the early 1950s was cushioned by high levels of national income, economic growth and agricultural subsidies. This meant that rural towns, highly dependent on agricultural activity, were able to diversify into other areas. In future, with harsher economic conditions, this seems less likely to be possible and the environmental reality of drought looks set to reassert itself as the true control of agricultural activity, greatly assisted by the proposed withdrawal of major agricultural subsidies by the end of the century.

8.5 THE IRRIGATION MOVEMENT —
THE HIGH PLAINS OF TEXAS (Figure 8.2)

The great drought of the 1930s brought home to many farmers of the Great Plains the immense difficulties of farming in a semi-arid environment. Many farmers left the area and moved westwards to California, while others turned towards technology for a solution to their problems. Underlying the Plains from Nebraska to Texas is a geological formation, known as the Ogallala, in which vast quantities of water are stored. Water had been withdrawn from the aquifer since the 1880s to water cattle, but the windpumps which achieved this were not capable of raising water from the depths at which it was found in sufficient volumes to permit the irrigation of large areas.

In the 1920s experiments with new pumping equipment were

Figure 8.2 High Plains of Texas

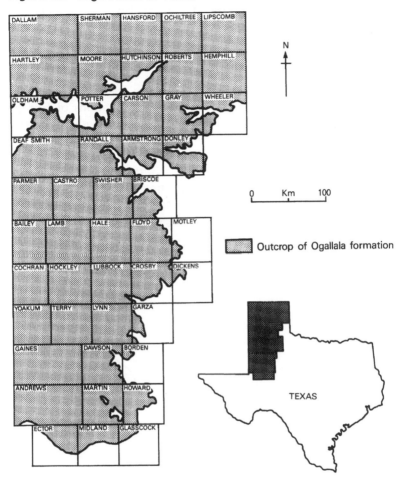

Source: Redrawn from Beaumont P., 1985, 'Irrigated agriculture and
ground-water mining on the High Plains of Texas, USA', *Environmental
Conservation*, v. 12, pp. 119–30. Reproduced by permission of the
Foundation for Environmental Conservation.

made in Texas, but capital and maintenance costs proved high,
and reliability was not good (Green, 1973). By the mid-1930s the
internal combustion engine had been refined and new and reliable
turbine pumps were available. Drilling operations were also
cheaper as a result of developments in the oil industry in California

315

and Texas. Thus, by the late 1930s, large-scale irrigation systems began to appear on the High Plains of Texas as a means of protecting the farmer from the vagaries of drought. The Second World War meant that drilling and pumping equipment was difficult to obtain, so few new schemes were constructed. However, once the war was over in 1945, the irrigation boom on the High Plains of Texas really began. It was greatly helped by the strong demand for food from the devastated nations of Europe and by an American agricultural policy which provided generous subsidies for farmers.

The irrigated area was initially concentrated around Lubbock where the water table was closest to the surface. By 1944, 154,701 ha were being irrigated, with the county of Hale accounting for a quarter of the total. From the mid-1940s the irrigation boom really got under way, with irrigation technology spreading out in all directions from the original heartland. In the 15-year

Figure 8.3 Growth of the irrigated area on the High Plains of Texas

Million hectares

Censuses of Agriculture — High Plains

12 'Heartland' counties

10 Northern counties

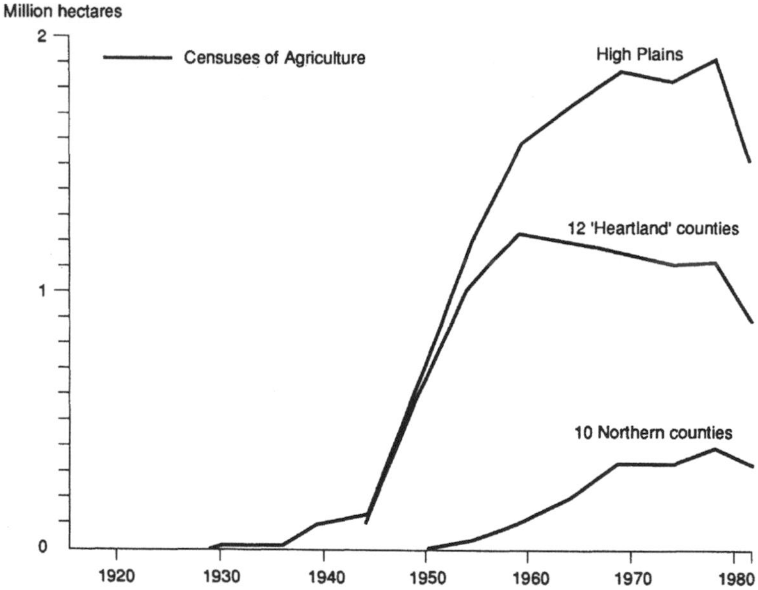

Source: Redrawn from Beaumont P., 1985, 'Irrigated agriculture and ground-water mining on the High Plains of Texas, USA', *Environmental Conservation*, v. 12, pp. 119–30. Reproduced by permission of the Foundation for Environmental Conservation.

period up to 1960 the irrigated area on the High Plains expanded to 1·6 million ha, and every county in the region now supported major irrigation systems (Figure 8.3) (Beaumont, 1985a). During the later 1960s and 1970s expansion of the irrigated area continued, but at a much slower rate than previously. From 1978 onwards it actually began to decline.

The water for the irrigation boom was provided almost exclusively from the Ogallala aquifer. As a result the pumped well is the most important item in the irrigation infrastructure of the High Plains, with each irrigation network being associated with at least one. From the early 1950s onwards it became obvious that the withdrawal of groundwater was causing a serious decline in water table levels. Prior to this time most of the farmers seem to have regarded the water resources as being limitless, so no consideration at all was given to conserving groundwater. The declining water table caused severe economic problems for the farmers. Firstly, it meant that a given pump size was producing less water and, therefore, the costs of each unit of water were increasing. To maintain a given water volume for their fields, farmers had to install a more powerful pump or to sink another well, both of which actions involved high capital costs.

From the middle 1960s it is interesting to note that the number of wells on the High Plains continued to grow substantially, at a time when the irrigated area was increasing only slowly and then even contracting. What is quite clear is that the area irrigated by each well has decreased, from around 45 ha in the early 1950s to about 25 ha in the late 1970s (ibid.). In the northern part of the High Plains, where irrigation development is most recent, individual wells still supply at least 80 ha of land.

With increasing water table decline conservation measures began to be introduced, including sprinkler systems, parallel terraces and tail-water return methods. The spread of sprinkler irrigation was particularly rapid after 1950 and now accounts for one-third of all irrigated land. Besides using less water, sprinkler systems also permitted the initiation of irrigation in sandy areas where surface water systems were not economic. In the older irrigated areas where heavy soils predominate, furrow irrigation still remains the most popular method of water application, though even here sprinkler systems have been gaining ground as a means of water conservation.

The irrigation boom on the High Plains has been sustained by the mining of groundwater from the Ogallala formation. What is now beyond any doubt is that this water is being depleted from reserves at a rate much greater than that of natural recharge (Texas Department of Water Resources, 1983). Indeed, it is estimated that abstraction is between 18 and 35 times the recharge rate. In some parts of the region a very serious hydrological situation exists. In 1974 the Texas Department of Water Resources estimated the groundwater reserves in the Brazos River basin. Using values of water abstraction, it can be shown that these resources will only last for between 18 and 31 years (Texas Department of Water Resources, 1979). A much more satisfactory position is found in the northern part of the High Plains, where it would seem that the rates of abstraction of the late 1970s can be sustained for at least 80 years. Most parts of the High Plains have reserves available for periods between the two extremes quoted above. Since the early 1970s it has become obvious that in many areas the volume of water withdrawn from the Ogallala aquifer has been declining. This has been the result of ever-increasing pumping lifts and rising energy prices, which have caused production costs to escalate (Sloggett, 1982). In order to remain profitable, farmers have been forced to reduce the amount of groundwater they use and to maximise the precipitation effectiveness through water conservation techniques (Swanson, 1985). This decline in water use naturally means that the groundwater reserves will last for longer periods than was previously believed.

In the mid-1980s, as irrigated areas decline, the importance of dry farming is increasing. In certain respects this is not surprising as the original occupation of the High Plains was based on dry farming methods. Even at the height of the irrigation boom, in the early 1960s, dry farming accounted for just over one-third of the harvested cropland. Looking back over the boom period it would seem that it reached its peak in the early 1960s when 63 per cent of the harvested cropland was irrigated (Figure 8.4). Since then the relative importance of dryland farming has increased steadily so that by 1982 irrigated lands only accounted for 46 per cent of the cropped area.

What, then, of the future? To a large degree this depends on decisions being taken away from the High Plains. To understand the nature of these decisions it is instructive to study how perceptions of the water resources of the High Plains have changed. From the late nineteenth century up until the late 1940s, the myth of unlimited groundwater resources underlying the High Plains prevailed amongst the populace of the region. This was

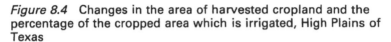

Figure 8.4 Changes in the area of harvested cropland and the percentage of the cropped area which is irrigated, High Plains of Texas

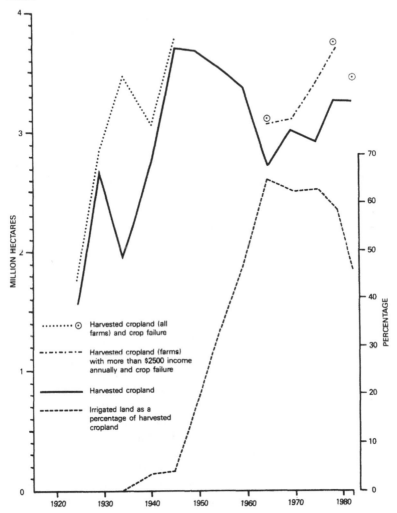

Source: Redrawn from Beaumont P., 1985, 'Irrigated agriculture and ground-water mining on the High Plains of Texas, USA', *Environmental Conservation*, v. 12, pp. 119–30. Reproduced by permission of the Foundation for Environmental Conservation.

despite reports from the Geological Survey in the early part of the twentieth century which stated:

Local precipitation is the source of the underground water of the High Plains . . .
Johnson estimates that not more than 3 or 4 inches annually soaks into the ground, an amount which would not saturate more than about 1 foot of sandy strata. This estimate of the amount of water absorbed seems rather small, but in all probability nor more than 6 inches of water are added to the groundwater each year. (Gould, 1906, pp. 35–6.)

In fact modern estimates suggest that the amount of recharge from precipitation is less than 0·5 cm per annum (Klemt, 1981). However, the volume of storage in the Ogallala aquifer was so large that it is perhaps not surprising that the early irrigation farmers felt that they could continue pumping indefinitely without any adverse effects. Concern about the falling water table was first expressed in the late 1930s and 1940s, but little was done about it. Indeed, attempts to introduce legislation to control the abstraction of groundwater failed in 1937, 1941 and 1947 (Thompson, 1966).

During the 1950s the irrigation farmers of the High Plains were opposed to any form of state control of groundwater. However, because of the growing worries of the people in urban areas and of conservationists, a District Groundwater Law was enacted in 1949. This bill permitted the establishment of underground water conservation districts, which could make and enforce certain rules about groundwater use. The districts were, however, powerless to prevent a landowner drilling a well on his or her property if it were so wished. A crucial point to realise here is that the laws applying to surface water and groundwater use in Texas are quite different. In law all surface water is regarded as public property, with the use to which it can be put administered by state agencies. In contrast, groundwater is regarded as private property, with a landowner having the right to use any water beneath his or her land as desired. On the High Plains, then, where groundwater is virtually the only water source, the state of Texas was completely unable to control water use. As a result landowners continued to exploit the Ogallala aquifer in an attempt to maximise their profits.

By the early 1960s, though, most farmers realised that the Ogallala aquifer was being seriously depleted and that long-term irrigation would only be able to continue if water were brought to the High Plains from elsewhere. Intense lobbying of the state government meant that when the Texas Water Plan was published in 1968, the objective of importing water to the High Plains was included in it (Texas Water Development Board, 1968). The main feature of the plan was the Trans-Texas canal which was expected by the year 2020 to be carrying almost 13,500 million cubic m of water each year from east to west Texas (Figure 8.5).

Figure 8.5 Texas Water Plan: the main features

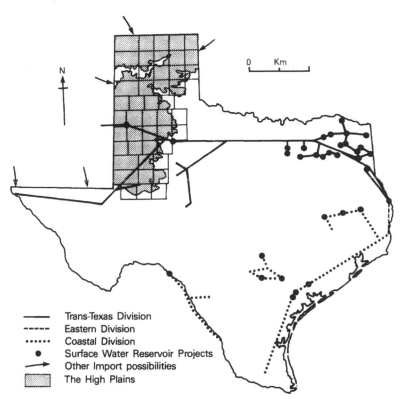

Source: Redrawn from Beaumont P., 1985, 'Irrigated agriculture and ground-water mining on the High Plains of Texas, USA', *Environmental Conservation*, v. 12, pp. 119–30. Reproduced by permission of the Foundation for Environmental Conservation.

Although in 1969 the Texas Water Development Board adopted the Water Plan as 'a flexible guide to State policy for the development of water resources in the State', the voters in the same year defeated a proposition which would have provided funds to build it. The High Plains farmers realised that they were not going to get extra water immediately and so during the 1970s an increasing emphasis was placed on water conservation measures and on crops which used little water. One point which it is interesting to note is that the farmers — who in the 1940s and 1950s had been against any kind of state control of groundwater — were now pressing the state to provide for their water needs. Water shortage was now perceived by them as a state rather than a local problem, and so state investment was needed to solve it.

Throughout the 1970s great improvements were made with regard to water conservation techniques for both irrigation and dryland farming: for example, certain crops grown by dry farming methods were able to produce yields of up to 75 per cent of those obtained by irrigation (Gardner, 1982). However, the major change was really one of attitude as irrigation began to be recognised as a supplement to precipitation and not the other way round. One of the main reasons for this was the massive increase in energy prices, with natural gas, the main pumping fuel for wells, rising by more than 400 per cent between 1976 and 1980 (Sloggett, 1982).

The quest for a solution to the water problems of the High Plains still continued and in 1981 the Texas legislature proposed a water assistance fund. With this fund it was hoped to construct certain parts of the Texas Water Plan as set out in 1968. Unfortunately for the farmers of the High Plains, the proposition was defeated by the voters and so the Texas Water Plan, once again, went back into cold storage. One of the main reasons for this defeat would seem to be that the voters in east Texas, where most of the population is concentrated, regarded the Texas Water Plan as a very expensive solution to the water needs of the High Plains, at a time when severe water shortages were not affecting the urban regions. The need for a state-wide plan, therefore, appeared to lack urgency. Following this defeat the Governor of Texas asked his Task Force on Water Resource Use and Conservation to reassess the water needs of the state. In 1982 the Task Force concluded that Texas must import water if its future needs were to be met. However, whence the water was to be obtained was left uncertain: the assumption remains that it would have to come from somewhere in the Mississippi basin.

Unless at some time in the future the Texas Water Plan is implemented, the farmers of the High Plains will be forced to rely more and more on dry farming techniques. This will not necessarily be because of lack of water, but rather owing to the ever-increasing cost of obtaining it. It will lead to a more extensive farming system than at present, but given the great improvement in dry farming techniques, profitability should be maintained. The fact that a major irrigation infrastructure exists means that irrigation could still be used for the supplemental addition of water as an insurance against dry farming crop failure in times of drought. Without doubt the peak of the irrigation boom on the High Plains of Texas is now well past, but it should not be expected that irrigation would disappear completely. Indeed, in the more favourable areas, such as where the aquifer is thick, the area of irrigation may well expand. However, overall the relative importance of irrigation now seems set to decline slowly as the move towards dry farming continues.

Although the High Plains of Texas provide the most spectacular example of irrigation growth, other parts of the Great Plains have also developed substantial irrigation systems based on the Ogallala aquifer. These include the states of Nebraska, Kansas, Colorado, Oklahoma and New Mexico (Bittinger and Green, 1980). Of these Nebraska alone possesses about one-third of the total surface outcrop of the Ogallala formation (Figure 8.6). However, in the eastern part of the state a younger formation overlies the Ogallala and this too provides valuable groundwater resources (Fricke and Pedersen, 1979).

Differences in irrigation development between the states over-lying the Ogallala formation are the result of many factors, including the hydrological characteristics of the aquifer, techno-logical advances and local historical factors. In the northern part of the region irrigation is dominated by 'centre-pivot' systems, which were developed in the early 1950s (Splinter, 1976). The intro-duction of these greatly reduced labour costs, though actual capital costs were high. The standard-sized system was geared to fit a quarter section of 64·75 ha (160 acres). However, the circular nature of the irrigation pattern meant that only between 52·61 and 54·63 ha (130–135 acres) in each quarter section were actually irrigated. During the 1970s centre-pivot systems spread rapidly through Nebraska and Kansas and became by far the most important irrigation technique in these areas. From the air they make an especially characteristic type of land use, quite unlike any other field pattern, with large circles superimposed on a uniform landscape. They also possessed the advantage that they allowed areas to be developed which were unsuitable for traditional irrigation methods on account of sandy soils or rolling terrain.

323

Figure 8.6 Outcrop of the Ogallala aquifer

One of the major influences on irrigation development has been the legal frameworks of the different states with regard to groundwater abstraction. Of all the states Texas undoubtedly has the most permissive system with the landowners being able to use any water under their land for whatever purpose they wish (Texas Department of Water Resources, 1983). In general the drier western states of Wyoming, Colorado and New Mexico have more stringent legislation and all regard water resources, both surface and groundwater, as being in public ownership. In the more humid states land and water rights go together.

In New Mexico the legislature passed a law in 1931 which still forms the basis of their groundwater legislation. The law stated that groundwater belonged to the public rather than to the owner of the land, and it gave specific powers to the state to control and regulate groundwater development. The regulations accepted that groundwater depletion was likely to occur and their objective was therefore to ensure that the resource would last long enough to provide adequate returns on the initial capital investment. In effect, this is protecting the interests of those who are already using the water in an analogous way to the prior appropriation rights applied to surface water.

A water law was passed in Kansas in 1945 which effectively gave the state the power to control water use, though the interests of existing water users had to be protected. Later, in 1957, new regulations were introduced to make all water users monitor and report the volume of water they were abstracting. However, it was not until 1978 that actual controls on the development of the Ogallala aquifer were implemented.

The first groundwater legislation in Nebraska was passed in 1963. It called for the registration of existing wells and of new ones as they were drilled, but no attempt was made for the state to take control of water use. In 1975 a tougher Ground Water Management Act was drawn up giving powers to districts under state supervision to prevent new drilling in certain areas.

8.6 IMPORTATION OF WATER

Today it is quite obvious that large-scale irrigation on the High Plains can only continue into the twenty-first century through the importation of water from other regions. As long ago as the 1960s grandiose schemes were being put forward to solve the water

Figure 8.7 The NAWAPA (North American Water and Power Alliance) Project

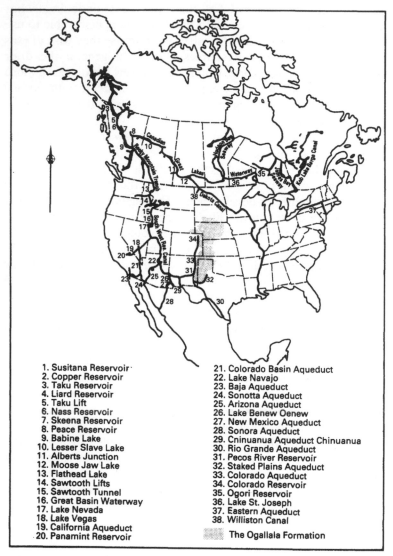

1. Susitana Reservoir
2. Copper Reservoir
3. Taku Reservoir
4. Liard Reservoir
5. Taku Lift
6. Nass Reservoir
7. Skeena Reservoir
8. Peace Reservoir
9. Babine Lake
10. Lesser Slave Lake
11. Alberts Junction
12. Moose Jaw Lake
13. Flathead Lake
14. Sawtooth Lifts
15. Sawtooth Tunnel
16. Great Basin Waterway
17. Lake Nevada
18. Lake Vegas
19. California Aqueduct
20. Panamint Reservoir

21. Colorado Basin Aqueduct
22. Lake Navajo
23. Baja Aqueduct
24. Sonotta Aqueduct
25. Arizona Aqueduct
26. Lake Benew Oenew
27. New Mexico Aqueduct
28. Sonora Aqueduct
29. Cninuanua Aqueduct Chinuanua
30. Rio Grande Aqueduct
31. Pecos River Reservoir
32. Staked Plains Aqueduct
33. Colorado Aqueduct
34. Colorado Reservoir
35. Ogori Reservoir
36. Lake St. Joseph
37. Eastern Aqueduct
38. Williston Canal

▓▓▓ The Ogallala Formation

problems of the High Plains and other water shortage zones within the USA.

The scheme which has received more publicity than any other is NAWAPA (North American Water and Power Alliance), proposed by the Ralph M. Parsons Company of southern California. The objective of the scheme was to collect water from rivers in Alaska and Canada, of which the major ones were the Yukon, Columbia and Fraser, and then divert this water southwards via the Rocky Mountain trench to the US border (Figure 8.7). It was hoped to be able to divert and collect about 194,890 million cubic m (158 maf) of water each year. Of this total approximately 71,542 million cubic m (58 maf) was destined for Canada and 98,678 million (80) for the USA. The remainder was for Mexico. The scheme itself was gigantic in its proportions, calling for the construction of 121 new lakes and reservoirs in Canada and 51 in the USA. Total costs were estimated at US$100 billion at 1964 prices. The NAWAPA project received considerable support from those areas which were to benefit from the water supplied. However, there were very serious concerns voiced about the environmental damage which was likely to occur, particularly in the area of the Rocky Mountain trench. Growing doubts in these areas, together with the staggering costs of the project, led to the plan being shelved.

Another scheme similar in concept to NAWAPA, but different in detail, was promulgated by Lewis G. Smith in the late 1960s. This used the Mackenzie River and a tributary in Canada as its main water sources. Water was brought southwards via the Rocky Mountain trench, as with the NAWAPA scheme. South of the US border, however, major differences were proposed. Water destined for the High Plains was to be stored in a massive reservoir in Montana's Centennial Valley. This lake, with a storage of 61,674 million cubic m (50 maf), was to supply canals carrying water to the western states (Figure 8.8). The 1968 cost of the project was estimated at US$75 billion for supplying 49,339 million cubic m (40 maf) of water each year.

It would appear that the sheer magnitude of this type of project meant that it was unlikely to be built without a strong commitment from the Federal government. For the government, though, the benefits were not always clear and chances of environmental disaster very real. Politically the risks were too great. The unacceptability of these major projects meant that other schemes had to be considered. Individual states evaluated various designs, but it was felt by many that what was still needed was some kind of integrated programme for the High Plains as a whole, though not on the same scale as NAWAPA.

In the early 1980s the Corps of Engineers conducted a study into

Figure 8.8 The Lewis Smith water diversion scheme

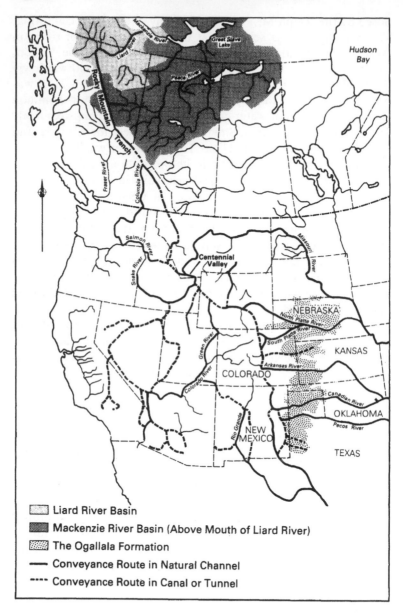

☐ Liard River Basin
■ Mackenzie River Basin (Above Mouth of Liard River)
▨ The Ogallala Formation
— Conveyance Route in Natural Channel
---- Conveyance Route in Canal or Tunnel

Source: Redrawn from Bittinger M.W. & Green E.G., 1980, *You never miss the water till . . . (the Ogallala Story)*, Water Resource Publications, Colorado. Reproduced by permission of M.W. Bittinger.

Figure 8.9 The Corps of Engineers plan for diverting water onto the Great Plains

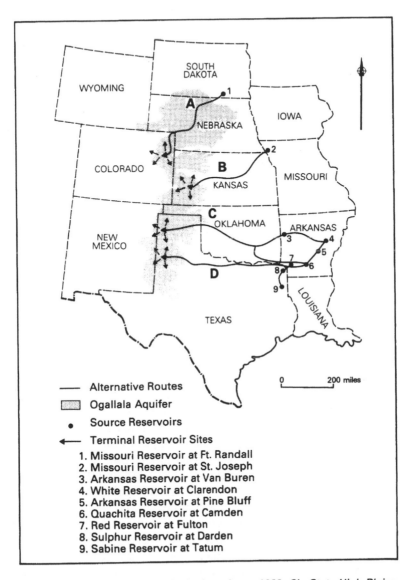

Legend:
— Alternative Routes
▨ Ogallala Aquifer
• Source Reservoirs
← Terminal Reservoir Sites
1. Missouri Reservoir at Ft. Randall
2. Missouri Reservoir at St. Joseph
3. Arkansas Reservoir at Van Buren
4. White Reservoir at Clarendon
5. Arkansas Reservoir at Pine Bluff
6. Ouachita Reservoir at Camden
7. Red Reservoir at Fulton
8. Sulphur Reservoir at Darden
9. Sabine Reservoir at Tatum

Source: Redrawn from High Plain Associates, 1982, *Six State High Plains, Ogallala Aquifer Regional Resources Study*, Austin, Texas.

the possibility of inter-basin transfers of water to reservoirs on the High Plains, as part of the High Plains–Ogallala Aquifer Regional Resources Study (High Plains Associates, 1982). The Missouri–Mississippi River system was to be the main source of supply. Four routes were chosen to carry water on to the High Plains (Figure 8.9). The two northern ones obtained their supply directly from the Missouri, while the two southern ones abstracted their water from a series of smaller rivers. It was expected that about 1·86 million ha (4·6 million acres) of land were to be supplied with irrigation water, of which 0·85 million (2·1 million) would be in Nebraska and 0·49 million (1·2 million) in Texas. Total water deliveries would be between 7,894 and 28,370 million cubic m (6·4 and 23 maf) per annum (Table 8.1).

Table 8.1: Range of water deliveries to the High Plains in the Corps of Engineers plan

Route	million m³/year	million acre feet/year
A	2,356 – 4,194	1·91 – 3·40
B	1,998 – 4,194	1·62 – 3·40
C	1,544 – 9,263	1·26 – 7·51
D	1,924 – 10,707	1·56 – 8·68
Total	7,833 – 28,358	6·35 – 22·99

Source: High Plains Associates, 1982.

The report concluded that from the engineering point of view, it was feasible to construct canal systems capable of supplying 11,101 million cubic m (9 maf) of water to the High Plains each year. However, the costs of building these projects would be high. With the cheapest scheme to deliver 1,998 million cubic m (1·62 maf) of water to Kansas the cost would be US$3·6 billion (1977), whereas the most expensive, to deliver 9,251 million cubic m (7·5 maf) to the Oklahoma and Texas Panhandles would cost almost $28 billion. The actual cost of delivered water was believed to vary from $227 to $569 (1977) per 1,233·5 cubic m (1 acre foot). Estimates of how much a farmer would be able to pay for water and still farm profitably suggested a maximum figure of about $120 per 1,233.5 cubic m (1 acre foot) (ibid.). This clearly indicates the very high level of subsidy which would have to be provided, presumably from Federal government funds, if the project were ever to be built.

What seems to be clear from all these studies is that the capital investment required to provide water for irrigation on the High Plains cannot be recovered and repaid solely from agricultural activities. If the Federal government decided to fund such a water project, even a small one such as that outlined by the Corps of Engineers, it would have to be thought of as a subsidised regional development programme, rather than a straight-forward economic investment. In effect, what the government would be saying is that the actual and potential problems of the High Plains were so great that it was in the country's interest to invest large sums of money to maintain a stable agricultural and social system. When one considers that even by the year 2020 the population of the High Plains will only be about 3 million people, such a decision does seem unlikely.

9

Rapid Economic Development
in the Gulf:
The Impact of Oil Revenues

9.1 HISTORICAL

The western coastal fringe of the Arabian/Persian Gulf is an extremely arid environment with few natural resources (Figure 9.1). Average annual precipitation is less than 100 mm, with the exception of the north-eastern part of Kuwait and parts of the Jebel Akhdar mountains of the United Arab Emirates and Oman. Nowhere, however, is agriculture possible without irrigation (Al-Feel, 1985; Dutton, 1985). When rainfall occurs it is usually associated with frontal systems of eastward-moving depressions. Rainfall can locally be intense and so flash flooding does occur. Along the Gulf coast, temperatures are high throughout the year. In July mean daily temperatures are usually in excess of 30°C and extremes of over 40°C are possible away from the immediate coastal fringe. During the winter months mean temperatures fall to between 13° and 18°C, though outbursts of cold air from Siberia can cause the temperature to fall rapidly for short periods in the northern parts of the Gulf. In coastal locations humidities are always high, making life unpleasant for the inhabitants (Babikir, 1986). In Bahrain, for example, it is rare for average relative humidities to drop below 60 per cent. Onshore humidities are high during the winter months, but are ameliorated during the summer by intensely hot and dry air moving out from the centre of the Arabian Peninsula.

The settlements which grew up in the area were located in favourable sites where fresh water was available and where defence against invaders was practical. This meant that almost all the settlements were either ports or inland oases (Petersen, 1986). Our knowledge of these early settlements is sparse. However,

Figure 9.1   The Gulf: major physical and economic features

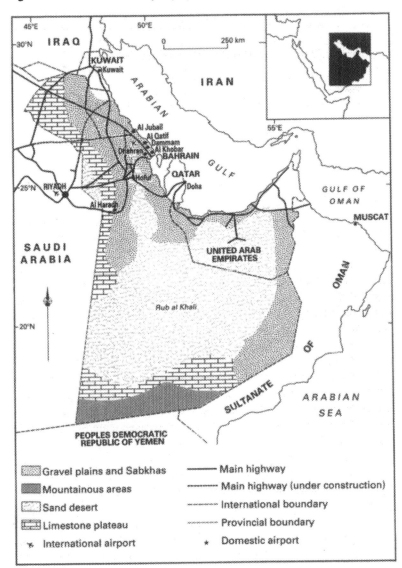

from the mid-part of the sixteenth century the influence of Europeans and in particular of the Portuguese, began to be felt in the Gulf region. The main interest of these Europeans was in securing shore bases to guard the maritime trade route between

Europe and the East. Little or no interest was shown in the countries themselves.

During the eighteenth century, tribal federations came into existence along the Gulf coast, and these settlements developed into the political units which exist at the present time (Petersen, 1977). With growing British influence in the Indian Ocean, in the late eighteenth and early nineteenth centuries, clashes occurred between local groups and the British, which were eventually settled by negotiation in 1820. The agreement led to the establishment of a British sphere of influence in the Gulf which was to continue until the mid-part of the twentieth century. This period of peace and stability gave rise to the development of economic activity, in particular of pearl fishing, which became the basis of the local economy. It also promoted the growth of the settlements, so that by the early years of the twentieth century a number of the ports had populations of between 10,000 and 30,000 people. In turn these towns included a large proportion of the total population of the states in which they were located. Dubai, for example, had 98 per cent of the people in the state of Dubai, and Kuwait, 70 per cent of the population of the state of Kuwait (El-Arifi, 1986). As these port settlements looked outwards for their livelihood it is not surprising that other culture groups, including Iranians, Indians, and peoples from East Africa were well represented within them. In the 1920s and 1930s the economies of the Gulf states suffered serious decline, largely as a result of competition from cultured pearls from Japan, which undercut the costs of the local fishermen. The urban areas declined in both wealth and population and were in a depressed state by the start of the Second World War.

9.2 OIL AND OIL REVENUES

The main changes in the urban fabric of the Gulf have taken place mainly in the post-Second World War period as a result of the huge oil revenues which have become available. What one is seeing today is an oil mining 'boom' similar to the gold and diamond 'booms' in the drylands of Australia and South Africa during the nineteenth centuries. The revenues from this 'boom' have resulted in a phase of urbanisation and development which has possibly been unequalled in world history and it has completely changed the urban fabric of these small port towns.

The settlements which had grown up along the Gulf coast had the characteristics of the traditional Islamic town. At the centre was the *suq* (market) and the harbour, with the mosque and the ruler's palace nearby. A fort was almost always present as well.

Figure 9.2 Oil production from the major oil producing states in the Gulf

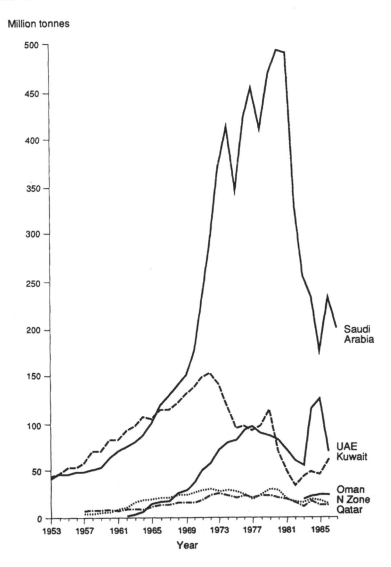

335

Residential quarters, consisting of traditional houses with thick walls insulating against the intense summer heat, were found on the periphery, and these were often strictly segregated on tribal lines. Streets were small and winding, giving a feeling of congestion and close packing.

The first discovery of oil on the eastern shore of the Gulf took place on the island of Bahrain in 1932, and this was the prelude to extensive and successful exploration in adjacent areas (Beaumont et al., 1988). In 1938 oil was discovered in Kuwait and at Dammam in Saudi Arabia, while in the following year a find was made in Qatar. The turmoil caused by the Second World War, however, meant that these fields were not developed until more stable economic conditions returned in the mid-1940s. Since this time production and exports rose rapidly as new fields were opened up. Kuwait, Saudi Arabia, Bahrain and Qatar were all significant oil producers by the early 1950s, while Abu Dhabi and Dubai, which were to become the United Arab Emirates, did not commence large-scale exports until a decade later. Throughout the 1950s, 1960s and 1970s many new oilfields were discovered onshore, while offshore exploration in the Gulf also revealed large oil and gas accumulations.

During the mid-1950s Kuwait and Saudi Arabia were the two main producers, each with an annual output of approximately 50 million tonnes (Figure 9.2). In the late 1950s and early 1960s Kuwait's production outstripped that of Saudi Arabia, but from 1966 onwards the position was reversed. Peak production in Kuwait was reached in 1972, followed by a marked decline. In contrast, Saudi Arabia has shown an almost constant increase in output to a peak of 493 million tonnes in 1980, but since then, this country too has registered a sharp decline. In relative terms the other producing states have been much less important than these two oil giants, with the exception of Abu Dhabi. Here oil production only commenced in 1962, but by the late 1970s it had reached 80 million tonnes per year. Thereafter this country also recorded a marked decline.

The key to understanding the rapid development of the region is money, in the form of huge oil revenues. In the early days of oil production, the oil companies held the dominant position and the individual states only received royalty payments. This situation proved extremely profitable to the oil companies, so it was not surprising that the emerging nation states should wish for a greater share of the rewards. At first the royalty payments were increased,

Figure 9.3 Oil production and oil revenues for Saudi Arabia

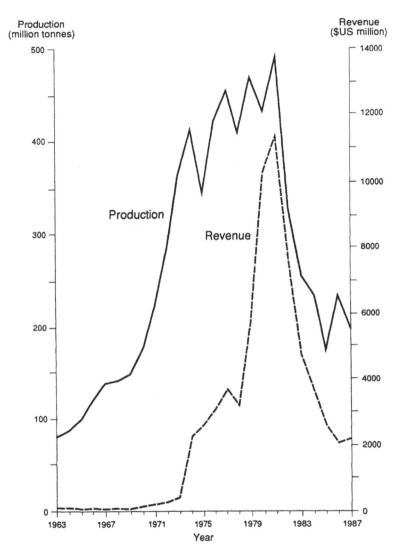

but by the early 1950s the states were demanding revenues which reflected the size of the oil company profits. Successful negotiations were carried out and a formula agreed which gave the producing countries 50 per cent of the net profits calculated on the basis of posted prices. As a result the income of the individual governments increased sharply.

By the late 1950s increasing oil production led to a reduction in posted prices, which in turn caused oil revenues to decline. This led to concern in the producer countries, many of which had embarked upon major development plans, and led to the formation of OPEC (the Organisation of Petroleum Exporting Countries) in 1960. The aim of this body was to protect the interests of the oil-producing nations in their negotiations with the international oil companies. One of its first achievements was to negotiate on changes regarding how revenues were calculated, so that in future they became a fixed payment on every barrel of oil produced. During the 1960s and 1970s further concessions from the oil companies were obtained by OPEC, and the trend to increasing control by government of the oil companies increased. This latter trend had been initiated in 1953 by Iran's nationalisation of the Anglo-Iranian Oil Company.

However, although these changes were important the most significant change was the quadrupling of oil revenues in 1973–4 by OPEC, following the October 1973 war between Egypt and Israel. This had a major impact on the world economy and in the Gulf gave rise to a spurt of development which has had a significant impact on the desert environment. The first major jump in revenues took place in 1974, to be followed by an even more spectacular one in 1979–81 (Figure 9.3). During the 1970s oil revenues in many states increased about twentyfold. In turn this necessitated a re-evaluation of all development plans to cope with the huge volumes of money which were now available.

9.3 THE URBANISATION PROCESS

The rapid urbanisation which has occurred has consumed large areas of land. This land has been viewed for the most part as worthless desert, and outside the urban centres themselves an extensive form of development has therefore taken place, with few real planning curbs being employed. The very great population growth along the Gulf coast over the last 40 years has been concentrated almost entirely in a small number of urban centres (Qutub, 1983) (Figure 9.4). Initially employment opportunities were to be found in the oilfields themselves and the related petrochemical plants. However, as the pace of development increased

Figure 9.4   Urban centres in the small Gulf States

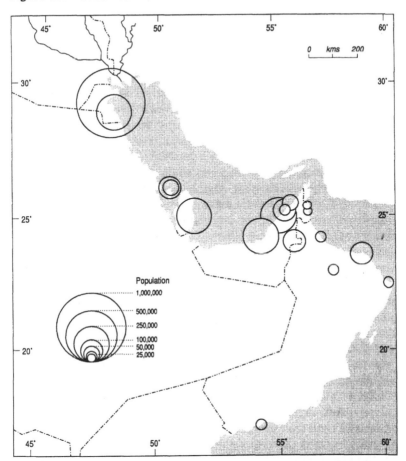

Source: Redrawn from El-Arifi S.A., 1986 'The nature of urbanisation in the Gulf countries', *GeoJournal*, v. 13, pp. 223–35. Copyright © 1986 by D. Reidel Publishing Company. Reprinted by permission.

and ever-larger sums of oil money were recycled into the general economy, more and more jobs were created in the construction industries associated with the rapidly growing urban centres and their communication infrastructures. Following this stage was the rapid growth of employment in the service industries. The ultimate form of this development seems likely to be large urban centres devoted to commercial and light industrial activities, connected to

339

each other and to large industrial complexes by very good road networks. This seems to be the favoured pattern of development up to the present and there seems little likelihood of it changing in the near future.

Between 1950 and 1980 the urban populations of the small states of the Gulf increased about thirteenfold to about 3 million people, with the greatest growth being registered in the capital cities. Today, both Kuwait City and Doha contain more than 80 per cent of the total population of their respective countries (El-Arifi, 1986). In contrast, in Oman, the capital, Muscat, accounts for less than 10 per cent of the country's population. The greatest overall increase has been registered in Kuwait City, which today has more than a quarter of the total population of the small Gulf states. Since 1945 urban growth rates have been very high, reaching values of over 14 per cent per annum in the UAE (United Arab Emirates) for the decade 1965–75. Elsewhere, rates in excess of 8 per cent have been registered in Kuwait, Qatar and Oman at various times.

Before 1940 the only settlements in the Eastern Province of Saudi Arabia with populations of more than 10,000 were the oases of Hofuf and Qatif. The population of the province at this time was probably less than 100,000. With oil development the population has increased to 1·5 million and urban population growth rates were in excess of 10 per cent between 1970 and 1980 (Barth and Quiel, 1986). By the 1980s the Dammam-Al Khobar-Dharan conurbation had a population of about 700,000, or almost 50 per cent of the total for the Eastern Province. Jubail, the industrial centre — which did not exist before 1972 — now has a population of 85,000 and is predicted to reach 350,000 by the turn of the century (Schliephake, 1985). By that time it is expected to cover an area in excess of 200 square km.

It should be remembered that the development process began at different times in the various countries. In Kuwait, Bahrain and Saudi Arabia major changes began to occur soon after the end of the Second World War (Al-Mallakh, 1980; Barker, 1982; Helms, 1981; Ismael, 1982; Khuri, 1982). In contrast, development did not get under way in the UAE until 1963, and in Oman, not until 1971 (Abdullah, 1977; Cottrell, 1980; Petersen, 1978; Townsend, 1977). However, with the decline in oil prices over the last few years the pace of development in all countries has definitely slackened.

The causes of the rapid population increase have been twofold. In the first place there have been high rates of natural increase.

Even more important, though, has been the migration of people into the urban areas to provide the necessary labour for the major construction projects. Some of this has come from local rural areas, but in the main it has been met by the immigration of foreigners, particularly from other Arab countries and from the Indian sub-continent. So great has been this influx that in the UAE and Qatar, non-nationals account for almost three-quarters of the total population. In Kuwait the figure is about 57 per cent. In Saudi Arabia there were 1·07 million foreign workers in 1985, making up about 41 per cent of the total work-force. In certain areas where specialist skills are in short supply the figure is even higher. In Jubail, for example, foreigners make up at least 60 per cent of the employees and it seems likely that these high figures will be maintained until at least the end of the century.

Kuwait provides an ideal example of this growth. Population estimates for the early to mid-1940s suggest a population of between 70,000 and 100,000 (Hill, 1972). At this time oil development was just beginning and labour was being sucked into the country. By the first census in 1957, the population had reached 206,000. A mere four years later 321,000 people were recorded and in 1965, 467,000. From about 1961 onwards non-Kuwaitis have outnumbered native born Kuwaitis. This extremely rapid growth rate continued in the late 1960s and early 1970s, but over the last few years it has declined sharply to about 3·4 per cent. The current population of 1·9 million (1985), therefore, represents an almost twentyfold increase over a 50-year period (Population Reference Bureau, 1985).

In many of the larger cities of the Gulf region over half of the population growth which has occurred can be accounted for by foreign immigration. It has meant that the cities now have many different culture groups within them, with views very different from those of the indigenous people. Although it is possible to distinguish between nationals and non-nationals in the cities, it should be realised that the indigenous peoples are not necessarily culturally homogeneous. Thus, for example, there are often great differences in outlook between the traditional urban dwellers and the newcomers such as nomads, oasis farmers, and fishermen (Cordes and Scholz, 1980).

Although in terms of absolute numbers the people from the West are relatively small, the influence which they have had on the economy and the environment has been very great. These are the people who have planned and overseen the expansion of the cities and the construction of the new industrial complexes. They have brought Western ideas with them and advocated horizontal urban development, with emphasis placed on single-house dwellings. They have also permitted the motor car to control city design, with roads providing the framework into which buildings are placed. The great wealth of the region has also meant that in many cities there is one car for every two individuals.

The growth of the cities has required that enormous sums have been invested in infrastructure provision for services such as water, sewage and electricity. Since 1950 and the advent of modern air conditioning systems there has been a massive increase in electricity consumption, particularly during the late 1960s and 1970s. In Saudi Arabia the electricity generating capacity increased twentyfold from 1·8 billion kW-hours in 1970 to 40 billion kW-hours in 1984 (Kingdom of Saudi Arabia, Ministry of Planning, 1985). By the end of the century the generating capacity of the Gulf region will have doubled compared with the 1986 figure (Table 9.1). In the Gulf region consumption of electricity reveals a

Table 9.1: Installed electricity generating capacity (MW) of the Gulf states, 1986–2000

	1986	2000
Bahrain	992	1,640
Kuwait	5,086	9,580
Oman	953	1,960
Qatar	1,095	1,670
Saudi Arabia	14,761	31,750
UAE	3,933	8,040
Total	26,820	54,640

Source: Wilkinson, 1987.

marked seasonal nature, with maximum demand occurring in the long hot summer period from April through to September (Figure 9.5). During June, July and August electricity demand is normally double the value for winter months. In all the Gulf states, electricity costs are highly subsidised by governments to minimise costs to consumers. This is one of the factors which has produced the large increase in demand. In order to reduce demand Saudi Arabia has tried to reduce the subsidy, but such moves have proved very unpopular. As a result industrial consumers only pay

Figure 9.5 Electrical power consumption in megawatts for the State of Qatar (1983 and 1984)

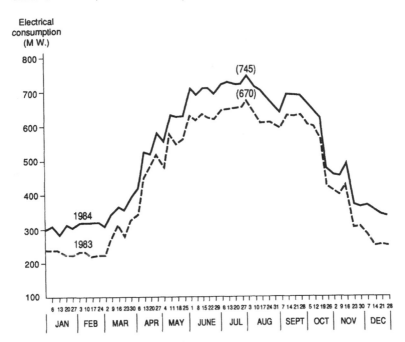

Source: Redrawn from Babikir, A.A.A., 1986, 'Some aspects of climate and economic activities in the Arab Gulf States', *GeoJournal*, v. 13, pp. 211–22. Copyright © 1986 by D. Reidel Publishing Company. Reprinted by permission.

about 40 per cent of production costs, while domestic consumers pay between 50 and 100 per cent, depending on the amount consumed. In Kuwait certain consumers are paying charges which account for only 7·5 per cent of actual production costs (Wilkinson, 1987). In such circumstances it is no wonder that demand continues to escalate.

With electricity demand still rising, one of the most ambitious proposed development projects is the construction of an electricity grid to link the states of the Gulf (ibid.). This will be 1,200 km in length and will eventually connect the electricity networks of all the states between Kuwait and Oman (Figure 9.6). The scheme will be expensive, around US$2,000 million, but will greatly reduce idle capacity and be able to cope with emergency failures. A major problem is that Saudi Arabia uses a 60-Hz electricity

Figure 9.6 Proposed Gulf power grid

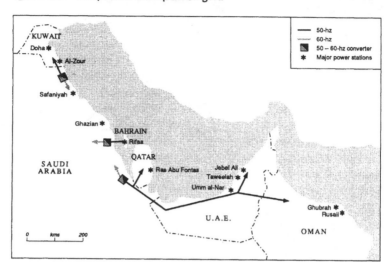

Source: Redrawn from 'Electrifying the Gulf', *Middle East Economic Digest*, 6th May 1987. Reproduced by permission of the Middle East Economic Digest Ltd.

system, while 50-Hz is used throughout the rest of the Gulf. If the project is implemented the first stage will link Kuwait, Bahrain, Saudi Arabia and Qatar. At a later date UAE and Oman will be connected to the system. Currently neither UAE nor Oman possess large integrated national grids of their own.

One of the really spectacular changes which has occurred along the Gulf coast has been the advent of high technology modern agricultural systems (Beaumont, 1985b). In former times the intense aridity meant that arable farming was only practised in oasis settlements such as Al Hassa, and at places where springs and shallow wells have permitted the cultivation of date groves and gardens. Farming methods were traditional and the total cultivated area was very small indeed. The massive oil revenues which the states of the Gulf received in the late 1960s and 1970s meant that large investments could be made in agricultural infrastructures. Wells could be sunk to tap the deep aquifers and new concrete irrigation systems constructed. Huge milk and beef units have been introduced, with animals always being kept indoors under zero grazing conditions. Similar complexes for egg and poultry production have also become common. Investment in controlled environment vegetable growth under polythene and glass has been considerable, and on a scale probably unmatched anywhere in the world.

The objective behind recent developments has been food production for strategic reasons, so that a port embargo would not immediately precipitate a political crisis. Although in many of the small countries along the Gulf this aim has been realised with regard to certain foodstuffs, it has meant that actual food costs produced by these methods have been extremely high and governments have consequently had to introduce large subsidies in different forms to enable their people to purchase the internally produced food.

9.4 ENVIRONMENTAL IMPACT

The environmental impact of all this rapid development along the Gulf coast is, as yet, difficult to assess with any accuracy. Perhaps the most obvious change has been the takeover of large areas of desert and semi-desert land for new activities (Figure 9.7). These include urban and industrial expansion as well as road and other infrastructure provision. Everywhere, this land has been perceived as having low value for alternative uses, resulting in a very wasteful policy of allotment. In turn, areas of land outside a project area have been greatly disturbed through a very casual form of constructional activity. Such areas of environmental blight commonly stretch for more than 100 m on either side of major highways and a similar zone often covering hundreds of metres exists on the outskirts of towns, criss-crossed by vehicles taking short cuts and dumping materials no longer needed.

The impact on water resources has also been considerable. The meagre existing resources prior to development permitted limited oasis agriculture at favoured locations as well as the growth of small urban centres. Modern development has only been possible by the desalination of sea water and the exploitation of the deep aquifers. This latter activity has led to excessive pumping from what are fossil water reserves and it has drawn poorer quality groundwater to the surface (Burdon, 1982). Declining water quality seems likely to occur at many localities as pumping continues. Even the desalination plants have not been without their problems. The highly concentrated brines which are a by-product

345

Figure 9.7 LANDSAT images showing changes and developments
in the Al Khobar-Dhahran-Dammam-Al Qatif agglomeration
between 1972 and 1981

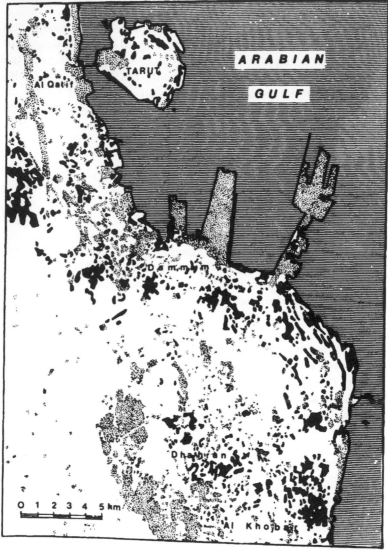

Note: White: no change from 1972 to 1981. Grey: quarrying, landfilling
and clearing for construction. Black: urban expansion, new industrial sites
and agricultural reclamation areas.
Source: Redrawn from Barth H.K. and Qviel F., 1986, 'Development and
changes in the Eastern Province of Saudi Arabia', *GeoJournal*, v. 13, pp. 251–9.
Copyright © by D. Reidel Publishing Company. Reprinted by permission.

of water purification have considerably disrupted ecosystems when dumped into the sea.

The disposal of urban and industrial waste waters is already giving rise for concern. Shallow aquifers are in danger of being polluted by pathogenic bacteria, while industrial effluents may introduce a range of new pollutants into the groundwater. In the Gulf waters themselves oil pollution has been a potential threat since exports first began. Coastal waters are polluted from spills from loading facilities and refineries, whilst the threat of petroleum leakage from offshore fields is always present. More recently there has been the danger of tankers being sunk or holed as a result of the Gulf War (Riad, 1983). To date little is known of the effects of this pollution on the Gulf ecosystem, though everywhere along the Gulf coast, balls of petroleum tar on the beaches testify to the presence of pollutants.

9.5 SAUDI ARABIA — EASTERN PROVINCE

Although the Eastern Province of Saudi Arabia has followed a similar pattern of development to that of the other Gulf states, its status as part of a much larger nation has meant that the scale of development has often been greater than elsewhere (Idris, 1986). A feature of the Eastern Province has been a programme of massive industrialisation and related urbanisation based on the petroleum industry (Barth and Quiel, 1986; Maadhah et al., 1985; Mirghani, 1986). It is this industry which provides a cheap energy source and also feedstock for many different chemical products.

Initial development was dependent on the oil industry itself, and it basically consisted of what were company oil towns and massive infrastructure provision to get the petroleum out of the ground and to its export markets. This involved the construction of well heads, pumping stations, pipelines and a large oil terminal at Ras Tanura. From the late 1960s, however, the emphasis changed from mere oil and gas production to a comprehensive programme of industrialisation and urban development. A key feature of this new policy was the building of a large gas grid — 'the Master Gas System' (MSG) — which was begun in 1975. Its objective was to provide industry with a cheap fuel and a raw material for chemical production. The first stage went into use in 1980. A major part of this strategy was the construction of a new industrial city at Jubail on the Gulf coast (Figure 9.8). The location was selected in 1972

347

and construction began in 1979. All the major industrial complexes in Jubail, with the exception of a steel plant, use petrochemicals for raw materials. The main products are chemicals such as ethylene, ethylene glycol and methanol, plastics and fertilisers. Secondary industries have also been established in Jubail and nearby Dammam to service the primary industries and the local population.

Figure 9.8 The new town and industrial site of Jubail, Saudi Arabia

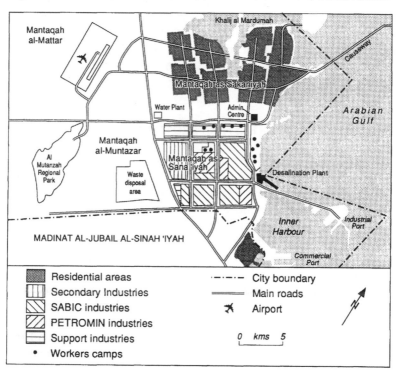

The infrastructure development in the Eastern Province has been remarkable. Even as recently as 1950 travel in the region was difficult and the major mode of transport was by camel. The first metalled road followed the Trans Arabian Pipeline (TAP-line) and then new roads were quickly built from Dammam, which became the centre of the oil industry, to Hofuf and Riyadh. Today a high quality road network services the region with connections to Kuwait, Qatar, and the UAE. Prior to the expansion of the petroleum industry ports were small and devoted to fishing and trade. With oil exploration Ras Tanurah was developed as the main export terminal, with Dammam being used for the import of goods required for industrial growth. Recently port facilities have been established at Jubail for the import of iron ore for the steel works and for the export of chemicals and related products. A new oil terminal has also been constructed at Ju'aymah.

The impact of oil revenues is also revealed in terms of agricultural activities. The low rainfall has always meant that agriculture is only possible with irrigation. Traditionally there were a number of oasis settlements such as Hofuf and Qatif, which utilised groundwater for the cultivation of date groves and other crops. With the abundant oil revenues Saudi Arabia invested heavily in agricultural projects. These included new irrigation schemes using water from deep wells, as well as the modernisation of already-existing irrigation systems. In all these projects the emphasis has been on high technology and little attention has been paid to the cost-effective nature of the scheme. Perhaps the most impressive results are to be seen in Saudi Arabia with regard to wheat production. In 1980 output was only 140,000 tonnes, yet by 1986 it had passed two and a quarter million tonnes (Table 9.2).

Table 9.2: Wheat production in Saudi Arabia (thousand tonnes)

1980	142
1981	199
1982	417
1983	817
1984	1,402
1985	2,047
1986	2,285

Source: *Middle East Economic Digest*, 1987b, p. 35.

As a result of developments the area under cultivation in the Eastern Province has increased from 30,000 ha in 1965 to 50,000 ha in 1985, with the main emphasis on wheat as a winter crop. The Eastern Province now contributes about 15 per cent of the total wheat crop of Saudi Arabia. In other cases large investments have been made in intensive livestock units for dairying and poultry production. With nearly all of the projects large government subsidies of one form or another have been involved. With wheat production, for example, the government

349

guaranteed the farmer a price per tonne which was more than three times the price on the world market. The results have been impressive, with the area under cultivation rising from 150,000 ha in 1975 to 2·3 million ha in 1984 for the country as a whole.

9.6 WATER

Throughout the Gulf, rapid urbanisation has led to serious water problems as the meagre available resources have been called upon to supply increasing numbers of people. Kuwait was one of the first states to tackle these difficulties. Until the early part of the twentieth century domestic water for the town of Kuwait was obtained from shallow wells along the coastal region. As the population grew, the available water resources proved inadequate, leading in 1925 to the import of water in barrels from the fresh waters of the Shatt al Arab. This project proved successful and in 1939 a company was formed to operate a fleet of dhows to bring Shatt al Arab water to the town to be stored in reservoirs near the port. At this time distribution of water took place by hand or donkey cart, in small containers. By 1948 about 364 cubic m per day were being transported.

In the late 1940s the boom in oil production led the Kuwait Oil Company to construct a small water desalination plant at its refinery. It was also agreed that this plant would supply the town, some 40 km away, with 364 cubic m per day. This venture proved so successful that the government committed itself to a major programme of desalinated water production. From the early 1950s onwards the increasing living standards of the inhabitants led to a major increase in the use of air conditioners. In turn this produced a rise in electricity consumption, and it was therefore fortunate that the increased need for water was paralleled by an increased demand for electricity. It was soon realised that the large quantities of waste heat associated with electricity generation could be utilised for desalination if electricity plants and desalination units were built side by side. The first such plant was built at Shuwaikh in the late 1950s with a capacity of 4,546 cubic m per day, and this was then doubled within a year to over 9,000 (Beaumont, 1977a). From this time onwards Kuwait embarked upon a major policy of desalinated water production.

Another major plan to obtain extra water for Kuwait was to construct a pipeline into Iraq so that fresh water could be abstracted from the Shatt al Arab. General agreement was reached with Iraq about this project, but owing to various difficulties, construction has never begun. Recently the scheme has been revived with a view to Iraq providing 0·91 million cubic m per day of drinking water and 0·45 million for irrigation. Until extra water is available Kuwait has been forced to make the best use of the resources it possessed. Considerable development of the shallow aquifers has taken place. At Raudhatain water-bearing sands and gravels were discovered in 1960 and connected to Kuwait City by pipeline. Later a similar aquifer was developed at Umm Al Aish.

The most important aquifer for Kuwait, though, is the Dammam Limestone, and numerous wells have been sunk into it to exploit its water. Water quality in the aquifer deteriorates in a north-easterly direction. Water from the aquifer is used for industrial purposes and irrigation, and some is added to the supply obtained from desalination. One of the major well fields tapping the aquifer is found at Sulibiyyah about 15 km to the south east of the city. The water, which is brackish, is connected by pipeline to reservoirs in the city. Close by is the Abduliya well field which supplies Al Ahmadi and the major oil installations with water. Similar brackish groundwater underlies most of the south-western part of the country.

The question of the cost of desalinated water is an interesting one which bears examination. As the flash distillation plants are always associated with electricity-generating stations, there is ample scope for allocating costs in different ways. The major recurrent cost is for energy, usually in the form of natural gas. This in itself can receive direct subsidies from the government in terms of the prices which are charged. More importantly, though, is the fact that it is a book-keeping exercise to decide whether the major charges should be allocated to the electricity or water side of the production process. It is possible, for example, to allocate most of the costs to electricity generation and, therefore, be able to produce cheap desalinated water, or vice versa. In general the Kuwaiti government has followed a policy of cheap water, and the government has also treated the desalination industry very favourably by charging low interest rates on capital investment for the plants. The distribution of water in the city is now achieved through a dual distribution system for fresh and brackish water. Potable water is used for drinking, cooking and personal hygiene, whilst brackish water is utilised for toilet flushing, washing, garden irrigation and similar activities. The dual supply system allows Kuwait to make the most efficient use of its water resources.

The question of future policy is an important one. It is possible for the government to continue constructing desalination plants to supply ever-larger amounts of water to the growing population. However, besides increase in numbers there is the problem that *per capita* use is increasing as standards of living rise. If sufficient capital were available it would be possible to supply growing needs by building more and more desalination plants. The Kuwaiti government has now realised that this could lead to an open-ended financial commitment which even it might have difficulty in funding. At the moment the government has no plans for new distillation plant. Current capacity (1986) is 0·98 million cubic m per day. A peak daily capacity of 1·13 million will be reached in 1988, before falling to 1·05 million in 1991 as old plants are closed or their output reduced (Middle East Economic Digest, 1987a). It therefore seems inevitable that the government will have to introduce a demand management policy, probably based on the cost mechanism, before the end of the century.

The water situation in Saudi Arabia is very similar to that of Kuwait, in that water demand is met from a combination of desalinated water and groundwater supplies. Given its huge financial resources it is not surprising that Saudi Arabia turned to desalination as a means of supplying the water needs of its expanding urban areas. Although ARAMCO had constructed a number of small desalination plants for its own use, the government did not construct any desalination plants for itself until 1970 when the first major plant was commissioned at Jiddah. This put it well over a decade behind developments in Kuwait.

The basic strategy of the government was to develop two areas of desalination capacity — one along the Red Sea and the other along the Gulf coast (Beaumont, 1977b). It was hoped that eventually the major plants in each of these two development areas would be connected by pipeline. About two-thirds of the installed capacity would be on the Gulf coast when all projects were completed. On the Gulf coast the first desalination plant was commissioned at Al Khobar in 1974 with an installed capacity of 28,500 cubic m per day, and other desalination complexes have since been built on the coast at Ras'al Khafji, Al Jubail and Al Uqayr. Currently the combined capacity of the Al Khobar, Jubail and Khafji plants is more than 1·3 million cubic m per day.

The other sources of water in the Eastern Province of Saudi

Arabia are the large aquifer systems which underlie the region (Figure 1.13). Although these aquifers store enormous volumes it is thought that the water in them is fossil in origin, with major recharge having occurred during the wetter parts of the Quaternary (Bakiewicz et al., 1982). Any water which is used is effectively being mined in the same way as oil. In the east the Cretaceous aquifer system is composed of the Wasia and Biyadh Formations, which are located at considerable depth. It is a confined aquifer system for the most part, though there does appear to be some sub-surface transfer of water into the overlying aquifers. Exploitation of this aquifer system has only been possible since the introduction of deep well construction techniques.

The Eocene aquifer system is also for the most part a confined aquifer system. However, it does outcrop at the surface in places, giving rise to oases at Al Hofuf and Al Qatif as well as springs on Bahrain Island. Evaporation losses from sabkhas along the coast fed from the Eocene aquifer system are thought to be about 500 million cubic m per annum (Pike, 1970). The most important unit within the system is the Umm ar Radhuma Formation. Pumping from this aquifer system is now considerable and water levels are

Table 9.3: Water budget for Eastern Province, Saudi Arabia

	1980	1990	2000
		(million m³/year)	
Water supply			
Groundwater	1,000	1,000	1,000
Desalination	11	169	169
Recycling	0	25	50
Total	1,011	1,194	1,219
Water demand			
Cities and industries	68	231	402
Irrigated agriculture	367	550	575
Rural areas	1	1	1
Total	436	782	978
Surplus	575	412	241

Source: Kingdom of Saudi Arabia, Ministry of Planning, 1981.

353

falling in a number of areas. Water quality is also declining in many boreholes.

The future use of water in Eastern Province obviously depends on the speed of industrial growth as well as any developments in irrigated agriculture. Government estimates suggest that urban and industrial demand for water will increase from 68 million cubic m in 1980 to 402 million in 2000. This represents an almost sixfold increase in just 20 years (Table 9.3). Similarly, irrigated agriculture will show an increase in demand, though not as great as that in the urban/industrial sector. Despite these rapid growth rates it is still predicted that there will be a water surplus of 240 million cubic m per annum by the year 2000.

The question of whence the water is to come is also revealing. Government estimates suggest that 1,000 million cubic m a year will continue to be obtained from groundwater by the year 2000. This appears rather optimistic in view of the fact that water levels in many boreholes are falling and that water quality is also deteriorating. Desalinated water capacity shows a remarkable growth over the planning period, with a fifteenfold increase. This represents a tremendous capital investment. Even here, however, Saudi Arabia, like Kuwait before it, will find that it cannot go on pouring funds into what would be an open-ended commitment. Before long a demand management policy will have to be introduced.

If the estimates of groundwater availability do prove optimistic, it is likely that the urban/industrial complexes of the Eastern Province will be facing severe difficulties before the end of the century. As elsewhere this would inevitably mean that water will be transferred from low- to high-value uses, such as from irrigated agriculture to urban/industrial usage. If this were to occur it would seem that there would be no major water shortages for at least this century, though the agricultural policy of high food production would have to be revised.

9.7 THE FUTURE

Since the Second World War the countries of the Gulf have experienced a spate of development which is probably unprecedented in world history in terms of the amount of capital that has been invested in major projects. As a result the eastern coast of the Gulf possesses an excellent road infrastructure which is continually being improved. The latest addition to this network is the opening of the spectacular causeway linking Bahrain Island to the mainland. The new parts of the cities which have expanded to consume so much land are, for the most part, nondescript in their architecture. Yet within these cities one finds some of the most innovative buildings in the world. Standards of services, such as water and electricity, are already high and within a few years will be comparable with any city in the West. Industrial development has also proceeded apace in countries like Saudi Arabia, and some of these new projects seem set to challenge similar activities in other parts of the world (Barth and Quiel, 1986; Schliephake, 1985). With policies of self-sufficiency in certain foodstuffs, agriculture has also boomed through government infrastructure provision and guaranteed food prices to the farmers. Some of the results, particularly the growth in wheat production in Saudi Arabia, have been quite spectacular.

A feature of all this development has been that the arid nature of the environment in which the development has taken place has been ignored. If specific problems have arisen — for example, shortage of water — these have been overcome by the injection of capital. The type of development in the future may well be different from that of the 1970s and early 1980s. The reason for this could be a fall in the oil price. In July 1986, after a decade and a half of almost continually rising oil prices, the price of a barrel of oil fell to less than US$10. This event had a tremendous psychological impact on the world oil market, as well as a financial one on the development plans of the oil-exporting countries.

In many respects the countries of the Gulf have been fortunate in the timing of the fall in the price of oil, in so far as many of them have almost completed their basic infrastructures for transport and services. Elsewhere in the economy the position is not so favourable, for in many cases large-scale industrial and agricultural production is dependent on large financial subsidies of one form or another. As far as industry is concerned it is the provision of cheap energy and cheap raw materials. In agriculture it is a policy of guaranteed crop prices, often many times the world market price. Even the urban dwellers receive cheap water supplies through heavy capital expenditure and low interest rates for desalination plant construction.

The inevitable consequences of falling oil revenues would seem to imply a downwards revision of development plan objectives. The 'boom' might not yet be over, but it is certainly past the peak.

It seems unlikely that many current agricultural objectives can be maintained, though industrial expansion may continue, if at a reduced rate. As a series of urban/industrial complexes the region seems destined to prosper in the future, though the pace of development will only be a fraction of that experienced in the recent past.

10

Israel: Integrated
Water Development

10.1 INTRODUCTION

The state of Israel was established in 1948 and since that time has
been at war with almost all of its Arab neighbours. As such the
country provides an interesting case study, permitting insights into
the development of water policies and methods of water management. Given its war footing, water had to play an important
strategic and ideological role in encouraging the settlement of the
countryside. At the same time the country was changing from a
traditional society to a materialistic, urban-based society of the
1980s. This change was accompanied by rapid population growth.
Part of this growth was the result of natural increase, but even
more important was the massive immigration of Jews from the
countries of war-torn Europe and more recently from other
Middle Eastern states. This meant that water policies had to cope
with a change in emphasis from a country in which agriculture was
a key element in the national economy, to one where the urban/
industrial lifestyle (with its high standard of living and growing
consciousness of pollution issues) has become the dominant one.
As if these changes were not in themselves enough, it has to be
remembered that Israel is an arid country with a limited and finite
resource base. In recent years this has posed increasingly difficult
choices as to which water management policies should be pursued.
   During the early years of the state's existence, immigration was
encouraged for political and idealistic reasons, with the result that
during the first 20 years of Israel's existence, about 1·25 million
Jews entered the country. Partly as a result of this immigration,
the population of the nation grew from 873,000 people in 1948

(Blake, 1972) to about 3·1 million in 1973 (Population Reference Bureau, 1973). Currently the population is 4·2 million, and it is expected to reach almost five and a half million by the end of the century (Population Reference Bureau, 1983). In 1948 a large proportion of the Jewish population was concentrated in Tel Aviv and the surrounding region. Much of the south of the country lay virtually empty, while the better-watered northern parts possessed an Arab majority. The new state therefore felt it essential that an important part of its development must include a viable and strategically safe form of rural settlement based on intensive agriculture.

10.2 CLIMATE

Most of northern Israel has a typical Mediterranean climate with summer drought and a winter rainy season. Southwards, however, annual precipitation decreases until true desert is found to the south of Beersheba. Between the two climates of the Mediterranean and the desert, a transitional dry steppe regime can be recognised. August is normally the hottest month with mean temperatures almost everywhere above 24°C. The highest mean temperatures, often in excess of 30°C, are registered along the Dead Sea Lowlands. In the coldest month (January) mean temperatures fall to below 15°C almost everywhere. At this time of year the warmest temperatures are found along the coastal fringe of the Mediterranean, as well as along the trough of the Dead Sea Lowlands. Throughout the uplands frosts are common.

In Israel precipitation declines in both an easterly and southerly direction (Figure 10.1). However, the decrease inland is masked by a tendency for precipitation to increase with altitude. Combining these factors means that the northern highlands receive the highest precipitation totals, in excess of 800 mm per annum, while in the extreme south-east at Eilat, annual rainfall totals are less than 50 mm. Almost all of the country north of a line from Tel-Aviv to Jerusalem receives at least 500 mm per annum and it is in this region that the major centres of population and agricultural activity are concentrated. The southern part of Israel, the Negev, has annual precipitation totals of less than 100 mm. The total amount of precipitation falling on Israel (pre-1967 boundaries) has been estimated at 6,000 million cubic m per annum. A further 4,000 million cubic m of water per annum drain into Israel from

Figure 10.1 Average annual precipitation for Israel (1931–60)

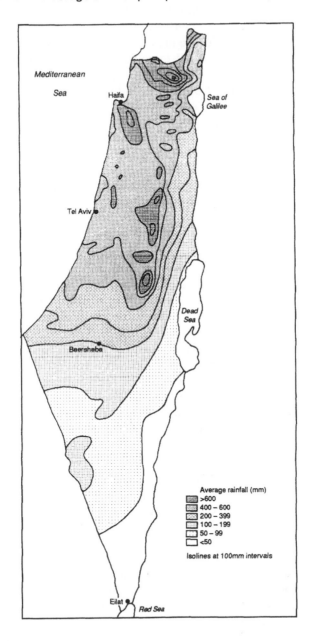

Mediterranean

Sea

Haifa

Sea of
Galilee

Tel Aviv

Dead
Sea

Beersheba

Average rainfall (mm)
>600
400 – 600
200 – 399
100 – 199
50 – 99
<50
Isolines at 100mm intervals

Eilat

Red Sea

(millimetres)

territories outside its legal jurisdiction to provide a total 10,000 million cubic m per annum (Orni and Efrat, 1971).

With the hot summer temperatures experienced over much of the country, evaporation figures are high. In July evaporation from an open water surface exceeds 24 cm along much of the structural trough of the Dead Sea Lowlands to the south of Lake Tiberias. On an annual basis, the highest evaporation totals, in excess of 270 cm, are observed around Eilat, while over much of the northern part of the state average values fall between 130 and 180 cm. Only in a narrow strip along the Mediterranean Sea does annual evaporation from an open water surface fall below 130 cm per annum. It has been estimated that in total between 60 and 70 per cent of the country's annual water supply from precipitation and river flow is lost by evapotranspiration.

10.3 HYDROLOGY

The drainage of Israel can be divided into two major zones, separated by a watershed running north to south approximately parallel to the Mediterranean coast. To the west are a series of relatively small basins with streams which flow directly from the uplands to the Mediterranean. Only four of the major streams — the Na'aman, Qishon, Tanninim, and Yarkon — are perennial. In all of the catchments, very great variations in discharge occur from year to year, dependent upon precipitation amounts. In drought conditions no storm runoff may take place during a rainy season, while at other times (following severe storms) a large proportion of the mean annual discharge can occur within a few days. East of the watershed is the inland drainage system of the Dead Sea Lowlands. Here by far the most important river is the Jordan, which rises in the well-watered areas to the north of the Israeli border in Lebanon and Syria. South of the Dead Sea little runoff flows into the structural depression except under very isolated storm conditions.

The River Jordan forms the key to understanding water resource development in Israel. It has its catchment divided amongst the four countries of Lebanon, Israel, Syria and Jordan. Although well known in the Western world because of biblical references, the Jordan is only a small river with a total length of 93 km from its source in southern Lebanon to its mouth on the Dead Sea and with a drainage basin area of 18,300 square km.

Figure 10.2 Water balance diagram for the Jordan River

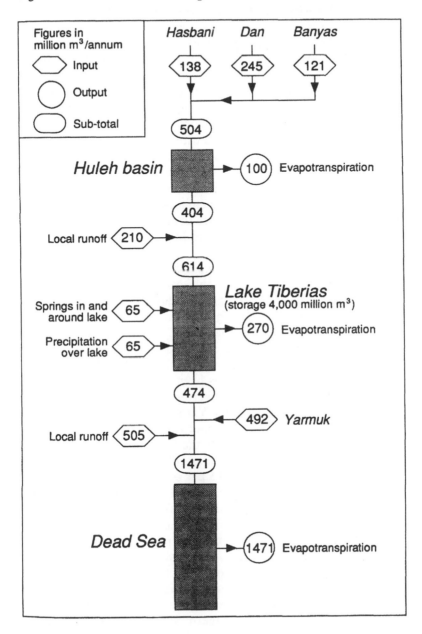

(million cubic metres per annum)

Three streams, which unite in northern Israel, form the head-waters of the River Jordan. The Hasbani rises from a series of springs north of the Israeli border in southern Lebanon, and its flow is further augmented just north of the border by discharge from the Ghajar springs. The mean annual flow of this stream is 138 million cubic m, although a wide range of discharges have been observed (Smith, 1966). To the south-east of the Hasbani the springs of the Dan provide the largest and the most reliable contribution to the waters of the Jordan with a mean annual flow of 245 million cubic m. The third tributary is the Banias rising on the southern slopes of Mount Hermon just within Syrian territory, although on land occupied by the Israelis since 1967. This has a mean annual discharge of 121 million cubic m, but it too, like the Hasbani, reveals considerable fluctuations from year to year. The waters of these three streams were dammed by a basalt sill in the southern part of the Huleh basin to form the lake and swamps of the same name (Neuman, 1955).

Between Lake Huleh and Lake Tiberias the bed of the River Jordan falls 270 m, with most of this drop concentrated in a 14 km rocky gorge cut through a basalt sill. This section of the river has always been viewed as having high potential for hydro-electric power generation. Lake Tiberias, which acts as a natural reservoir on the upper Jordan, covers an area of 165 square km and possesses an average storage capacity of 4,000 million cubic m. The discharge from the lake has been controlled since 1932 by a dam constructed as part of a hydro-electric project built at the confluence of the Jordan and the Yarmouk. Operation of this dam permits a controlled storage of about 1,000 million cubic m (Smith, 1966). The mean annual flow into the lake is considered to be 760 million cubic m, of which 615 million cubic m (81 per cent) is water from the Jordan; 80 million (10·5 per cent) is local runoff together with spring discharge from the floor of the lake; and 65 million (8·5 per cent) comes from precipitation over the lake (Figure 10.2) (Neuman, 1953). Evaporation from the lake is high and thought to be of the order of 270 million cubic m per annum.

South of Lake Tiberias the Jordan falls a further 180 m before reaching the Dead Sea. In this stretch significant additions to discharge are received. The largest single source, of 492 million cubic m per annum, is provided by the River Yarmouk draining north-eastern Jordan and southern Syria. Winter floods are a common feature of this stream. Smaller wadis and springs furnish a further 505 million cubic m per annum, giving a total flow into the Dead Sea of 1,471 million cubic m. The Dead Sea itself has exhibited considerable changes in level since the beginning of the nineteenth century, from which time reasonably accurate historical records are available. Between 1807 and 1897, a rise in level of about 11 m was recorded, with most of this rise occurring after 1884 (Klein, 1965). During the twentieth century, a fall of approximately 6 m took place up to the mid-1960s. Climatological changes within the basin of the River Jordan are generally accepted as the cause of these observed fluctuations.

There are a number of major aquifers in Israel which, in recent years, have become widely used for groundwater abstraction. In the limestone uplands, springs fed by groundwater are numerous and play an important role in the agricultural economy by providing an assured water supply throughout the year. All of the perennial streams receive waters from at least one major spring. Many villages also obtain their domestic supplies from wells dug into the limestone aquifers. The other major aquifer is the Plio-Pleistocene system of the coastal plain, consisting of alluvial material forming large fan systems sloping from the uplands to the Mediterranean Sea. Total discharge from the aquifer in the period 1933–5, prior to the large-scale exploitation of the groundwater reserves, was estimated at 190 million cubic m per annum (Eriksson and Khunakasem, 1969). Since 1948 groundwater has been widely used to supply irrigation and domestic demands. The result has been that many aquifers have been overpumped and groundwater quality measured in terms of chloride content has fallen rapidly in a number of areas. In the coastal zone excessive water abstraction has led to sea-water penetration into the aquifer.

Large quantities of groundwater (estimated as several hundred billion cubic m) are known to exist within the Upper Nubian Sandstone aquifer of central Sinai and the Negev in areas where present precipitation totals are mostly below 100 mm per annum (Issar *et al.*, 1972). Most of the water appears to be fossil in character, with carbon 14 dating indicating ages from 13,000 to more than 30,000 years BP. Research suggests that the Nubian Sandstone forms a massive interconnected aquifer system of at least 200 m in thickness which was recharged during colder and wetter environmental conditions during the latter part of the Pleistocene. Present-day recharge of the Nubian Sandstone aquifer is believed to be low, totalling only 3 million cubic m per annum (ibid.). It is thought that water could be mined from this aquifer to supply projects in the Negev.

10.4 WATER RESOURCE DEVELOPMENT

During the early period of the British Mandate, water supplies in Palestine were developed on a local basis, with numerous wells tapping shallow aquifers. Later, more complex schemes were embarked upon as larger organisations such as the Palestine Water Company and the Mekorot Water Company were established. After 1948 and the formation of the state of Israel, the new government decided to undertake a comprehensive plan of water resource development based on the ideas outlined in Lowdermilk's *Palestine: land of promise* (Lowdermilk, 1944). Following much discussion and advice from international hydrological experts, a Master Plan incorporating a national water system was finally agreed in 1956. At the time when the plan was conceived two factors greatly influenced the course and type of development. These were the shortage of capital within the newly formed state and the necessity to obtain immediate results as waves of new immigrants poured into the country. It was therefore decided that emphasis should be placed on maximum water conservation and on strict water allocations, together with the design of a highly integrated water system throughout the country (Weiner, 1972).

The first projects undertaken were low-cost schemes which provided quick returns. By far the most popular method of water exploitation proved to be the drilled well, which allowed the rapid irrigation of newly colonised land in the coastal plain and northern Negev. Unfortunately the extraction of groundwater by these pumped wells was so great that sea water penetrated into the coastal aquifer.

The Coastal Plain aquifer (of Plio-Pleistocene age) did not suffer major exploitation until the late 1940s when pumping began on a large scale. From this date the yearly output increased to a maximum in 1962 when 480 million cubic m of water were produced from it, representing 45 per cent of Israel's total water supply at that time (Aberbach and Sellinger, 1967). By the early 1960s water production from the aquifer was thought to be approximately two and a half times the safe yield. As a result of this groundwater mining falls of 15–20 m in the water table were recorded over a period of 25 years.

In the Tel Aviv region water supply for the conurbation was obtained from local wells in the coastal aquifer system until the late 1950s. Over-pumping was considerable with an annual production of almost 80 million cubic m, which was estimated to be almost five times the safe yield (ibid.). Water levels were falling rapidly and the intrusion of sea water up to two km inland had already caused 25 wells to be abandoned. To combat these difficulties a programme of artificial recharge was initiated in 1964, accompanied by the import of water from other regions as well as by a reduced pumping rate from the aquifer. At first, producing wells in which the intrusion of sea water was greatest were utilised for groundwater recharge, supplemented at a later date by newly drilled wells. The water used for recharge was passed though the network of the Tel Aviv municipal water supply system and so had to conform with drinking water standards. Recharge took place during winter months when excess water was available, but the operation of the scheme meant that the intrusion of sea water was checked, so protecting the inland portion of the aquifer from further contamination.

The medium-term development programmes were based on a number of important criteria. Schemes were selected which provided the minimum investment per unit volume of water supplied, were not technically complex, and were capable of having the investment income divided into a number of stages. It was also considered essential that, in planning the project, sufficient thought should be given to the fact that the data on which decisions were taken might not be accurate and, therefore, that the project design should be sufficiently flexible to permit changes to be made in the future (Weiner, 1972). At the same time the idea evolved that every project within the country, no matter what its size, should be capable of eventual integration into a nation-wide hierarchical system, so that should a single project fail, any water deficiency could be met by water transfers from another more successful scheme.

Attention was also focused on a number of long-term projects of regional rather than local significance. The largest of these was the Yarkon–Negev Project, which diverts water from the River Yarkon near Tel Aviv southwards towards the Negev. The first stage was completed in 1955, when a 106 km pipeline was commissioned. The supply from the Yarkon, together with the water produced from numerous wells integrated into the project, was 270 million cubic m per annum (Prushansky, 1967). At a later stage a second pipeline was built to carry water from the Yarkon, together with water supplied by the National Water Carrier from Lake Tiberias.

Another major project at this time was the drainage of the

Huleh marshes, which destroyed one of the most important wetlands in the Middle East. Work began in 1949 on a small drainage project for lands which were already cultivated on the northern margins of the marshes, while the main project to drain the Huleh marshes began in 1951 and was completed in four stages (Karmon, 1960). A small area of the former marsh was preserved as the Lake Huleh Nature Reserve. The drainage of the marshes removed an important water storage area from the river system and so accentuated the potential flood problem of the lower reaches of the river. Associated with the drainage of the marshes was a change in agricultural land use from a cereal-producing and grazing economy to one which relied on intensive sprinkler irrigation methods and modern scientific farming techniques. What this has also meant was a much greater pollution risk from fertilisers and pesticides.

One of the aims of draining the Huleh marshes was to try to increase the total water volume flowing into Lake Tiberias. Although accurate data are difficult to obtain this does not seem to have occurred. By 1967 the total area of fishponds in the upper Jordan valley was 15 square km — an area slightly in excess of that of the former lake itself. The total area of cultivated land rose from 32 square km in 1952 to 120 in 1967. Assuming a mean irrigation rate of 0·85 million cubic m per square km, this is an added water demand of 75 million cubic m per annum, of which perhaps one third percolates back to the river (Shick and Inbar, 1972). When one compares this figure with an estimated saving of 45 million cubic m per annum by evaporation reduction from the lake and marsh areas compared with the same drained area, it will be noted that the overall balance is, if anything, a negative one with increased water use.

A result of the drainage of Lake Huleh has been a significant change in the movement of sediment within the upper part of the Jordan catchment. Until its drainage Lake Huleh acted as a settling basin for most of the sediment which was carried into it by the Jordan and the other streams. Further large quantities of easily erodible sediment were made available to the River Jordan as a result of the drainage scheme itself when the major canals and drainage ditches were under construction. Almost all of this sediment was transported to and deposited in Lake Tiberias. In the early 1970s this corresponded to a sedimentation rate of 0·5 to 1·0 cm per annum, compared with an estimate of 0·2 from palynological records over the previous 18,000 years (ibid.). Most of the sediment reaching the lake is coarse-grained and tends to concentrate in the delta area.

10.5 NATIONAL WATER CARRIER

The largest water project in Israel is the National Water Carrier, which transports the waters of the River Jordan southwards to the agricultural and industrial centres along the coastal plain (Figure 10.3). This scheme stems from earlier ideas and concepts for the unified development of the water resources of the Jordan valley. One of the earliest proposals was put forward in the opening decade of the twentieth century when an engineer named Wilbush planned to increase the water resources of the Jordan basin by diverting the River Litani into the Hasbani (Brawer, 1968). Following the First World War, when the details for the mandated areas of the eastern Mediterranean were being discussed between Great Britain and France, the British pressed strongly for the water resources of the Jordan to be located exclusively within the boundaries of Palestine to allow the development of large-scale irrigation projects (Lowdermilk, 1944; Prushansky, 1967).

Of the later plans, perhaps the best known is that of Lowdermilk, who proposed that a large part of the flow of the Jordan should be diverted to the northern Negev for irrigation usage (Brawer, 1968). He also suggested that the height difference of 460 m between the Mediterranean and Dead Seas would permit the generation of hydro-electric power if sea water were channelled into the Dead Sea Lowlands. This water would also replace the water lost from the Jordan by irrigation. Another project, the Hays-Savage scheme, was similar in overall concept to the Lowdermilk proposal, but did include Wilbush's earlier idea of diverting the waters of the Litani into the Jordan valley (Hays, 1948).

After the establishment of the state of Israel in 1948, it became obvious that these schemes for the unified development of the water resources of the basin as a whole were not feasible and both Jordan and Israel made their own plans for water usage. As a result of Israel's drainage work in the Huleh basin, beginning in 1951, clashes between Israeli and Syrian armed forces took place which prompted the United Nations to try to get the combatants to agree on a plan for the unified development of the waters of the Jordan basin (Smith, 1966). The plan, which was drawn up for the

Figure 10.3 National Water Carrier and related water networks in Israel

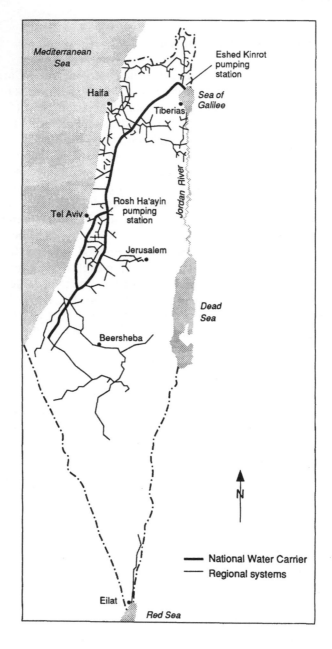

United Nations, is usually referred to as the 'Main Plan' (Garbell, 1965; Main, 1952). It formed the basis of discussions between Israel and her Arab neighbours held from 1953 to 1955, with Eric Johnston, the special representative of the United States government, acting as mediator. After prolonged discussions, modifications to the original plan were agreed and the new version became known as the 'Johnston Plan'. This plan gave Israel 36 per cent of the utilisable waters of the Jordan, Jordan 52 per cent, Syria 7 per cent, and Lebanon 3 per cent. For the purposes of the plan the total amount of water available was considered to be 1,378 million cubic m per annum. It is generally assumed that the technical experts approved the details of this plan, but that the governments rejected it for political reasons.

With the failure of these negotiations, both Israel and Jordan decided to proceed with water projects situated entirely within their own boundaries. In 1958 Jordan commenced work on the East Ghor Canal, which would carry the diverted waters of the Yarmouk to irrigate the terrace lands of the Jordan valley to the east of the mainstream. In the same year Israel began work on the National Water Carrier, designed to carry Jordan water abstracted from Lake Tiberias south to the margins of the Negev desert.

Lake Tiberias forms the major storage capacity and also the starting point for the National Water Carrier. The amount of live storage needed to be kept in Lake Tiberias to supply the National Water Carrier is about 1,000 million cubic m. This amount, which represents about one-quarter of the total volume of the lake, can be provided by permitting the lake surface to fluctuate between an upper level of 209 m, and a lower level of 216 m below sea level (Schick and Inbar, 1972). Water is abstracted from the north-west margin of the lake at a height 212 m below sea level at Eshed Kinrot and pumped through pipes to an altitude of 39 m above sea level. From here it flows in a gravity canal to a reservoir at Tsalmon, where a second pumping station is located. The water is then raised to an altitude of 145 m, whence it flows by canal to a large storage reservoir, with a capacity of about 455 million cubic m, at Beit Netofa. This reservoir is designed to supply 80 per cent of maximum irrigation demand of the coastal region for a period of up to 14 days, should the pumping system up to the reservoir fail for any reason (Smith, 1966). To the south of the Beit Netofa reservoir the water is transported in a pipeline to the starting point of the Yarkon–Negev distribution system at Rosh Ha'ayin.

With the completion of the National Water Carrier, Israel was able to embark upon a more sophisticated water management strategy which enabled the co-ordinated use of surface and groundwater resources. In particular advanced techniques of groundwater management have played a crucial role in the development of Israel's water resources with a policy aimed at transferring up to one-half of the surface water supplies into groundwater storage by the use of artificial recharge techniques. (Weiner, 1972). The objective of the Israeli water supply system was to make the most effective use of the scarce resources of the nation. This was to be achieved by transfers of water between a complex system of storage, both surface and underground.

Short-term seasonal water storage was achieved by the use of a large number of small reservoirs often associated with flood conservation projects. Medium-term storage, which is almost exclusively underground, utilises limestone strata along the coastal strip, which possess large storage capacities combined with relatively high flow-velocities. Such storage permits artificial recharge in wet years and water withdrawal during periods of drought. The aim of long-term storage is to accumulate cyclic surpluses of surface water as underground water reserves for use in prolonged dry spells lasting a number of years. For this purpose the coastal plain aquifer is used as the main storage capacity. Movement of water from one storage system to another is achieved mainly by transmission through the National Water Carrier and other systems, although in some cases the aquifers themselves can be used as the transporting medium.

Given the growing importance of groundwater to the economy of the new state, it is not surprising that Israel invested considerable time and money in studying aquifer characteristics and, in particular, how they responded to water abstraction. In central Israel the limestone–dolomite aquifer system covers 3,500 square km and is between 500 and 1,000 m in thickness. Most of the system operates as a confined aquifer, with natural recharge occurring as a result of winter rains falling on the eastern outcrops. Major discharges from the aquifer are to the River Tanninim in the north and to the River Yarkon in the south.

Since 1950 many new wells have been sunk to extract water from the aquifer. This caused water abstraction to increase more than tenfold between 1951 and 1963 to 280 million cubic m (Harpaz and Schwarz, 1967). As a result of this heavy exploitation the outflow of the Tanninim and Yarkon springs was progressively reduced. Once it was realised that the safe-yield of this aquifer was likely to be exceeded if unrestricted development continued, efforts were made to determine the feasibility of artificial recharge operations. Research established that the long-term average replenishment of the aquifer was about 317 million cubic m per annum; a figure which agreed well with the mean annual discharges from the Tanninim and Yarkon springs (ibid.). Using this figure as a base, various operational procedures for the aquifer were studied, including its use as a storage reservoir for the regional water systems.

10.6 POLLUTION AND ENVIRONMENTAL ISSUES

The 1960s mark a change in views with regard to environmental issues in Israel. During the 1950s technological progress, economic growth and the survival of the state were the prime aims of the Israeli government and people. Environmental issues raised little interest or attention, as is witnessed by the fact that the draining of the Huleh Marshes, one of the great wetlands of the Middle East, raised few protests in the country. Indeed, if anything it was seen as an example of how humanity 'conquers' nature and changes 'wilderness' areas into productive farmland. However, in the 1960s it became apparent that the very rapid rate of water resource development was producing negative environmental impacts.

One of the major problems which Israel faces is that of deteriorating water quality, particularly with regard to its aquifer systems. This is largely because of the high rate of groundwater use relative to the natural recharge rate of the aquifers. In the coastal aquifer dissolved salt levels, mainly chlorides, have been rising for many years. Although salt build-up is a naturally occurring phenomenon in arid regions, in Israel it has been exacerbated by irrigation. What happens is that irrigation return waters with high salt contents percolate into the ground and so contaminate groundwater. Since 1964 this problem has been intensified in coastal areas by the use of water from the National Water Carrier, which has been delivering high salt content water from Lake Kinneret for irrigation. The overall result is that in some parts of the coastal aquifer, chloride concentrations are rising by 1·15 milligrams (mg)/year (Mercado, 1980). If this continues unchecked to the end of the present century the water in many of the wells will be too salty for agricultural use. Currently acceptable levels of chloride in irrigation water vary from 170 mg/litre (1) in the north of the country to 250 mg/l in the south. If the aquifer becomes seriously polluted, then much agricultural land will either have to obtain higher quality water from elsewhere or cease production of crops which are not salt-tolerant. In future it would seem inevitable that water use from the coastal aquifer will decline either in an attempt to limit groundwater exploitation or as a result of the water being too saline to use.

Nitrate pollution of groundwater in Israel has been largely the result of the growing use of fertilisers in agriculture. Before the Second World War the average nitrate concentration in the coastal aquifer was between 0 and 10 mg/l (ibid.). Following the establishment of the state of Israel and the use of modern farming techniques the mean value rose rapidly to reach 51 mg/l in 1972. Although nitrogen fertilisers contribute almost two-thirds of the total nitrate input, sewage wastes from urban centres are locally important. High nitrate levels of over 45 mg/l are potentially dangerous when used as drinking water sources as they can cause methemoglobinemia in small children and may also be carcinogenic in adults. However, since 1974 the Israeli Ministry of Health has relaxed its nitrate standard from 45 to 90 mg/l, mainly as a result of tests which suggested that risks were minimal (Shuval, 1980b). In many instances it would also have proved exceedingly difficult for the original standards to have been adhered to. In the long-term the nitrate problem will only be solved by lowering the amounts of fertiliser use and by the more efficient treatment of sewage return waters in terms of nitrate removal.

The ecological effects of human activity on the watershed of the upper Jordan are more difficult to define precisely. One feature which is well known, however, is the eutrophication of the waters of the Jordan brought about by the drainage of agricultural fertilisers into the river, the decomposition of peats from around the former Lake Huleh, and effluent discharge from sewage works. Increasing tourism with its threat of added nutrients, together with possible oil pollution from motor boating and car use, remain potential problems for the future. By the late 1960s the pollution problem had become so serious that a special committee was established to combat it (Christian Science Monitor, 1970). An unusual, though extremely serious, pollution problem occurred in the upper Jordan catchment in May 1969 when the Trans-Arabia pipeline (TAP-line), which crosses the watershed area north of the Israeli border, was blown up. This caused some 8,000 tonnes of crude oil to enter the river system and to form a huge slick over much of the surface of Lake Tiberias (Schick and Inbar, 1972). One pollution problem which has been alleviated is that of the growing salinity of the waters of Lake Tiberias. In the 1950s and 1960s the salinity of the water had risen to about 400 ppm and was causing concern amongst water users. However, the saline springs around the lake which were causing the problem were diverted into a special conduit and then into the lower Jordan River.

The importance of water quality in Israeli agriculture is shown by the increase in irrigation water requirements locally as water quality has declined: for example, with water containing 100 ppm chloride, citrus fruits require 7,710 cubic m per ha per annum (Matz, 1967). With increasing salinity, irrigation needs go up to 9,044 cubic m per ha per annum for water with a concentration of 200 ppm chloride; 10,872 at 300 ppm; and 13,838 at 400 ppm. These extra amounts of water are needed to flush accumulated salts from the plant root zone.

By the 1980s, Israel, like most other developed countries, has become more environmentally conscious and sensitive to the problems caused by resource development. However, the magnitude of its water supply difficulties probably means that pollution problems will continue to grow worse, at least in the short term.

10.7 RE-USE OF WASTE WATERS

A partial solution to the water supply problem, on which Israel has put considerable emphasis in recent years, is the reclamation and re-use of waste waters. Large volumes of such waters are already used for irrigation and plans exist for up to four-fifths of all waste waters generated by urban centres to be used for such purposes. However, there still remains the problem of what to do with these waste waters for over half the year when they are not required for irrigation purposes. To date they have been allowed to flow into nearby water courses where they have often produced pollution problems. Illegal use of sewage effluents for the irrigation of salad crops near Jerusalem in 1970 led to a major cholera epidemic which brought home to the government the dangers of using untreated sewage effluents.

As a result of this cholera epidemic the government approved a national plan for sewage aimed at improving sewage treatment works and constructing effluent storage reservoirs to hold flows

Figure 10.4 Dan Region Sewage Reclamation project (after Tahal 1977)

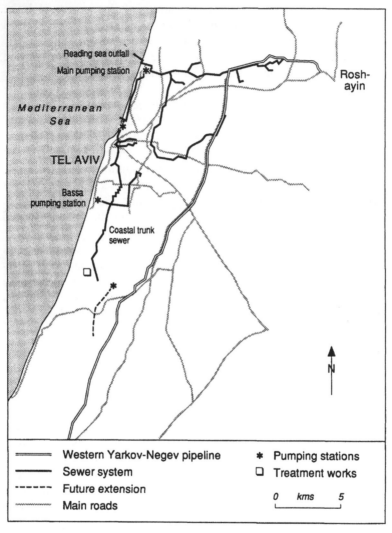

Source: Redrawn from Shuval H.I. (ed.), 1980, *Water quality management under conditions of scarcity — Israel as a case study*, Academic Press, London.

during the winter months when irrigation was not needed. The plan was also seen as an economically viable way of making extra water available to the consumer. It is interesting to note that by the early 1970s, all the major water supply construction projects, such as the National Water Carrier, had been constructed, and few other large projects were planned. It is not surprising, therefore, that the many water agencies welcomed the opportunity to support the National Sewerage Plan as an important new scheme for them to work on. It meant a change in emphasis from water supply to reclamation, but it did mean that lots of work was once more available.

The largest water reclamation scheme in Israel is the Dan Region Wastewater Reclamation Project which, as a water resource venture, is second in size only to the National Water Carrier itself (Figure 10.4). The aim of the project was to gather and reclaim the sewage from the Tel Aviv conurbation. During the 1960s when the project began it was hoped to convert the sewage into drinking water following treatment. However, technical considerations and worries over health meant that this objective had to be abandoned and replaced by the aim of producing water almost up to drinking standards, which could be used anywhere for agricultural purposes. Recent views suggest that these treated waters should be substituted for the high quality waters previously used for irrigation, which could then be diverted to urban regions. Considerable resistance to this view has been expressed by the agricultural lobby, which would like to see these waste waters being used to open up new irrigation areas in the south.

With the growth of Tel Aviv after 1948 new sewers were added and the resulting effluent was disposed of directly into the sea or the Yarkon River. Growing concern about the safety of the bathing beaches nearby led to the Ministry of Health imposing a standard for bathing water quality in 1950 (Shuval, 1980b). Complaints were also received concerning smells from the Yarkon River and adjacent streams. As early as 1950 as a result of public concern a plan was commissioned to deal with sewage in the area around Tel Aviv. This proposed the construction of a large activated sludge plant at the mouth of the Yarkon River, but construction at this time did not go ahead owing to the high costs.

The reason that action was finally taken about the sewage problem of Tel Aviv was the construction of the Yarkon–Negev pipeline. The prime aim of this was to transport all the perennial flow from the Yarkon south to the Negev. However, it was quickly realised that this would reduce the Yarkon to an open sewer unless something was done. The solution was the construction of an interceptor sewer along the Yarkon River to the coast. In the mid-

1960s, following continuing public anxiety about the polluted state of the beaches, another interceptor sewer was constructed along the coast to the south of the Yarkon River. This proved highly successful and led to the reopening of some of the beaches.

During the 1960s Tahal drew up a comprehensive plan for the re-use of the waste waters of the Tel Aviv region. It followed the objective, already employed throughout the rest of the country, that water should be of a quality suitable for drinking. It was initially envisaged that the water would be added to the western line of the Yarkon–Negev project for transmission to the south. However, during the early 1970s growing concern about the chemicals left in the water after treatment led to the abandonment of this objective in the mid-1970s. Eventually, it was decided that the water should be moved via the Yarkon–Negev pipeline, but that all communities obtaining drinking water from this source would now have to obtain alternative sources of supply. The first stage of this major project dealt with the sewage of cities to the south of Tel Aviv, but it was only partially successful and led to many complaints about smells. It did, however, effectively act as a pilot project and showed that a more sophisticated system was required if it was going to cope with the effluent of over 1·5 million people in the Tel Aviv conurbation by the end of the century.

10.8 WATER LEGISLATION AND POLICY

The ways in which water policies are formed and water management carried out depends on the legislative framework that a country establishes. In Israel a highly centralised system has been set up which is based on the Water Law of 1959, which effectively nationalised the water resources of the country. The law stated that: 'The water sources in the state are public property, subject to the control of the state and destined for the requirements of its inhabitants and for the development of the country.' The Ministry of Agriculture was given responsibility for carrying out the Water Law, a fact which indicates the prime importance of the farming and irrigation lobby in Israel's politics. A National Water Council was appointed by the government to advise the Ministry on water matters.

In Israel three bodies are involved in the management of water, with separate responsibilities for policing the law, constructing water projects, and water planning (Laster, 1980). The Water

Commission is a government agency located in the Ministry of Agriculture. It is headed by a Water Commissioner, appointed by the government, and has the duty to oversee all aspects of water management and in particular to control the operational provisions of the law. The Commission itself possesses four technical and four administrative divisions. The technical divisions are concerned with the efficient use of water in Israel, hydrology, drainage, and pollution prevention and control. In effect, the Commission is a regulatory agency with no long-term planning role.

The construction and implementation of water projects is the responsibility of the National Water Company. Originally an organisation known as Mekorot was established in 1937 to oversee water supply for both domestic and agricultural use. After the enactment of the Water Law of 1959 Mekorot became the National Water Company, and maintained responsibilities for constructional activities of all water projects.

The third body is Tahal. This is a corporation set up in 1952 by the government, the Jewish Agency and the Jewish National Fund. It was primarily a planning body designing water projects which were later built by Mekorot. When it was established it obtained the planning divisions of both the Ministry of Agriculture and Mekorot. It remains unusual in being a corporation outside the government, and it is therefore not subject to direct parliamentary control. As a result it has been criticised for being too independent and somewhat insensitive to the water needs of the public. It is certainly true that the Water Commission does not have the expertise to provide an adequate review of Tahal's plans or to suggest alternative schemes.

Galnoor (1980) put forward the view that water policy-making in Israel can be divided into two periods — namely, before and after 1964. In the first period from the establishment of the state in 1948 until 1964, he claims that there was not a basic shortage of water, but rather a problem of poor distribution. The real issue at this time was making these resources available to the users. It was achieved by the construction of a number of large projects, culminating with the completion of the National Water Carrier in 1964.

At this time the ideology underlying water resources development was not one based on economic considerations, but rather on Zionist ideals. With these the prime objective was to create a society in the new state of Israel which was based on rural agricultural development. As Galnoor (ibid., p. 293) states: 'During the first phase, ideology dictated water development. No

377

plan for a new agricultural settlement was ever abandoned only because the cost of supplying water was too high'. As a result of this ideological position it is not surprising that the Water Law of 1959 should be geared in particular to the provision of irrigation water for agriculture and that water policy-making was under the control of the Ministry of Agriculture. Mekorot, the organisation involved with constructional and supply activities, was the dominant water body at this time. With such a clear objective and the necessary agricultural control, the first phase (1948–64) was a considerable success in developing water resources for irrigation usage. The main instrument for achieving this was the National Water Carrier, which had been planned and built entirely by government agencies. During this period financial constraints were not limiting factors, given the strategic significance of water resources in the development of the state.

The second phase, from 1965 onwards, has been identified by Galnoor (ibid.) as a time when attention began to be focused on how to deal with the growing water shortage. Such water shortages obviously challenged the ideology that agriculture should have unrestricted access to water resources. However, this ideology does not appear to have been significantly changed: 'Instead, desperate efforts have been made to prove that the ideological objectives and the pragmatic limitations of water development can co-exist' (ibid.). Initially, attempts were made to provide water from other sources, such as by desalination, but eventually this was shelved owing to the high costs involved. Its abandonment showed that conflict between ideology and the operation of the system at maximum efficiency was bound to arise. It is interesting to note that it was the Ministry of Agriculture (in overall charge of water policy-making) which had to state that priority would be given to the water needs of urban/industrial areas. The implication of this was that at some time in the future, water might have to be transferred from agricultural use, so undermining the ideological framework of water planning as carried out in the first phase.

During a period of shortage it is the planning arm of the water industry, rather than the constructional and supply side, which becomes important, and this meant that the focus of attention switched form Mekorot to Tahal. However, given the great inertia of the system, Tahal has not been as effective as might have been expected in changing the approach to water policy planning. Indeed once the idea of a major desalination plant fell through, no long-range plan replaced it. Galnoor (ibid.) sees the desalination objective as a possible attempt to provide new sources of supply and so obviate the need to change the ideological framework of water planning. It should be noted, though, that it was the worry of the USA that the scheme was economically unsound which eventually led to its downfall.

10.9 WATER USE

Israel's renewable water resources are estimated at 1,500 million cubic m a year, with the major sources being the Jordan-Kinneret basin, the Yarkon–Tanninim aquifer and the Coastal Plain aquifer. Of this total approximately 570 million cubic m (38 per cent) are generated in the upper Jordan catchment and about 850 million (57 per cent) come from aquifer recharge. The remaining 80 million cubic m (5 per cent) comes from the collection and use of storm water runoff (Table 10.1). Since the mid-1960s these sources have been connected to each other and to the major demand centres by a transportation and distribution system, the largest element of which is the National Water Carrier. This system was extremely expensive to build and its operational costs are high. In the late 1970s it was estimated that between 15 and 20 per cent of the country's electricity production was used for the pumping of water (Shuval, 1980c, p. 333).

Because of a highly variable climate the annual runoff and groundwater recharge exhibits large fluctuations from year to year, necessitating that any distribution system should possess a large storage capacity. Indeed, it is estimated that for Israel a total

Table 10.1: Renewable non-saline water resources of Israel (million m³/year)

Fresh water resources	
Jordan River	570
Groundwater	850
Storm runoff interception	80
Sub-total	1,500
Reclaimed sewage effluents	300
Total	1,800

Source: Vardi, 1980.

of 4,500 million cubic m is required for the full inter-annual regulation of the available resources (Vardi, 1980). Given the relatively small size of Lake Kinneret, the main regulatory storage for the distribution system has to be the two major aquifers (Table 10.2).

Table 10.2: Major storage reservoirs in Israel

Reservoir	Operative water stock (1977) (million m³)		Dead storage (below the operating levels)
	Total	Out of which inter-annual	
Lake Kinneret	500	250	4,000
Yarkon–Tanninim	1,500	800	40,000
Coastal Plain	1,000	800	16,000
Total	3,000	1,850	60,000

Source: Vardi, 1980.

It will be noted that the total operative storage of 3,000 million cubic m is well below that required for full regulation of the system. It is also very small in relation to the total dead storage. However, it should be realised that the line which separates the operative storage from the dead storage is an arbitrary one, which has been fixed so as to minimise the risk of pollution to the aquifers. By changing the position of the line it would be possible to increase the usable storage capacity, but it might cause so much contamination as to render the water unsuitable for use. The Yarkon–Tanninim aquifer is especially prone to pollution, owing to the high rates of water movement through it. Although the dead storage is not directly utilised for water supply, the fact that it exists in such large volumes means that it has a major role to play in the dilution of any polluting inputs.

During the period from 1948 to the mid-1960s Israel's use of its renewable water resources rose from 17 per cent to almost 90 per cent of the average annual value. Over the same period the irrigated area increased from 30,000 to about 150,000 ha. At this time the national water grid connected all the major water resources and demand regions of the country, with the sole exception of a few desert areas of the south which were still fed only by regional projects. In wet years the surface water supplies were used as much as possible and any surplus runoff was used for groundwater recharge so that the underground reservoirs could be

Figure 10.5 Water use in Israel

Million cubic metres

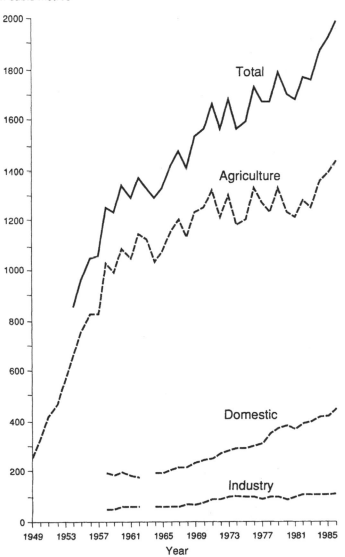

topped up. In contrast, in dry years when surface supplies were restricted, groundwater extraction had to be increased to supply the water demands. The total amount of water used in 1965 was 1,329 million cubic m. Of this total, 80·9 per cent was used for agriculture, 15 per cent for domestic supply and 4·1 per cent for industry. This date can be taken as the end of the era of easy access to water supplies.

Some 20 years later the position has changed markedly, for Israel reached the point at which all its renewable water resources were in use by the late 1960s. Since the mid-1970s Israel's total water use has been between 1,700 and 2,000 million cubic m each year (Figure 10.5). Agriculture still continues to be by far the largest single user, though absolute demand has remained relatively stable at between 1,200 and 1,400 million cubic m a year since the end of the 1960s. As a result the proportion of water devoted to agriculture has fallen in the early 1980s to only 72 per cent of the total, compared with a value of over 80 per cent in the period before the mid-1960s. Since the late 1950s both domestic and industrial water demands have doubled, but domestic demand (which is approximately four times greater than that for industrial needs) only accounts for about 22 per cent of total water demand in the early 1980s. Since the late 1960s the nation has only been able to survive by the mining of groundwater and the extensive re-use of waste waters. In the future the country's development can only continue by increasing the yields from the present system, transferring water from lower- to higher-value products, increasing the efficiency of water utilisation, desalinating saline waters, greater re-use of waste waters or a combination of these approaches.

10.10 THE FUTURE

Israel has already proved herself to be amongst the most efficient countries in the world in terms of obtaining maximum economic output from minimum water input. However, the problem of continued population growth, together with the increase in standards of living, will put even greater strains on the water supply network in the future. How long Israel can go on supplying these expanding needs from its present water resource base without some demand areas beginning to suffer considerably is a matter of conjecture. A feature of the last two decades has been the growth in demand of domestic supply as population numbers have increased. At the same time, *per capita* use has gone up as standards of living have risen. Agricultural use still remains the dominant one, though its relative importance has been falling steadily.

The cultivated area of Israel amounted to 437,000 ha in the early 1980s. The proportion of the land which is irrigated has increased dramatically since the state of Israel was established, rising from 18 per cent of the total cultivated area in 1948–9 to over 50 per cent in the early 1980s (Figure 10.6). The scarcity of water in Israel necessitated the adoption of efficient sprinkler methods of irrigation, which, even as early as the mid-1960s, were used to irrigate over 90 per cent of all irrigated crops (Goldberg, 1967). Use of this method has permitted accurate water dosage of crops, minimised drainage needs, and reduced salinity hazards and leaching requirements. Largely as a result of sprinkler use, the area of irrigated land rose by almost a third during the period 1958–65, without any increase in the amount of water used for irrigation purposes (Shumeli, 1967). A characteristic feature of Israeli agriculture has been the way in which water use per unit area of irrigated land has fallen since 1948 (Figure 10.6). By the early 1980s average use was about 0·6 million cubic m per 1,000 dunams (100 ha). This represents a reduction in water use of almost one-third compared with conditions when the state of Israel was established. What is quite surprising, at present, is that as of yet, the rate of reduction of water use does not appear to be showing signs of slowing down, and it might well prove possible to make yet further significant savings with the extension of drip and trickle irrigation systems.

It is a common practice in many countries to subsidise water used for agriculture, and Israel is no exception. Subsidies for irrigation water are provided from a number of sources. One of the most important is by the use of an Equalisation Fund which attempts to redistribute money from areas in which water costs are low to those where water distribution difficulties make water expensive (Vilentchuk, 1967). This fund also receives a direct subsidy from the Treasury to aid its work. A considerable government subsidy is also provided through the government-controlled Mekorot Water Company which supplies more than half of all water consumed in the country, by charging consumers a very low interest rate on its total investment. Finally, a cheap electricity tariff is also available to farmers for pumping irrigation water.

Figure 10.6 Irrigation water use and irrigated land in Israel

Whether this can continue in the future on such a large scale is a question that Israeli planners have been grappling with over the last few years. However, the political power of the agricultural lobby is still very large, and its interests are not going to be harmed

384

without considerable resistance being encountered. In the late 1970s Tahal did draw up a plan which recognised that urban/ industrial demand would increase and that the only realistic way to meet this need would be to reduce agricultural consumption by about 20 per cent between 1977 and the year 2000. As yet, though, the implications of such changes have not been accepted by water policy-makers in Israel and translated into a coherent plan.

Given a water resource base of around 1,500 million cubic m per annum, Israel will face ever-increasing problems in the future as demands for water continue to increase. By the end of the twentieth century the population in Israel is expected to reach at least 5 million and may go as high as 5·6 million. To serve this population annual urban water demand is thought likely to be between 995 and 1,195 million cubic m (Tahal, 1977). This alone accounts for at least two-thirds of the renewable water resources of the country. Likely total water demand by the year 2000 is estimated to be over 2,000 million cubic m a year (Table 10.3). This means that at least 500 million cubic m annually will have to be obtained from other sources, including re-use of water or the creation of fresh water by desalination techniques. Whether these extra water volumes can be provided, or whether it will be irrigated agriculture which is drastically cut back, still remains to be seen.

Table 10.3: Israel — water balance, 1977–2000

	1977	1990	2000
		(million m³/year)	
Supply			
Existing sources	1,565	1,495	1,495
Reclamation of sewage	30	250–90	325–425
Development of other sources	–	127	127–227
Overdraft	130	–	–
Total sources	1,725	1,872–912	1,947–2,147
Consumption			
Industry, domestic and miscellaneous	515	800–80	995–1,195
Total left for agriculture	1,210	1,032–72	952

Source: Tahal, 1977.

11

Soviet Central Asia:
Water Transfer and Irrigation
Development

11.1 INTRODUCTION

The arid and semi-arid lands of the USSR are located mainly in
Kazakhstan and Central Asia (Figure 11.1). They form an inverted
triangle with a north–south axis of 2,000 km and an east–west base

Figure 11.1 Arid and semi-arid lands of the USSR

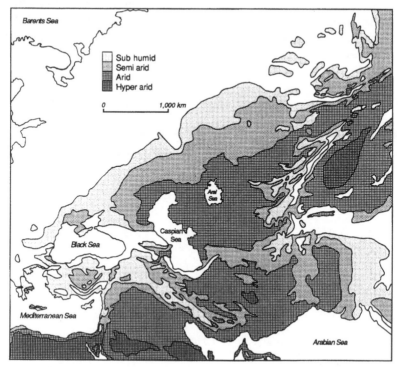

of 3,000 km. The area consists of widespread lowlands, with a highland fringe along the southern flanks. Given the favourable temperature conditions during the spring and summer periods, the agricultural potential is high. Water resources are, however, sparse and for the most part limited to streams draining the uplands to the south and east. These streams are fed largely by snowmelt and have peak discharges in late spring and early summer. Only one major river enters the region from the north — the Volga, which drains a large proportion of European Russia and discharges into the Caspian Sea.

If the arid and semi-arid lands of the USSR are to be further developed, additional water supplies will have to be obtained to supplement the already-strained water resources of the region. This will not be easy, for in the USSR most of the drainage is either northwards into the Arctic Ocean or westwards into the Pacific Ocean. Indeed, these two oceans receive 84 per cent of all the runoff from the USSR (Figure 11.2). River flow in the basins of the Caspian and Aral Seas, covering much of the arid zone, accounts for only 9 per cent of the total. Most of this comes from the Volga, draining a relatively well-watered area of north-west Europe. Another feature of the hydrology of the USSR is that a few very large rivers account for a large proportion of the total discharge of the country. Four rivers — the Amur, Lena, Yenesei and Ob — account for 41 per cent of all surface runoff. Even in the drylands of Central Asia a similar pattern prevails. Here the Volga (with an annual discharge of 250 cubic km is by far the largest river, with the next largest, the Amu Darya, only a quarter of the size (63 cubic km).

Potentially irrigable land in Kazakhstan and Central Asia is estimated at anywhere between 50 and 100 million ha (Voropaev and Velikanov, 1986, p. 81). At present the irrigated area in the USSR is 19.1 million ha (1983), of which about two-thirds is located in the dryland areas of Kazakhstan and Central Asia. Over the last 20 years the irrigated area has developed rapidly from a figure of only 9·5 million ha in 1962 (FAO, 1966). Before this time the irrigated area seems to have remained relatively stable. Between 1960 and 1975 irrigation water supply was increased from 8 to 15·5 cubic km per annum (Zonn, 1982, p. 82).

11.2 IRRIGATION DEVELOPMENT

It is estimated that there are 120 million ha of irrigable land in the

Figure 11.2 Major river basins in the USSR and average flow values

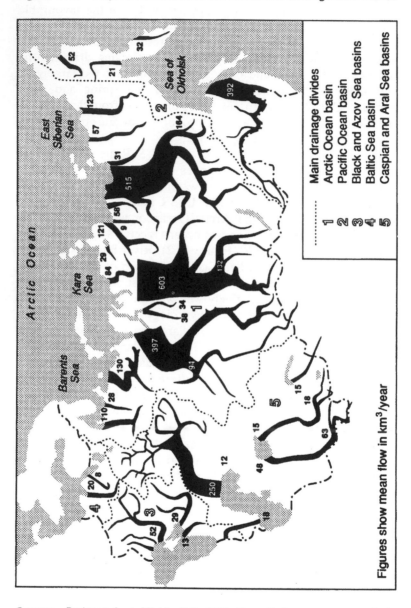

Source: Redrawn from Micklin P.P., 1981, 'A preliminary systems analysis of impacts of proposed Soviet river diversions on Arctic Sea Ice', *EOS*, v. 62, no. 19, pp. 489–93. Copyright by the American Geophysical Union. Reproduced by permission of the American Geophysical Union.

Soviet Union (Micklin, 1978a, p. 3). In the drier parts of the country 26 million ha of irrigable land lie in Central Asia and Kazakhstan, 10 million ha in the north Caucasus region and 8·2 million ha along the Volga. The crops grown on the present irrigated land are quite varied. Cereals and cotton account for about a fifth each of the area, with fodder crops making up a further quarter. Orchards and vegetables account for a further 6 to 8 per cent each. Irrigated cereals provide only 5 per cent of the total cereal crop, whereas with cotton and rice all production is by irrigation. About one-half of the vegetable crop and a quarter of the fruit production of the USSR comes from irrigated land.

Since its origin the Soviet state has always gone in for detailed planning, though not always successfully. In terms of land use planning the Virgin Lands Programme, 1954–60, did not produce the leap in food production that had been hoped for. As a result, from the mid-1960s onwards it led to agricultural policies geared to raising crop yields and to minimising fluctuations from year to year. In particular, considerable emphasis was placed on the extension of irrigated agriculture and land drainage. The expansion of the irrigated area has been the result of a conscious policy to improve food production for the growing population of the USSR. It has involved the opening up of new lands for irrigation and has relied on local water resources, though these have often had to be moved great distances.

In the arid zone of Central Asia and Kazakhstan most of the major projects have been associated with the Amudarya and Syrdarya drainage basins (Figure 11.3). The largest single project has been the Kara Kum Canal which carries water from the Amudarya westwards along the foothills of the Kopet Dagh. The canal is 1,400 km long and transports water almost to the Atrek River. When fully developed it will abstract 13·5 million cubic km of water annually and permit the irrigation of up to one million ha. The objectives behind the Kara Kum Canal were of a multi-purpose nature. One of the main aims was, of course, to provide irrigation water, but it was also intended to supply urban/industrial centres *en route* as well. The canal was constructed in a series of stages, with increasing amounts of water being withdrawn from the Amudarya. The first stage, which was put into operation in 1962, used 3·5 cubic km of water a year. In the second stage this was increased to 4·7 cubic km, and in the third stage to 8·3 cubic km per annum. When finished the total volume extracted annually will be 13·5 cubic km. In total, 1·5 million ha of land will eventually be

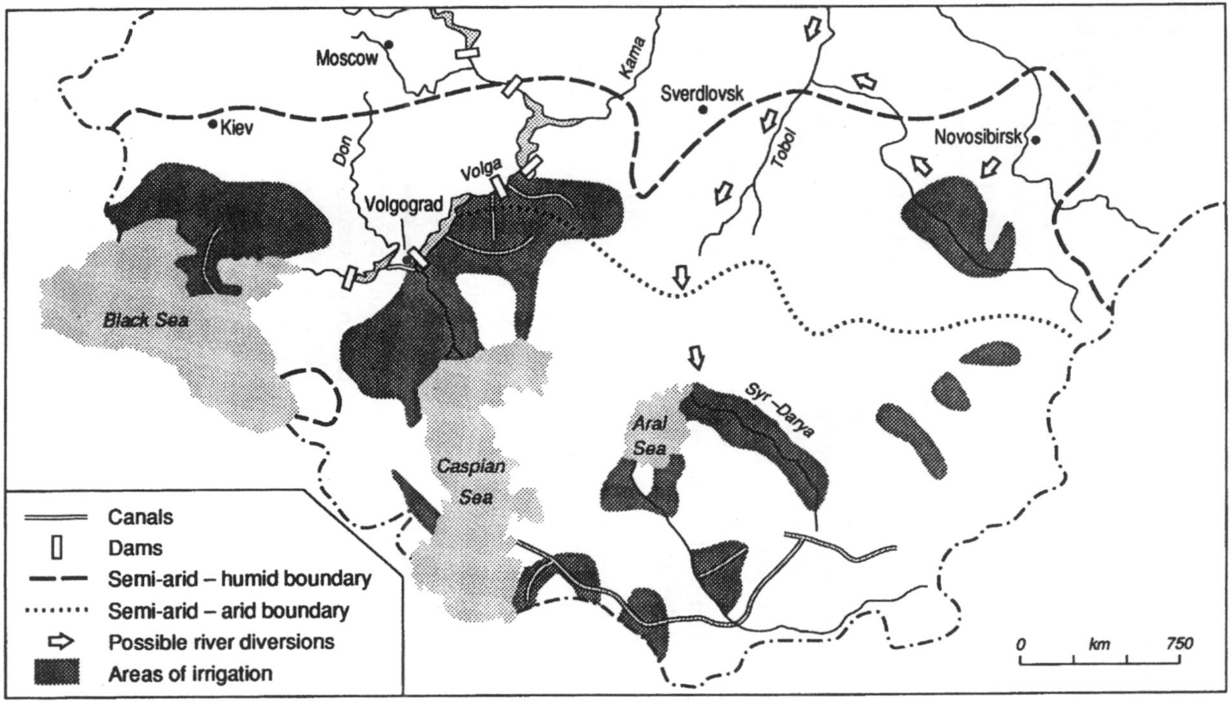

Figure 11.3 Irrigation projects in the Ukraine, the Lower Volga and Central Asia

irrigated along the foothills of the Kopet Dagh. In the upper part of the Amudarya watershed is the Karshi Steppe scheme. Work began on this in 1964 and ten years later 160,000 ha were receiving irrigation water. When fully completed 1 million ha will be irrigated.

On the Syrdarya the Golodnaye Steppe Project is one of the oldest of what might be termed modern schemes, dating back to 1913. With the construction of the Farkhad Dam in 1948 a major expansion of irrigation took place, so that by the late 1960s, 300,000 ha were receiving water. The full potential of the project is once again around one million ha. The Fergana valley is the most important cotton-growing area in the USSR. Major irrigation began here with the construction of the Great Fergana Canal in 1939 and has been extended by other structures. By the late 1960s, 800,000 ha were being irrigated.

Since 1945, by far the most impressive irrigation development has taken place in the southern Ukraine, fed mainly by waters from the Dneiper (Figure 11.3) In 1950 only 156,000 ha were being irrigated. By 1965 this had risen to 540,000 ha and ten years later in 1975 to 1·5 million. This has been largely on land classified as semi-arid or even sub-humid in terms of the UNESCO map of the arid zone (UNESCO, 1979). The most important project has been at Kakhovka, which in total has developed 1·165 million ha.

Almost equally spectacular growth has been witnessed along the middle and lower Volga, where new canal systems have permitted the growth of irrigated cereals. In 1965 only 165,000 ha were irrigated, but by the early 1980s this had risen to close to 2 million ha. A further 2 million ha of irrigated land is planned with the development of the Volga–Ural canal. However, the full utilisation of this project will require the import of waters from northern Europe. Finally, mention should be made of irrigation development in the northern Caucasus along the valleys of the Don, Kuban, Sulak and Terek. By the early 1970s a number of small schemes accounted for 1·3 million ha of irrigation.

A feature of irrigation projects in the USSR since 1945 has been organisational and technological failures. Projects have been finished late and have often been commissioned with key features, such as land levelling and drainage systems, incomplete. Even when finished the operational procedures have lacked sophistication, resulting in either too much or too little water being applied to the crops. In general water has been applied in excess and this — coupled with inadequate drainage systems — has led to widespread waterlogging and salinisation of the soil. So great was this problem in the early 1970s that the annual loss of irrigated land was 1·2 to 2·7 per cent of the total (Micklin, 1978a, p. 12).

Estimates suggest that irrigation water demand for a varied crop mix should be about 6,000 cubic m per ha in the cooler areas and perhaps 8,000 in Central Asia and Kazakhstan. The actual usage figures are, however, much higher, at between 10,000 and 12,000. In the older systems of Central Asia, where conveyance losses can be in excess of 50 per cent of water withdrawals, water usage rates can approach 20,000 cubic m per ha. Indeed, in the early 1970s total water withdrawals in the Amudarya basin were between 20,000 and 25,000 cubic m per ha.

This excessive water use has had serious effects, especially in those areas of Central Asia where saline soils are common. Gerasimov et al. (1976) estimated that strongly saline soils cover 70 per cent of the areas where new irrigation systems are planned. The high volumes of water used produced rising water tables which increased surface water evaporation and the deposition of salts. As a result of inadequate drainage, often by open ditches which were easily choked with weeds, large areas of irrigated land had to be abandoned. In the 1960s, in the oases of Central Asia, it is estimated that the area of land going out of production because of high salt content equalled the area of new land which was coming under irrigation. Fortunately, this position no longer prevails owing to better irrigation management practices, and the salt-affected land on irrigation projects has fallen from 40 per cent in 1964 to 20 per cent in 1975.

Today, the main water problems are occurring in the basins of the Amudarya and the Syrdarya. These rivers have a total annual discharge of about 127 cubic km. By the early 1980s it was estimated that water withdrawals for irrigation were around 90 per cent of the flow in average years, and considerably more in dry years. Any further developments in irrigation will require the import of water from elsewhere.

Increased water use for irrigation has produced environmental problems in the Azov, Caspian and Aral Seas. In the Sea of Azov the reduced flow of the Rivers Don and Kuban has already caused a salinity rise which has harmed the fisheries. In the inland seas of the Caspian and Aral, lower water inputs have caused reductions in water level and reduced areas of the water surface (Hollis, 1978). The Caspian Sea has fallen by 2·5 m since 1929, and its area reduced by 30,000 square km. In recent years the dominant cause of this has been increased irrigation activity along the lower Volga. Much the same picture is seen on the Aral sea where a fall of 3 m took place between 1960 and 1975, and a further fall of 5 m is predicted by the end of the century. In 1960 the area of the sea was 64,000 square km. With continued water loss it has been postulated that the area might be reduced to as little as 16,000 square km by the year 2000.

Since the 1960s the Russians have gone some way towards eliminating excessive irrigation water use by modernising irrigation networks, but much remains to be done. In the Aral Sea basin alone, it has been estimated that more efficient water delivery and monitoring systems could save up to 20 cubic km a year. On the Syrdarya and Amudarya, increased river regulation by the construction of new reservoirs might also provide an extra 20 cubic km per annum. To achieve this would probably require storage volumes of 26 cubic km on the Syrdarya and 22 cubic km on the Amudarya. The elimination of two large drainage lakes, which at times can reach areas of close to 1,000 square km, into which irrigation return waters flow might also save up to 2 cubic km per annum. Loss of water from riverine vegetation is believed to consume up to 10 cubic km each year through evapotranspiration. If this vegetation loss could be reduced, then considerable savings would accrue. It has even been proposed to use water from the Caspian Sea for irrigation purposes. The Caspian Sea has a salt content of 12,000 ppm, about one-third of that of sea water. Experiments have suggested that it might be used for the cultivation of cereals and alfafa.

11.3 WATER TRANSFER SCHEMES

By the 1960s it became obvious that the available water resources of Kazakhstan and Soviet Central Asia would be fully committed in a decade or so. At the same time, the need for food made the government conclude that it was essential to utilise the favourable climatic conditions of the region for crop production. It was obvious, though, that this could only be achieved by the import of water from outside the arid zone. The possible source areas for water are divided neatly by the Urals into European and Asiatic sections. In European Russia the rivers are small and frequent in their distribution, whereas in Asia three major rivers — the Ob,

Figure 11.4 Planned river diversions in the European part of the USSR

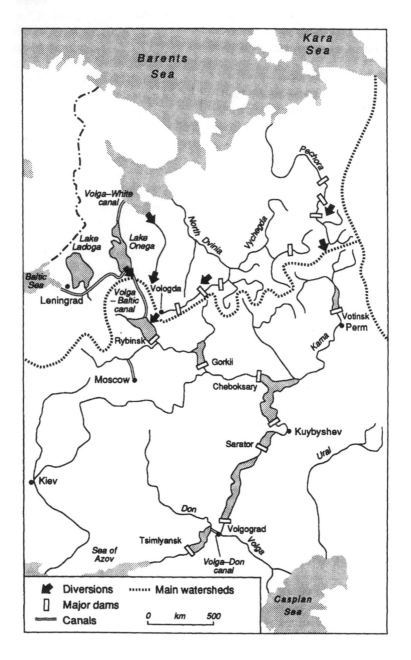

Yenesei and Lena — dominate the runoff pattern (Micklin, 1977, 1978c, 1984 and 1985).

In the European sector a large proportion of the drainage is already carried into the arid zone by the Volga River, and if water could therefore be diverted into the headwaters of the Volga, the task of water delivery to Central Asia would be far simpler (Micklin, 1983; Moiseev *et al.*, 1980) (Figure 11.4). Partly for this reason early attention was focused on obtaining water for Central Asia from the European sector. In the 1950s plans were drawn up to divert as much as 43 cubic km per annum from the Pechora and Vychegda Rivers, which flow into the Arctic, into the Volga basin. However, considerable opposition to the scheme was generated and it was finally dropped in the 1960s. As water problems intensified the idea of diverting waters from the north was revived in the 1970s, and in 1981 the 26th Communist Party Congress decided that the European scheme should be delivering water by 1990 at an annual rate of 19·1 cubic km. At the same time it was agreed to go ahead with a detailed study of Siberian diversion schemes aimed at transporting 27·2 cubic km per annum.

Work on the European diversion was scheduled to begin in 1986 with the first water being delivered by 1990. Initially this will be at an annual rate of 5·8 cubic km and will be fed into the upper part of the Volga network. The water sources for this phase will be the Sukhona River and Lakes Lacha, Vozhe and Kubena. During the 1990s other diversions will be completed which will raise the total annual import of water into the Volga system to 19·1 cubic km. This will be achieved by moving 3·5 cubic km per annum from Lake Onega along the already-existing diversion routes. The rest, amounting to 9·8 cubic km per annum, will be from a new diversion, this time of the Pechora River into the upper Kama, a tributary of the Volga.

By the early 1980s environmental problems were becoming obvious in Central Asia: for example, both the Caspian and Azov Seas were experiencing severe reductions in inflows. In the Caspian Sea, inflow volumes were 35–40 cubic km (12–13 per cent) below natural flow levels, while for the Sea of Azov the annual figure was 13 cubic km (30 per cent) (Hollis, 1978) (Table 11.1). Estimates for the Caspian basin suggest that inflows might be reduced by up to 60 cubic km per annum in the year 2000, and by as much as 88 cubic km by 2030, as a result of withdrawals for irrigation and other uses (Micklin, 1986). In itself, this may cause severe damage to the fisheries of the Caspian Sea, as the main

Table 11.1: Water resources of the Syrdarya and Amudarya Rivers

	Water resources where rivers pass from mountains to arid lands		Inflow into Aral Sea	
	Syrdarya	Amudarya	Syrdarya	Amudarya
		(km³/year)		
1926–50	37·3	70·0	13·0	41·3
1951–60	43·7	74·5	15·2	40·1
1961–8	36·2	68·9	8·2	30·2
1969	65·0	109·5	16·6	56·3
1970	43·3	73·3	9·3	28·4

Source: Hollis, 1978.

breeding grounds of the fish are in the shallow northern section. If the water level of the sea falls below −28·5 m, the breeding of the fish could be severely disrupted. As a result it is essential that extra water supplies are obtained to prevent this fall in water level from reaching critical proportions. Of the 19·1 cubic km which will be transferred into the Volga basin by the late 1990s, about 5 million cubic km per annum will be transferred to the Don River and then to the Sea of Azov, leaving 14·1 cubic km per annum to flow into the Caspian Sea. It is hoped that the extra water provided will permit an increase in the irrigated area of some 4 million ha, as well as producing 3·5 billion kW-hours of electricity in the hydro-electric power stations along the Volga. Even allowing for the pumping requirements of the diversion schemes, this should still provide a net benefit of 2 billion kW-hours. It will also improve navigation and water quality along the Volga and its tributary, the Kama.

Beyond the year 2000 it is possible that further diversions may be made. It has already been suggested that an extra 10·2 cubic km per annum might be available from the Sukhona and 3·6 from Lake Onega. In total this would provide almost 33 cubic km of water. Yet further schemes propose an extra 20 cubic km a year from the lower Pechora basin, and as much as 40 cubic km per annum from the damming of the Onega Gulf. If these were to go ahead it would seem that they would be able to meet the water needs of the Caspian Sea and Sea of Azov basins well into the twenty-first century.

The Siberian water diversion schemes are designed to come on-stream after the European ones. They are much bigger projects, in

Figure 11.5 Planned river diversions in the Siberian part of the USSR

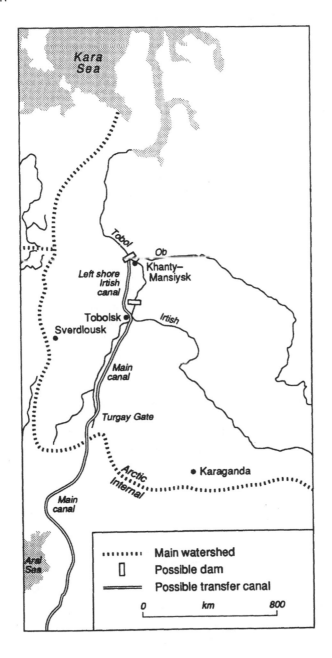

that the actual transportation route associated with the diversion is over 2,000 km long (Figure 11.5). The reason for this is that in Siberia all the major streams flow northwards into the Arctic Ocean. What is perhaps surprising, though, is that the pumping lift for the project is only 113 m, even though the water is being moved large distances against the natural slope of the land. Phase one of the river scheme calls for the diversion of 27·2 cubic km from the River Ob and its main tributary the Irtysh. The actual method of diversion along the lower Irtysh remains uncertain, with either the construction of a canal or a reversal of the river's flow using small dams being possible. From Tobolsk on the Irtysh the water will be fed into a major canal which will transport the water 2,200 km southwards to the Amudarya. Water will have to be pumped up to the main watershed, but will then be able to flow under the influence of gravity to the Aral Sea.

The canal itself will be trapezoidal in form, with a top width of between 108 and 212 m and a depth of 12–15 m. Along most of its length it will be unlined, so losses are expected to be high and average about 2·5 cubic km per annum. It is hoped, however, that about two-thirds of these losses will be able to be reclaimed by the use of interceptor drains. Of the total water delivered, about 5 cubic km will be used for urban/industrial purposes, leaving the rest for irrigation. As a result it is hoped that an extra 2·3 million ha can be brought under irrigation. Construction is likely to begin in 1988 but water will not be delivered until the late 1990s.

By the mid 1970s about 9 million ha were being irrigated in Central Asia and growing worries were being expressed about the available water supplies (Cole, 1984). The first stage of the Siberian water diversion project would merely reduce existing pressures and lead to relatively little new land being irrigated. It would certainly do little to help overcome the hydrological

Table 11.2: Predicted decrease in inflow of surface waters to the Caspian and Aral Seas as a result of industrial and agricultural activity

| | Natural inflow to sea | Decrease in inflow (km³/year) | | | | |
		1971–5	1976–80	1981–5	1986–90	1991–2000
Caspian	300	31	41	44	53	74
Aral	54	10·5	21	30	39	51

Source: Hollis, 1978, after Voskresenskiy *et al.*, 1973.

imbalance of the Aral sea which has registered a fall of 8 m between 1960 and the early 1980s, and a corresponding decrease in area of 20,000 square km. These changes have been almost entirely due to increasing water withdrawals from the Syrdarya and the Amudarya Rivers for irrigation (Table 11.2).

A second-phase project in Siberia is also proposed, by which diversions will be increased to 60 cubic km per annum, and for this project water will again be abstracted from the Ob. At this time water may also be diverted from the Yenesei into the Ob to make up for the large volumes of water being abstracted. In the very long term, future diversions from the Ob and the Yenesei are envisaged which could lead to total annual water transfers of up to 200 cubic km. Such projects would not, however, take place until well into the twenty-first century.

11.4 IMPACT OF THE WATER TRANSFERS

A crucial feature of these diversions is the environmental and economic impacts they will have on both the water-exporting and water-importing regions (Micklin, 1978b, 1986). It is quite obvious that the benefits will not be uniformly distributed and it is the water-importing areas which are likely to benefit most. In particular the hydrological imbalances of the Caspian Sea, Aral Sea and the Sea of Azov can be corrected, while the expansion of irrigation is likely to increase economic activity in general. Along the Volga the extra water will help to reduce pollution as well as providing electricity.

Of the two projects it would seem that the Siberian diversions will cause the greater problems. With the European schemes, at least during the first phase when up to 19·1 cubic km are being exported, likely environmental effects are thought to be relatively minor (Pedersen and Moller, 1981). However, with any major project there is always damage caused during the construction stage when heavy machinery is being employed and large-scale earth-moving operations are in progress.

Water circulation in lakes from which water is being abstracted may be reduced, so causing oxygen deficiencies in the hypolimnion. It will also increase eutrophication rates if any polluted water is entering the lakes. Lower water levels along the exporting rivers may affect fisheries. On the Onega River, for example, flow will be reduced by over a half at the abstraction point and by 12 per cent at the mouth. A similar problem is envisaged on the Sukhona River with flows being reduced by 43 per cent. Additional likely problems here are thought to be a spate of downcutting along the tributary streams as local base levels are reduced, as well as sedimentation along the main stream.

During the second phase of the European projects it seems likely that greater problems will be experienced, particularly along the Pechora River. Here a large reservoir of over 2,000 square km will have to be constructed in the upper part of the basin, and flow below the dam will be reduced by three-quarters in a low-flow year. This will have deleterious effects on navigation and lower water quality, which in turn may harm fisheries. High water-tables alongside the reservoir will cause drainage problems, though downstream from the dam, drainage will be improved. Tributary downcutting and sedimentation may also occur.

The effects of the Siberian schemes on the environment are believed to be much greater, partly because of the huge distances over which the water is to be transported (Mikhaylov *et al.*, 1977). In the first stage 27·2 cubic km and in the second stage 60 cubic km will be exported each year from the Ob catchment. Besides this, though, it is expected that water use in the upper part of the Ob will reach 25 cubic km by the end of the century. Taken together, this means that in the first stage the discharge of the Ob at its mouth will be reduced by 13 per cent under normal flow conditions. During the second stage this will increase to 22 per cent, unless compensatory flows are made from the Yenesei. On the Irtysh below Tobolsk, from where water withdrawals would be taking place, discharge would be reduced by as much as 60 per cent during the first stage. This will not increase during the second stage as no further water will be abstracted at this point. Navigation on both the Ob and the Irtysh seems bound to suffer, as too do breeding conditions for certain fish species. On the positive side, the lower water levels along the rivers will improve drainage on the floodplains in spring and summer, as well as reducing the area of flooding.

One of the great debates about the Siberian schemes has centred on the effects of the water diversions on ice cover in the Gulf of the Ob and the Arctic Ocean (Kelly and Gribbin, 1979; Micklin, 1981). Estimates suggest that the thermal inputs into the Ob Gulf will be lowered by 15 per cent in the first stage and by as much as 30 per cent in the second. Once the second stage is in operation, it is expected to increase ice cover by 10 per cent and to delay spring melting by 8 to 10 days. The freshwater interface with the sea is also likely to move inland by one degree of latitude and water quality will decline. However, major changes in the heat balance of the Arctic Ocean seem unlikely.

Perhaps the greatest problems will occur along the line of the diversion canal, which will be transporting water towards areas of increasing aridity. Water losses through the bed of the unlined canal, though minimised by interceptor drains, seem set to raise water tables substantially. Given the high summer temperatures and arid conditions, it seems inevitable that this will lead to salination of the soils in a belt along the canal. The canal itself is so large that erosion and sedimentation problems may be generated by water moving along it. Over parts of its length it traverses sandy terrain and, under high wind conditions, this sand may be drifted into the canal and so limit the conveyance capacity.

Finally, there are the environmental effects which will occur in Central Asia. The chief impact here seems likely to take place as the result of too much water being applied for irrigation. Already, excessive water use, probably related to a cheap water pricing policy, has led to enormous areas becoming saline since the 1950s. With further expansion of the irrigation network this problem seems set to get worse. The soils themselves are naturally heavy and contain a high salt content. Drainage is difficult and the soil structure can soon become damaged. By the end of the present century it is not inconceivable that large areas of formerly irrigated land will have been withdrawn from cultivation as crop yields decline in response to increasing salinity.

12

The Los Angeles Conurbation:
Problems of Environmental Management
and Resource Provision

12.1 INTRODUCTION

Los Angeles is the second largest urban system in the USA, and amongst the world's largest city regions (Nelson, 1983; Steiner, 1981). In the mid-1980s it has a population of 11 million and still shows signs of continued growth. From the time of the gold rush in the mid-nineteenth century until the late 1960s, southern California was viewed by many as the 'promised land', because of its mild winters, fertile soils, sunshine, and outdoor lifestyle (Baur, 1959). Today that image has been tarnished and replaced by one which sees pollution, over-crowding and crime as the main attributes.

Unlike the eastern cities of the USA, Los Angeles did not experience major growth through immigration in the latter years of the nineteenth century (Bowman, 1974; McWilliams, 1973). Thus by 1900 the population of the city of Los Angeles was 100,000 and the entire urban region only boasted about one-third of a million people. In the 1920s Hollywood became the centre of the motion picture industry, adding further glamour to the already very positive image of southern California. Perhaps even more important at this time, however, was the discovery of petroleum just north of the downtown area, which provided an energy source for the growing numbers of automobiles. The 1920s were also a period of great agricultural productivity and prosperity based on the orange groves which bordered the city to the north (Zierer, 1934). Given this general economic prosperity the city began to grow, commencing an urban sprawl of single-family dwellings which, over the years, has overwhelmed the orange groves and prosperous agricultural regions (Fogelson, 1967). During the

1920s and 1930s, population grew steadily and by 1940 the Los Angeles metropolitan area contained about three and a half million people.

It is since the end of the Second World War, however, that the explosive growth of Los Angeles has occurred, for between 1940 and the mid-1980s an extra 6 million people were added to the conurbation, with almost half of this total being made up of inward-moving migrants (Weaver, 1980). In the 1950s and 1960s most of these newcomers were from other parts of the USA and Mexico, but since the 1970s they have been joined by new waves of people entering Los Angeles from places such as Vietnam, Iran and other countries of South-east Asia. As a result minority groups are found in large numbers. Today in Los Angeles there are 3 million people of Hispanic origin, over 1·1 million blacks and more than 150,000 Japanese. With the rapidly growing population, lifestyles are also changing. Single-family dwellings, although still common in the affluent suburbs, are being physically replaced by apartment complexes in the older urban areas. So great has been this change in recent years that by 1980, 60 per cent of the families in the city of Los Angeles were living in multi-occupied property.

12.2 THE ENVIRONMENT

A key feature of Los Angeles is its site (Figure 12.1). The old core of the city is situated on a coastal plain composed of coalescing alluvial fans deposited by the Los Angeles, San Gabriel and Santa Anna Rivers (Jahns, 1954). The original town was on the Los Angeles River at a point where solid rock just beneath the surface ensured perennial surface flow and so provided an assured water supply. The Los Angeles Plain is about 65 km across and 24 km deep. It is roughly semi-circular, bounded to the north-west by the relatively low Santa Monica Mountains and to the east and south by the Santa Anna Mountains. The coastline is made up of two crescentic bays — Santa Monica and San Pedro — separated by the small highland massif of the Palos Verdes hills. Along this coastline there are 100 km of sandy beaches, while in San Pedro Bay the artificial harbours of Long Beach and Los Angeles were created by the construction of a breakwater completed in 1899.

During the twentieth century the city, which was located well away from the coast, has expanded into three major inland valleys as well. On the landward side of the Los Angeles Plain, rivers have

Plate 12.1   Major flood channel in Los Angeles.

Figure 12.1 Site of Los Angeles

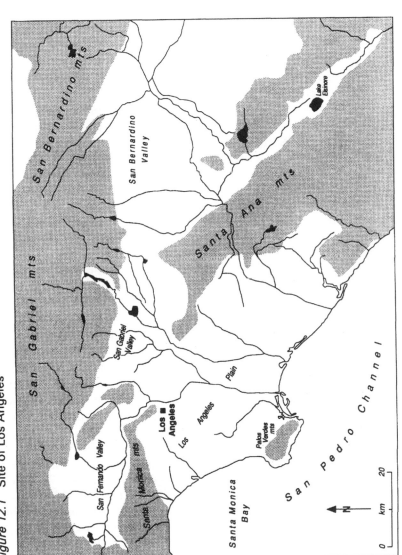

broken through the coastal ranges to permit drainage from the San Fernando, San Gabriel and San Bernadino valleys. These valleys are essentially outwash plains from the adjacent highlands and slope down from about 450 m above sea level to around 150 m, where they merge into the Los Angeles Plain. The valleys and the coastal plain cover a total area of approximately 5,000 square km. Behind the valleys high mountains provide a significant barrier to the north and west. The San Gabriel Mountains attain heights of between 900 and 3,000 m, while in the San Bernadino Mountains, the highest peak, Mount San Gorgonio, reaches a height of 3,501 m (Robinson, 1977). These mountains, which have few easy routes through them, today provide important recreational areas for the citizens of the metropolis.

The climate of Los Angeles can best be described as semi-arid, with mild winters in which precipitation occurs, and hot summers (Bailey, 1966; Keith, 1980). However, the very large relief contrasts over short distances mean that a variety of often very different micro-climates exist throughout the metropolitan region. This is especially well seen in summer when average maximum temperatures in the San Bernadino valley can be 5·5°C above those of downtown Los Angeles, and as much as 13°C greater than in coastal locations such as Santa Monica. Much smaller variations than these are recorded in winter.

In the Los Angeles downtown area the average rainfall is around 380 mm per annum, though variations from about 125 to 950 mm have been recorded. In the foothills of the Santa Monica and Santa Anna Mountains annual totals rise to between 380 and 635 mm, whilst in the higher mountains precipitation can reach up to 1,000 mm per annum. As in all semi-arid regions, although rainfall does not occur often, it can be intense, and Los Angeles has recorded a fall of 125 mm in 24 hours. Such storms often produce devastating effects in terms of flooding and landslips. Rainfall is also very variable from one year to another (Figure 12.2).

The proximity of the ocean means that the maritime influence is considerable. During the day sea breezes can reach into the inland valleys, while at night a cool reverse land breeze results. In early summer, sea fog often drifts inland with the breeze. Although at low levels the fog is often dissipated, high fog layers can prevail in the coastal areas, giving cool, sunless days. Even when all the fog is evaporated the cool marine layer moving inland often forms a temperature inversion against the warmer and drier air above.

Figure 12.2 Annual precipitation variations in Los Angeles

Rainfall (inches)

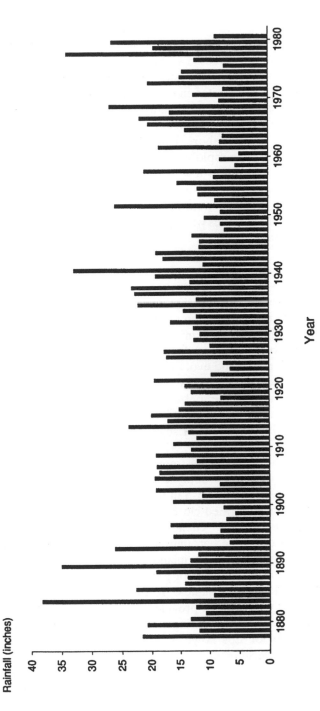

Year

Source: Redrawn from Nelson H.J., 1983, *The Los Angeles Metropolis*, Kencall/Hunt Publishing Company, Iowa. Reprinted by permission of Kendall/Hunt Publishing Company.

Such conditions are ideal for generating the photochemical smog for which Los Angeles has become infamous, by preventing the escape of pollutants created within the urban area.

A final climatic characteristic worth mentioning is the Santa Anna winds of the winter months. These are strong, warm and very dry winds from the north and north-east which blow when a high pressure system is situated to the north-west and a low pressure zone exists offshore. Under these conditions dry arctic air is drawn in, warming rapidly as it moves southwards and also as it loses altitude. Extremely low relative humidity values of less than 10 per cent can be recorded and fire danger is high. As the winds enter the Los Angeles area they are funnelled down valleys and canyons, which increases windspeed and often causes considerable property damage.

12.3 WATER SUPPLY

Water supply to a population of 11 million people enjoying a high standard of living in a semi-arid climate is obviously a major logistical problem. In Los Angeles it has been solved over the years by the growth and evolution of a water supply system which has been forced to obtain its water resources from ever-greater distances away from the metropolitan area. At present adequate supplies can be maintained, but if future urban growth continues it seems likely that ever greater supply difficulties will result (Kahrl *et al.*, 1979).

The location of the earliest non-Indian settlements in the area was dictated by water supply considerations. The San Gabriel Mission, built by the Spanish in the heart of what is now modern Los Angeles, was located close to a perennial section of the Los Angeles River in the Glendale Narrows, where underlying impervious rocks forced groundwater flow to the surface. Within the alluvial deposits of the inland valleys and the Los Angeles Plain are extensive volumes of groundwater recharged by runoff from the adjacent highlands. It has been this groundwater, abstracted by wells, which provided much of the necessary water for the development of Los Angeles in the late nineteenth century. When the city of Los Angeles was incorporated, in 1850, the perpetual and permanent rights to the waters of the Los Angeles River given by the Spanish were confirmed. Later disputes did arise over water rights, but in 1899 the State Supreme Court

Figure 12.3 Water diversion from the Owens Valley and the
Colorado River to Los Angeles

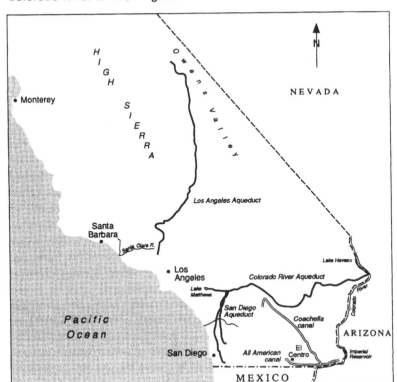

confirmed the right of Los Angeles to all the water it needed from
the Los Angeles River, even if it meant that the farmers of the San
Fernando valley — in which the river had its source — were not
able to pump water from their wells.

By the beginning of the twentieth century the waters of the Los
Angeles River were already fully utilised and large volumes of
groundwater were also being abstracted. A major conservation
measure taken at this time by the newly established Los Angeles
Water Department was the installation of water meters for all its
customers in 1902. This had the marked effect of reducing daily
water consumption from 909 litres (240 US gallons) *per capita*
daily to as low as 511 litres (135 US gallons) in 1907. However,
such measures were not enough to provide the extra water needed
for the growing city and so other sources had to be found. The

409

source which was chosen was water from the Owens valley (Figure 12.3). This was situated on the eastern side of the Sierras with the water of the Owens River draining into a lake of inland drainage. The feasibility of the project was assessed in the first few years of the twentieth century, and in 1907 funds to build an aqueduct to bring water to Los Angeles were agreed. Construction began the following year and by 1914 water was being delivered by gravity flow along the 360 km aqueduct to Los Angeles. With the implementation of this project Los Angeles obtained about five times as much water as had been previously available to it (Hoffman, 1981; Nadeau, 1974).

This extra water was looked upon enviously by adjacent communities. However, it was decided — after considerable debate — that the Owens valley water would not be delivered to townships outside Los Angeles unless they were willing to be annexed by the city and to pay an appropriate share of the costs of building and operating the distribution system. In 1915 the people of the San Fernando valley were the first to vote to join Los Angeles to gain access to the Owens valley water. In the following years many other districts were also annexed. Spare water resources proved an extremely valuable asset in the growth of the city of Los Angeles and annexation of new areas to obtain extra water resources became definite city policy. Many residents in the Owens valley resented the way in which their water resources were being taken over by the city of Los Angeles, but the city continued to purchase land and water rights. Today it controls over 98 per cent of the land on the valley floor.

In the 1920s drought conditions revealed that there was insufficient water in the valley to fill the aqueduct to capacity, and so it was decided in 1930 to extend the system by 160 km to tap water from the Mono basin to the north. This permitted total water deliveries to be increased by about 25 per cent. Much later, in the 1960s, it became obvious that water from the Owens valley was of higher quality and also cheaper than water which could be obtained from elsewhere. Because under average rainfall conditions more water was available in the valley than could be transported along the aqueduct, it was decided to construct a second transmission line to increase water transport by as much as 50 per cent. This new system was completed in 1970.

Although the water from the Owens valley provided the lifeline for the growth of Los Angeles in the early years of the twentieth century, it soon became obvious that even more water would be needed as the city expanded. Attention was focused on the Colorado River as a possible source, as in 1921 the US Congress had passed a bill authorising the construction of the Hoover (Boulder) Dam to provide flood control and also to make available irrigation water (Nadeau, 1974). In 1924 the city of Los Angeles applied for water rights for 42·5 cubic m (1,500 cubic feet) per second of water from the Colorado River. This led to a meeting of the states with interests in the Colorado River and the signing of the Colorado River Compact in which the rights for river water were allocated amongst themselves. The state of Arizona, however, refused to agree the terms and it would not sign the Compact. Nevertheless construction of the Boulder Dam went ahead in 1931, and the right of Los Angeles to obtain water from the Colorado was confirmed.

To deal with the Colorado water the California State legislature created the Metropolitan Water District. This was to construct and operate the aqueduct and to act as wholesaler for the water in Los Angeles and adjacent districts. The scheme was to be financed from property taxes and the sale of water. Water was to be taken from the Colorado River at Parker Dam and then delivered along the 390 km Colorado Aqueduct (Figure 12.3). In 1933, 13 cities voted to make US$220 million available to construct the project. The first water deliveries were made in 1941. At first there were considerable worries about the viability of the project, owing largely to the high costs of operation as a result of the large amount of pumping required.

In 1964 an important Supreme Court decision concerning the waters of the Colorado went against California and as a result, more water was allocated to Arizona. This means that when the Central Arizona Project is finally completed in the late 1980s, the current Metropolitan Water District entitlement to 1,480 million cubic m (1·2 maf) will be approximately halved.

State-wide problems with water supply during the 1950s led in 1960 to the decision to construct the California State Water Project. The idea behind the project was to bring water from the wetter northern part of the state to the San Francisco Bay area, to the Central Valley for irrigation purposes and to the urban regions of southern California. The scheme was superimposed on to the Central Valley Project which had been constructed during the 1930s. A new dam at Oroville on the Feather River was constructed and this, together with the already-existing Shasta Dam, would provide water storage and regulate the flow along the Sacramento River.

411

Figure 12.4   California State Water Project

Water released from the two dams flows down the Sacramento River to the delta, where it joins the waters of the San Joaquin River. A pumping station then lifts water out of the delta and feeds it into the 715 km long California Aqueduct (Figure 12.4). Along the Central Valley, flow in the aqueduct is by gravity flow, but to reach the Los Angeles region, water has to be pumped 610 m up to cross the San Gabriel Mountains at the Tehachapi Pass. As delivered water prices reflect actual water transmission costs, any water crossing the San Gabriel Mountains is very expensive indeed (Figure 12.5). The California State Water Project was designed to supply about 5,427 million cubic m (4·4 maf) of water each year, with approximately half of this being delivered along the Californian Aqueduct to southern California.

Figure 12.5 Water costs along the California Aqueduct

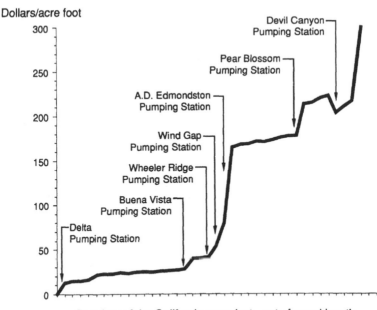

Reaches of the California aqueduct - not of equal length

The water from the project, as with that from the Colorado River Aqueduct, is heavily dependent on pumping for its operation. As a consequence it is inevitable that water prices will continue to rise steeply in the long term as energy costs go up.

During the 1960s, when water supply for the rapidly growing southern California area looked as if it would continue to be a major problem, a number of novel ideas were suggested as solutions. One of these was iceberg mining. The aim was to tow icebergs from Antarctica northwards to the Californian coast where they could be mined and so provide a fresh water source (Weeks and Campbell, 1973). Despite considerable discussion nothing has yet been achieved with this plan. Another suggestion was the construction of an undersea aqueduct along the Californian coast to bring fresh water from the high discharge rivers of the northern part of the state. Once again, though, such a scheme has never reached even the feasibility stage. The most ambitious scheme of all, however, was NAWAPA (North American Water and Power Alliance), which entailed bringing water southwards from the Canadian arctic via the Colorado and other rivers to supply the needs of water deficient areas of North America (figure

Figure 12.6 Sources of water for the Los Angeles Municipal Water District under dry year conditions

Million acre ft/yr

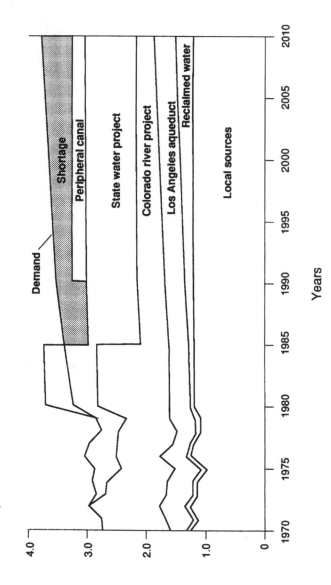

8.7). This grandiose scheme has been much discussed, but the sheer magnitude and cost of it has meant that it is never likely to go ahead.

How, then, will water be supplied to the Los Angeles region in the future? (Figure 12.6). It is quite obvious that no new sources of water are easily and cheaply available. Given the coastal location the idea of desalinating water has been mooted, but the costs have always looked prohibitive. However, such techniques do produce high quality water and, as water quality becomes an ever more important issue, it seems possible that at least some desalination capacity might be installed.

The more likely key to the solution of the water problems of Los Angeles seems to be demand management by use of the price mechanism. Currently, large quantities of water are being used in the urban area for the maintenance of plants in gardens and public places. As water costs rise it would seem possible that people could switch cultivation practices to grow more native species which have lower water requirements. This has already occurred in Tucson, Arizona, and seems to have worked well. It would mean, however, that the 'green' oasis image which Los Angeles conveys to the person arriving by air would change: it would be replaced by the much browner and arid view of Tucson. What is beyond any doubt is that the cheap and abundant water era, which lasted from about 1915 to the late 1960s, is over (Englebert and Scheuring, 1982). Water is now correctly perceived as a precious and expensive resource in the Los Angeles region, and future usage patterns will increasingly reflect this reality.

It is therefore possible to recognise a pattern of water use in Los Angeles. To begin with, local water supplies were exploited, but when these were exhausted the city turned to ever more distant water sources. By the late 1970s, however, the situation had been reached whereby the costs of delivering water from these distant sources was becoming so high that the range of uses to which it could be put was greatly reduced.

12.4 SEWAGE COLLECTION, TREATMENT AND DISPOSAL

12.4.1 Introduction

The existence of 11 million people in a relatively small area requires an efficient sewage system if disease is to be prevented. In Los Angeles this is achieved by three major and one minor sewage systems which all terminate in ocean outfalls. Together these four plants account for about 99 per cent of all waste waters discharged into the Pacific Ocean. Sewage collection and removal is basically a transport problem, while sewage treatment is a processing operation. Every day each citizen in the Los Angeles urban area uses on average 663 litres (175 US gallons) of water. Given that there are approximately 11 million people in the conurbation, some 7·293 million tonnes of water have to be collected and disposed of each day. By any standards this is an enormous quantity of water to handle. To minimise costs it obviously has to be transported by gravity flow and so this fact determines that natural drainage basins play a crucial role in sewer system design.

12.4.2 Sewage treatment techniques

To understand the environmental problems associated with sewage collection, treatment and disposal, it is essential to know what sewage is and how it can be dealt with. Sewage is about 99·5 per cent water. It contains considerable quantities of organic matter, in both dissolved and particulate form, as well as a huge range of dissolved chemicals. If storm water from road drainage is also fed into the sewage system it will contain grit, together with varying amounts of chemicals associated with the automobile.

In general it is the organic matter which causes the most immediate problems. This organic matter comprises human waste products, together with many food wastes produced from garbage disposers. This organic matter, whether particulate or dissolved, is biodegradable and so can be broken down by bacteria to more stable chemical forms. The nature of the problem is that these bacteria are aerobic, meaning that they need free oxygen to survive. As they biodegrade the organic matter, they consume oxygen from the water. If large amounts of sewage are dumped into a water body, the bacteria will multiply to consume the organic matter and, as a result, all the oxygen will be consumed. When this happens, an anaerobic environment, in which no free oxygen remains, is produced. This type of environment is extremely unpleasant and prone to smells. It also causes all aerobic organisms, such as fish, to die. It is therefore essential that water bodies are protected from becoming anaerobic. The main aim of a sewage works is to achieve this.

Sewage works remove as much solid organic matter as possible and convert the rest into more stable and, therefore, less oxygen-demanding forms. If this is achieved the sewage effluent will have a low oxygen demand which can be met by the normal re-oxygenation rate of the water body which is receiving it. This means that the system remains aerobic and that fish and other aerobic life forms will not be killed.

Sewage treatment processes are relatively simple. Primary treatment consists of sedimentation, whereby the coarser organic material sinks to the bottom of a tank from which it can be drawn off as sewage sludge. Although the solids are more concentrated than in the incoming sewage they still account for only about 4 per cent of the total volume of the sewage sludge. This sludge can then be digested in an anaerobic system to produce a lower oxygen demanding product, still basically in liquid form, which has to be disposed of. This can be done by drying through use of a centrifuge or in drying beds. This dry organic matter can then be used as a soil conditioner/fertiliser or for landfill, and occasionally it is incinerated. Sewage sludge handling and treatment is usually the main problem in any sewage works. This is because the anaerobic digestion process is a slow one and it is therefore difficult to deal with large amounts. Any drying process is also very expensive, and as a result sewage sludge is often dumped without any processing, if this is possible. The oxygen demand of untreated sewage sludge is, however, very high and if dumped into a small water body will almost certainly lead to anaerobic conditions.

Following sedimentation the remaining liquid effluent can then be either discharged into a water body, with its oxygen demand lowered, or else it can be subjected to secondary treatment. Secondary treatment involves an engineered system whereby bacteria are encouraged to consume the organic matter, which is dissolved in the water or is in suspension, in an environment in which adequate oxygen is available. In most urban sewage works today secondary treatment is achieved by the activated sludge process, in which bacteria and oxygen are continually mixed into the incoming effluent. Such secondary treatment can reduce oxygen demand by as much as 33 per cent. Effluents following secondary treatment can then usually be discharged into water bodies without problems.

In most sewage works only primary and secondary treatment is used. However, such techniques have little impact on dissolved chemicals within the waters. One of the biggest problems here is the growing amount of nutrients which get into sewage through use of household and related chemicals. In general nutrients are beneficial to the growth of plants and animals, but they can unfortunately cause major algal growths in water bodies if present in excessive amounts. When these algae die the bacteria breaking them down consume large quantities of oxygen, which in extreme cases can cause the environment to become anaerobic. Tertiary treatment methods are designed to remove nutrients such as nitrogen and phosphorus. At present such techniques are still relatively complex and costly, so they are only used in plants where special water problems exist.

In recent years a whole range of dissolved chemicals, many of which are toxic, have been added to sewage lines from both household and industrial sources, including pesticides, herbicides and heavy metals. The problem here is that no type of sewage treatment has a significant effect on these chemicals. This means that they are disposed of either within sewage sludge or effluent and are capable of damaging any life forms with which they come into contact. As this problem has increased during the twentieth century it has meant that increasingly, sewage sludge has not been able to be used as a soil conditioner for fear of any toxic chemicals being transferred to food crops, and in turn to human beings. Similar factors restrict the use of effluent for irrigation, and have also hindered the development of water reclamation projects which have considered recycling sewage effluents after treatment back to the consumer. In the USA to date, it has been felt to be too much of a risk to attempt such recycling on a large scale, but increased water shortage in the future might force it to occur.

12.4.3 The sewage systems

The residents of the city of Los Angeles and people and industry within the catchment of the Los Angeles River are linked by a large sewage system to a treatment plant at Hyperion, situated on Santa Monica Bay, just to the north of the International Airport (Figure 12.7). Today this plant serves in excess of three million people (Los Angeles, Department of Public Works, Bureau of Engineering, 1982).

Figure 12.7 Hyperion sewage works and collector systems, Los
Angeles

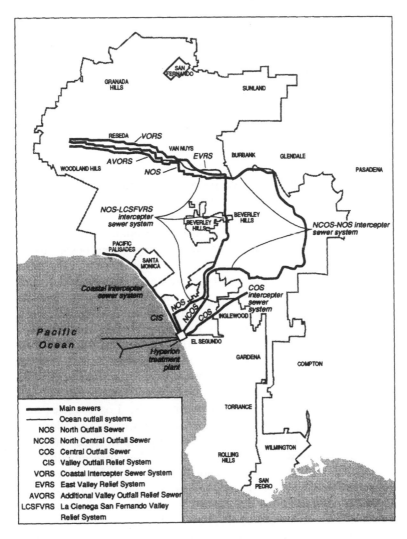

In the early days of the city of Los Angeles in the mid-
nineteenth century sewage disposal was primitive. Human wastes
were fed into pits in the ground, while water from baths and sinks
was drained into nearby irrigation ditches. The first public sewer
was constructed in 1876 and served a population of about 7,000. It

419

transported the water-borne wastes to the southern part of the city, where some of the waters were used for irrigation purposes.

The population boom of the late nineteenth century, together with the desire to preserve the image of Los Angeles as an important health centre, led to the drawing up of a major plan for sewage disposal. This consisted of three large interceptor sewers within the Los Angeles River basin, which were joined along the line of the river to an outfall on Santa Monica Bay. Despite concern being expressed that dumping water into the ocean was a wasteful procedure, an ocean outfall solution to the sewage problem was eventually agreed by the voters and the scheme was completed in 1894. A second outfall to cope with growth in the region was established in 1908. When finished the project solved the city's immediate sewage problem, but the dumping of raw sewage into the Pacific Ocean was to transfer the problem to another area.

A few years after the original outfall was completed complaints about the state of the beach around Hyperion began to be received. In 1915 the State Board of Health ordered the city of Los Angeles to do something about the problem. However, it was not until 1924 that a new screening plant was constructed at Hyperion and a 1·61 km long submarine outfall built. At this time, too, other cities in the Los Angeles River basin were installing new sewer systems and finding that the easiest solution to their difficulties was to connect their trunk sewers to the Los Angeles main sewer and so send all their wastes down to the Pacific.

Given the increasing volumes of sewage which were being handled by the Hyperion plant as population in the Los Angeles area grew, pollution of the beaches steadily grew worse. So serious did this become that in 1943 the State Board of Health closed 16 km of beach to the south of the outfall. It was therefore decided that something had to be done and the solution chosen was the construction of a new plant at Hyperion which would provide secondary treatment of the sewage by use of the activated sludge process. This major facility was built by 1951, but three years later sewage volumes had increased so much that the plant was already working at full capacity. Once this occurred additional waste waters were not able to be given secondary treatment and had to be directly discharged along the submarine outfall. The basic aim of the plant had been to stop this situation from happening.

When the new plant had been opened it was intended that the sewage sludge, when dried, would be sold. However, the high costs and the large volume of sludge produced meant that the process was uneconomic. The only practicable disposal for the sludge was to send it along the submarine pipeline into the ocean. The necessity to dispose of the sludge and also to cope adequately with the growing volumes of sewage being supplied led to a decision to upgrade the plant in 1957. Major modifications made at this time included the construction of two large submarine outfalls. One of these was entirely for the disposal of sewage sludge. It was 11 km long and terminated at the head of a submarine canyon. The other, for waste waters, was 8 km long and disposed of the effluent through a series of diffusers at a depth of 58 m. During the 1960s and the 1970s increasing loads on the Hyperion plant meant that an ever-reducing amount of sewage was capable of being given the full treatment of which the plant was capable. All sewage in excess of plant capacity was merely screened to remove the larger material and then dumped into the sea.

An even bigger sewage system than that of the city of Los Angeles is operated by the Sanitation District of Los Angeles County (Orange and Los Angeles County Water Reuse Study, 1982). This covers 72 cities and 45 unincorporated communities and serves, in total, about four million people, and is known as the Joint Outfall System. Its limits are largely those of the San Gabriel River basin, together with about 1,550 square km of the coastal plain to the south of Los Angeles. Within the system there are a number of treatment plants, but by far the largest is on the coast at Carson, which handles almost 80 per cent of the total discharge. The Carson plant was originally built in the late 1920s. By the early 1980s it was handling in excess of 1·23 million cubic m (325 million gallons) a day, though it was only providing primary treatment. All effluent from the plant is transported to the coast via two 10 km-long pipelines, and then discharged by two submarine outfalls 1·75 miles off the Palos Verdes Peninsula at a depth of 60 m. The sewage sludge produced from the primary treatment is anaerobically digested, dewatered by centrifuge and the final drying is achieved by drying beds. The end product is then used for soil conditioning or for land fill.

The third of the major sewage systems is run by the Orange County Sanitation District, and operates largely within the confines of the Santa Anna River basin. There are two large interconnected sewage plants on this system. The first is located about 6 km inland at Fountain valley and the second near the mouth of the Santa Anna River itself. Total daily capacity is about

0·861 million cubic m (230 million gallons). Most of the sewage handled by these plants receives only primary treatment. The effluent is then chlorinated to kill bacteria and to destroy viruses before being pumped 8 km out to sea via a submarine pipeline.

Within the Los Angeles urban area there are a number of small sewage treatment plants which produce effluents to relatively high standards for specific purposes, such as for cooling, irrigation and groundwater recharge. The Orange County Sanitation District, for example, operates a facility which produces high quality water by a variety of techniques (including reverse osmosis) for injection into the ground to prevent sea water intrusion into the aquifer. Similarly, the cities of Los Angeles and Glendale have a joint plant near the boundary of the two cities which utilises primary, secondary and tertiary treatment methods. Part of this high quality water is utilised by Glendale for cooling purposes in a power plant and the rest is used by Los Angeles for park irrigation and for groundwater recharge.

To date the costs of producing high quality water by sewage treatment are great. In addition, it is currently illegal to connect reclaimed water to domestic supplies and little incentive therefore exists to develop sophisticated processes which are capable of supplying water to drinking quality standards. The main worry here still remains the problem of certain chemicals present, often only in small quantities, but which may in the long term produce health risks in humans. However, the outlook for non-domestic users looks likely to improve as demand rises. Water from the Colorado already has a dissolved solids content of over 700 ppm, and many of the local water sources record values between 300 and 600 ppm. It is only the waters from Owens valley and the California State Water Project which possess total dissolved solids values of less than 300 ppm. As water from the State Water Project is already very expensive, it seems likely that reclaimed water may prove increasingly attractive for industrial and certain agricultural uses where the highest standards are not required.

In recent years the major problems faced by the three main sewage systems have been how to cope with ever stiffer state and Federal legislation on pollution. Part of the difficulty has undoubtedly been that all three authorities have tended to accept minimal treatment methods in the past in the face of ever-growing discharges. All too often the preferred solution has been simply to dump everything as far out into the Pacific Ocean as possible, and then just hope for the best.

To comply with the ever-more stringent pollution legislation, the City of Los Angeles produced a Wastewater Facilities Plan in the late 1970s to upgrade its wastewater treatment works, and other authorities drew up similar plans. The most important national legislation was the Federal Water Pollution Control Act, with amendments of 1972 and in 1977 through the Clean Water Act (1977). The basic aim of the legislation was to restore and preserve the water quality of the nation's water sources. A number of objectives were stated, of which the most important was the elimination of polluting discharges into navigable waters by 1985.

At the state level the critical legislation is the Porter-Colone Water Quality Act, first enacted in 1969 and amended in 1979. The aim was to maintain the highest water quality which was reasonable and also to attempt to reclaim wastewater for beneficial uses. This legislation was complemented by two plans, known as the Oceans Plan and the Bays and Estuaries Plan, which were designed to protect waters by, amongst other things, prescribing effluent quality requirements for discharges.

As a result of these laws and plans the California State Water Resources Control Board and the Environmental Protection Agency ruled that the City of Los Angeles must stop discharging sewage sludge into the ocean, and that full secondary treatment must be given to all effluents discharged into any receiving waters. The City, however, applied for a waiver with regard to full secondary treatment for its Hyperion Plant, and this was effectively granted by a 1977 amendment to the Federal Water Pollution Control Act which permitted less than full secondary treatment for effluents discharged into deep ocean waters.

The chief problem still facing the City of Los Angeles was the disposal of sewage sludge which could no longer be dumped into the sea after 1985. Various options were studied and it was eventually decided that thermal processing of dewatered sludge was the best method which could be used. The proposal is that the digested liquid sludge be mechanically dewatered and the cake then processed through a thermal system. The residual ash is used for landfill (Figure 12.8). An innovative feature of the process chosen for Hyperion was in the thermal evaporation process which permitted the generation of electricity in a combined plant. In this way the plant is able to supply its own energy needs and those of the rest of the Hyperion plant.

As a result of enormous capital investment and research, the City of Los Angeles and adjacent areas have been able to produce working wastewater treatment plants which have greatly reduced the pollution problems of southern California. It is true that these improvements have been forced on the authorities as a result of legislation, but the speed and efficiency of the technical response has been impressive. The still unsolved problem is the reclamation

Figure 12.8 Thermal processing of sludge, Hyperion sewage works, Los Angeles

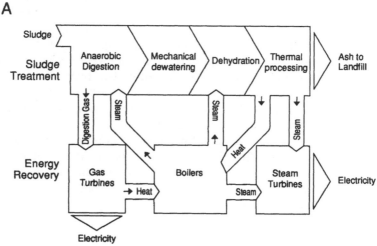

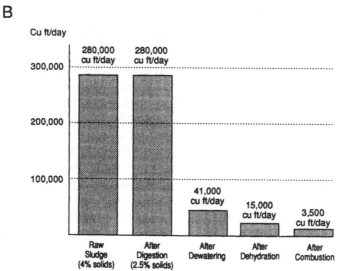

of these wastewaters for beneficial uses. Some work has already proved successful, but much more still needs to be done to recover the millions of tonnes of water each day which are dumped into the Pacific Ocean. This undoubtedly represents the biggest challenge for the future.

12.5 ENVIRONMENTAL PROBLEMS CAUSED BY TOO MUCH WATER

It is ironic that a conurbation so short of water should suffer serious problems from water in excess, yet such is the lot of Los Angeles. The causes of this excess water are violent winter storms tracking eastwards across southern California. It should not be thought that these severe storms occur every year, for they do not, but about every ten years or so a significant flood event is registered. The effects of such storms are twofold. In the mountain and foothill areas landsliding and slumping are common, while along the river valleys and on the alluvial plains widespread flooding can take place.

Landslides and mudflows are almost completely confined to the mountain and foothill regions. Here slopes are steep and this, coupled with deeply weathered, unconsolidated material, can mean that during intense rainfall slope failure occurs. Human activity, by undermining or overloading slopes and by alteration to drainage, can exacerbate the problem, though in general one is dealing with a naturally occurring phenomenon. Every year slides and mudflows occur after heavy rains. For the most part the individual slides are small, but the cumulative effect in terms of damage is very high. It was after a winter of severe slope failures in 1951–2 that the City of Los Angeles introduced strict regulations for building and development work on hillslopes. Permits were required before construction could begin, and stringent standards had to be complied with when any changes to the angles of slopes were contemplated. These regulations were further strengthened in 1956 when the City demanded that geological reports on a site had to be made before any development permit would be issued. Even more stringent rules on cut and fill operations, compaction of fill and drainage lines were enacted in 1963. The result of all this legislation is that the problems caused by landsliding and mudflows have been substantially reduced, though not, of course, eliminated (Cooke, 1984). The situation now is that most of the human causes

425

of slope failure have been removed, but natural slippages still continue.

The heavy rains which promote slope failure also cause flash flooding in the semi-arid environment of Los Angeles and over the years a number of major flood events have been documented. One of the earliest recorded floods was in 1861–2 when all the lowlands between Los Angeles and the sea to the west and south were covered in water. The year 1884 was another bad flood year when downtown Los Angeles recorded an annual total of 1,000 mm of rainfall. At the time of both these floods, though, the city was relatively small and so only minimal damage occurred. However, by the early years of the twentieth century the city had grown considerably and so when the next major flood event took place, in February 1914, the damage was extensive. During a four-day storm, 483 mm of rain fell over the San Gabriel Mountains on to ground that was already saturated by earlier rainfall. The result was massive flooding along both the Los Angeles and San Gabriel Rivers, causing damage in excess of $10 million. So great was the disruption and damage that the elected representatives in the city created a County Board of Engineers on Flood Control and charged them with solving the flood problem (Bigger, 1959).

Before looking at what they achieved, it is essential to assess the true nature of the flood problem in the metropolitan area. Here the most important factors are topographical, combined of course with climatic ones. The Los Angeles basin is surrounded by high mountains, rising to over 3,000 m, which are all within 100 km or so of the sea. This means that the channel slopes are exceedingly steep, especially in the mountain and foothill regions. As a result any runoff is concentrated rapidly and flows quickly down on to the alluvial fans and floodplain zones, often with very high sediment concentrations.

The lowlands in which the urban area of Los Angeles is built have been formed almost entirely by recent fluvial deposition. On the alluvial fans and plains each new flood event deposits sediment and slightly raises the bed of the stream, so that the next flood will almost certainly flow along another line as it crosses the alluvial fan zone. In such a manner the whole fan area is alluviated (i.e. built up). Across the broad and gently sloping alluvial plains the rivers do not flow in well-defined channels, but instead follow washes which might be anything up to 3 km in width. For most of the year such washes are dry, but during flood conditions they can become raging torrents. These observations show that under naturally occurring conditions, large areas of the Los Angeles basin have been prone to flooding. The construction of houses and other buildings did not in any way change this situation, except — if anything — to make it worse. This was achieved by the increased area of impervious land, which further accelerated the speed of water runoff and so augmented local flood discharges.

Superimposed on these physiographic conditions was a semi-arid climate, with an extremely variable climatic regime. Average annual rainfall throughout the region varied from 250 mm in the coastal lowlands to as high as 1,000 mm in the mountains. At individual stations, particularly those with low average annual totals, variations from year to year are very great. At Los Angeles, for example, the 1975–6 rainfall was only 48 per cent of the long-term average, while two years later in 1977–8 it was 223 per cent of the same figure. Individual storm events can also produce high rainfall totals. During one storm a station in the Santa Anna Canyon recorded a fall of 660 mm in 24 hours. Under such conditions it is not surprising that widespread flash flooding can occur.

Having established a Los Angeles County Flood Control District, an election took place in 1917 which agreed a US$5 million bond issue for flood protection works. From this time until the passing of the Federal Flood Control Act in 1936, all flood control measures were undertaken by local bodies. However, following the 1936 Act the protection of urban areas from flood damage became the responsibility of the Federal government. Since then most of the flood control structures in the Los Angeles area have been built by the US Corps of Engineers.

There are two chief objectives of the Flood Control programme. The first concerns the highland phase. Here the basic aim was to reduce the speed of runoff and to cause less material to be eroded. In the lowland section the aim was to store as much water as possible in reservoirs on a temporary basis, and move the rest of the water as quickly as possible into the sea.

In the highlands it was recognised that vegetation cover helped to reduce erosion. Accordingly, attempts were made to protect the area from fires and to maintain as dense a vegetation cover as possible. Along the larger gullies, check dams were constructed to collect debris and so prevent it from reaching the lowlands. After each flood this debris has to be removed by earth-moving equipment at high cost.

Within the Los Angeles County Flood Control District, some 20 large dams have been constructed since the Devil's Gate Dam in

Figure 12.9 Flood control systems for Los Angeles

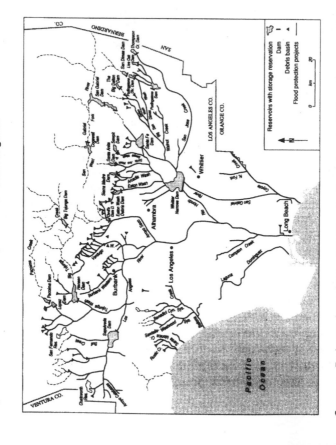

Source: Redrawn from State of California, Department of Water Resources Bulletin 199, *California Flood Management* — an evaluation of flood damage prevention programs, 1980. Reproduced by permission of the Department of Water Resources, State of California.

the Arroyo Seco was opened in 1920 (Figure 12.9). All of the early structures were built in the mountain canyons, but the later ones such as Hansen, Sepulveda, Santa Fe and Whittier Narrows were constructed further downstream. These downstream structures had the advantage of controlling runoff from a larger proportion of the total watershed. To date they have been remarkably successful in containing and controlling the flood flows across the lowland parts of the urban area (Department of Water Resources, State of California, 1980).

Once the waters left the mountain valleys, they flowed across the lowlands in a series of major washes up to 3 km in width. These washes were not permanent channels and often moved their position in response to sediment deposition. With the growth of the urban area this situation could no longer be tolerated and so the major drainage lines have had to be stabilised. This has entailed the straightening, narrowing and deepening of the channel. To facilitate further the rapid removal of water, over 725 km of the banks and beds of these channels have been concreted. The net result of all this activity is that the lowland parts of the metropolitan area have been successfully protected from the heavy flood losses which characterised the region prior to the 1920s. In the mountains and the foothills the nature of the terrain has meant that it has been impossible to provide total protection from the flood menace, and here heavy losses can still occur.

A severe test of the efficiency of the flood control system which had been developed since the 1920s was provided by major storms between 18–26 January 1969. During this period as much as 1,209 mm of precipitation was recorded in some of the mountain stations. In the lowland areas the flood control facilities worked well and it is estimated that they prevented US$1·2 billion of damage (Rantz, 1970, p. 8). Even during the severest period of runoff none of the flood control reservoirs were filled to capacity, and their impact on downstream river discharges was considerable. At the Prado flood control basin on the Santa Anna River, for example, the peak inflow was 2,123 cubic m (75,000 cubic feet) per second on 25 January, but the peak outflow was controlled to a mere 142 cubic m (5,800 cubic feet) per second as a result of water storage.

In the same storm the debris basins in Los Angeles County collected an estimated 1·53 million cubic m (2 million cubic yards) of sediment (ibid., p. 9). Of 61 basins only three were filled to such an extent that eroded material was permitted to move downstream.

429

Although much damage was prevented, losses were still high, with 92 people killed and more than 10,000 people having to leave their homes. Total damage caused by the storm was estimated at US$62 million. The greatest damage occurred in the headwaters of the San Gabriel and Los Angeles Rivers, in the mountain and foothill zones in the northern and western parts of the metropolitan area. Here the streams in the small canyons overflowed their banks and, together with mudslides, caused extensive damage. Further east, in the Santa Anna River basin, conditions were not as severe and damage was restricted to a number of upland tributaries.

Today Los Angeles has a complete flood control system which is an integrated combination of projects built by both the Los Angeles County Flood Control District and the United States Corps of Engineers (Figure 12.9). It is centred on the Los Angeles and San Gabriel Rivers, but also includes numerous small structures along headwater streams. In the lowlands there are four major flood control basins — at Hansen, Sepulveda, Santa Fe and Whittier Narrows — which are capable of regulating flows. All the major channels have been shaped and concreted to ensure the most rapid controlled runoff that is possible. The statistics of the system are impressive. There are 14 major dams, 125 debris basins, 757 km of improved open channels, 2,900 km of storm drains, 55,000 catch basins, 29 groundwater recharge spreading basins, 23 pump plants and three sea water intrusion barrier projects.

Besides the structural facilities provided for flood control, a policy of flood/plain management is also practised by Los Angeles County. This is closely linked to the National Flood Insurance Programme which was introduced by the Federal government between 1968 and 1973, when it became obvious that structural measures alone were not enough. In order to gain the benefits of the programme local authorities had to join the insurance scheme. The County of Los Angeles joined in 1980.

The basic planning tool of the programme is a Flood Insurance Rate Map (FIRM) which delimits the floodplain and flood-prone areas with regard to a flood event with a 100-year return period. Once this map has been produced for an area, new properties can only be established on the floodplain provided that the ground floor level is above the height of the 100-year flood. Moreover, a flood-way free of buildings must be maintained to ensure adequate drainage of the area. By such co-ordinated planning activity the damage caused by flood events has been substantially reduced. As a result of both structural and behavioural measures the recent major storm and flood event in February and March 1983, which produced 203 mm of rain in downtown Los Angeles and 660 mm in the San Gabriel Mountains, caused relatively little flood damage to property (Los Angeles County Flood Control District, 1983). Damage to the flood control works themselves did, however, amount to almost US$9 million (Figure 12.10).

Figure 12.10 Storm damaged areas in Los Angeles in 1983

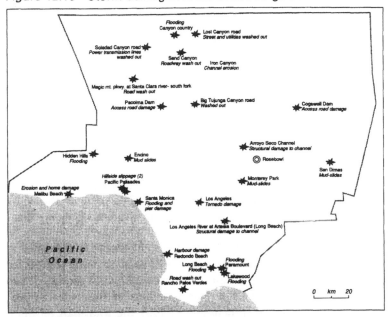

Source: Redrawn from Los Angeles County Flood Control District, 1983, *Storm Report on Los Angeles County*. Reproduced by permission of the County of Los Angeles, Department of Public Works.

12.6 AIR POLLUTION

12.6.1 Introduction

Today one of the most common images people have of Los Angeles is of smog generated by the large number of automobiles which are found within the city. The problem of photochemical

431

smog is, however, a surprisingly recent one, only dating back to July 1943 when a thick smoky cloud reduced visibility over downtown Los Angeles to less than 5 km. Before then Los Angeles had been noted for the health-promoting characteristics of its atmosphere. The blame for the smog of 1943 was originally put on a plant making synthetic rubber close to the downtown area, but it was soon obvious that other sources must also have been responsible. This first outbreak of serious smog did have the remarkable effect of galvanising the legislators into action, and in 1945 the City of Los Angeles passed a regulation limiting the amount of smoke which could be produced from a stationary single point source. Soon afterwards about half the nearby cities introduced similar laws to reduce pollution. Even the *Los Angeles Times* became involved, leading a citizens' campaign against the smog menace, which included commissioning a specialist's report on what needed to be done to reduce the pollution risks. A recommendation of this report was that air pollution control districts should be established. In 1947 this idea was taken up by the Board of Supervisors in a bill which established air pollution control districts in each county. These organisations, which were charged with making rules and regulations to control air pollution, came into operation early in 1948. Thus, within five years of the smog problem first being recognised, an administrative framework for controlling it had been established.

The Los Angeles County Air Pollution Control District quickly introduced regulations to limit the most serious point sources of pollution, such as oil refineries and waste disposal dumps. Over the years stricter controls on the emission of fumes and particulates were introduced, including in 1957 the total ban on open fires in gardens. During this period, industry invested heavily in air pollution control measures, with the result that by the late 1950s emissions from industry had been greatly reduced. It was not until 1950, however, that the true nature of the smog was recognised — namely, that it was largely caused by motor exhaust gases being subjected to photochemical reactions and change in the atmosphere. Despite this research finding it was almost ten years before it was widely accepted, in part because of the resistance of the automobile manufacturers and the petroleum industry.

In recognition of the nature of the problem the state of California established a motor vehicle control board in 1960 to set emission standards for automobiles and to approve emission control devices. The following year the board drew up regulations for the reduction of hydrocarbons by 80 per cent and carbon monoxide by 60 per cent from automobiles which were to apply from the 1966 model year. Oxides of nitrogen were to be reduced from 1971 and by the mid 1970s these stricter regulations necessitated the fitting of catalytic converters on all cars. One of the unforeseen early effects of these control measures was to reduce hydrocarbons, but at the same time to increase the oxides of nitrogen. This meant that the conditions for certain photo-chemical reactions were made even more favourable.

12.6.2 Nature of the smog problem

The Los Angeles smog is caused by three main factors: firstly, the prevailing climatic conditions; secondly, the partially enclosed nature of the Los Angeles basin; and thirdly, the very large output of pollutants. The chief cause is undoubtedly the favourable climatic conditions which trap pollutants within the basin. Over Los Angeles the atmosphere is mostly composed of two layers. The lowest is a mass of cool maritime air which drifts inland from the Pacific, whilst overlying this is hot dry air. This cool air below with hot air above forms a temperature inversion, which blocks the upwards movement of air and so allows pollutants to accumulate (by lack of natural ventilation).

This situation is made worse by the barrier of hills to the north-west and south-west, which effectively block off air movement at night and in the early morning. During the heat of the day the upper slopes of the mountains become hot and act as vents, sucking pollution from the basin. However, this effect is relatively small compared with the size of the basin and under inversion conditions the smog is only removed by strong winds, which unfortunately do not occur very often. An exception which does sweep away the smog is the Santa Anna winds which blow from the north. The general daytime pattern is for a sea breeze to drift the pollution inland and for a land breeze to blow in the opposite direction at night-time. If the sea breeze is strong enough the pollutants can be blown out of the basin in an easterly direction. This often happens in spring and early summer, but in late summer and winter calmer conditions prevail, permitting the pollutants to build up.

The pollutants which exist in the air above Los Angeles are cate-gorised as primary, secondary, and tertiary. Primary pollutants are those released directly from pollution sources, either static or mobile, while secondary and tertiary pollutants are formed successively in the atmosphere as a result of photochemical reactions. The chief primary pollutants are carbon monoxide, hydrocarbons, oxides of nitrogen, sulphur dioxide and particulates. Of these more than half the carbon monoxide, hydrocarbons and oxides of nitrogen are generated from vehicles of one form or another (Table 12.1).

Table 12.1: Air pollution sources in the Los Angeles basin

Pollutant	Tons per day	Motor vehicles (% of total	Other sources (% of total)
Carbon monoxide	7,876	75·8	24·2
Reactive organic gases (mainly hydrocarbons)	1,512	51·6	48·4
Oxides of nitrogen	1,337	57·2	42·8
Oxides of sulphur	276	15·8	84·2
Suspended particulates	621	14·0	86·0

Source: South Coast Air Quality Management District (SCAQMD), 1981.

Carbon monoxide is a colourless and odourless gas which is toxic. It is formed by the incomplete combustion of fuels and almost three-quarters of this gas is generated from motor vehicles. Sulphur dioxide is also colourless, but it possesses a strong irritating smell. This gas is produced whenever any fuel containing sulphur is burnt, so it can be emitted from both stationary and mobile sources. In the presence of moisture, the sulphur dioxide will lead to the formation of droplets of sulphuric acid which are highly corrosive. Particulate material such as dust, soot and mist, can be generated by natural causes as well as from agricultural land and industrial activities. Their effect is to produce haze and hence reduced visibility.

Of the primary pollutants the two critical ones are the hydrocarbons and the oxides of nitrogen, both of which can be altered by photochemical reactions in the atmosphere to produce secondary and tertiary products, with deleterious effects. Many different types of hydrocarbon are present, although the olefins are the most reactive and, therefore, subject to greatest change. The hydrocarbons, when only partially oxidised, produce numerous compounds, almost all of which are undesirable.

Nitric oxide is a colourless and odourless gas which is created from nitrogen and oxygen in the atmosphere during combustion. Particularly large amounts are formed when combustion occurs at high temperatures and pressures, such as occurs in an internal combustion engine. The oxides of nitrogen emitted into the atmosphere react to form the yellow brown gas nitrogen dioxide. Nitrogen dioxide is itself reactive, and after being subjected to ultraviolet light it changes to produce tertiary pollutants. Nitrogen dioxide concentrations are greatest in winter, as at this time the less intense sunlight means that the photochemical reaction is unable to go beyond the nitrogen dioxide stage. In summer, in contrast, the reaction can continue to produce ozone and other components. The most important of the tertiary pollutants is ozone, which is a faintly bluish gas, with a slightly chlorine-like smell. Ozone itself is unstable and continuously forms and decays in the complex atmospheric reactions which are occurring. Ozone cannot begin to accumulate in the atmosphere until all the nitric oxide has been converted into nitrogen dioxide. On a diurnal basis ozone begins to form after sunrise, but does not reach high levels until all the nitrogen dioxide which has been created following the morning rush hour has been used up. As a result peak levels usually occur between noon and 2 p.m.

The spatial distribution of all the pollutants other than the primary ones is difficult to predict with any accuracy. In general the primary products reveal maximum concentrations in the source areas and in the plume downwind (Figure 12.11). The main stationary pollution sources — such as power stations, refineries and large factories — are located near the coast and in the eastern end of the San Fernando valley. Motor vehicles are found everywhere within the basin, with a concentration in the central areas. While secondary pollutants, such as nitrogen dioxide, do show a correlation with the sources from which they are derived, they also reveal a downwind movement (Figure 12.12). This tendency is even better developed with the tertiary products, such as ozone, where the highest concentrations may be kilometres away from the sources of the original pollutants (Figure 12.13). It should not be forgotten that there are temporal variations of both a seasonal and diurnal nature in the output of pollutants. As it is colder in winter there is increased fuel consumption at this time, while on a diurnal basis it is the morning rush hour which emits the highest pollution levels.

One of the main problems with photochemical smog is that it is a health hazard, particularly for people who suffer from respiratory

Figure 12.11   A. Los Angeles — Carbon monoxide: 1984.
Number of days on which Federal standard was exceeded (8-HR CO
> 9.3 ppm)
B. Los Angeles — Total suspended particulate: 1984. Annual
Geometric Mean, $\mu g/m^3$

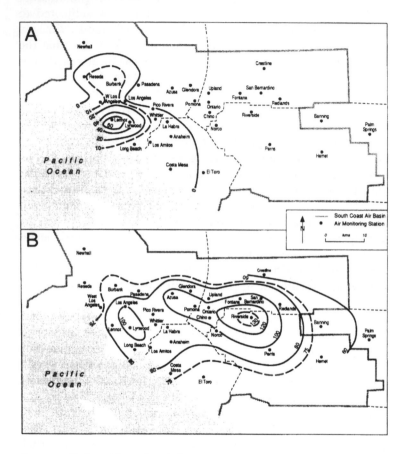

Source: Redrawn from South Coast Air Quality Management District, 1985,
Air quality trends in the South Coast Air Basin 1975–1984. Reproduced by
permission of the South Coast Air Quality Management District, El Monte,
California.

and heart diseases. However, it should be remembered that plants
are also sensitive to it. In San Bernardino National Forest,
Ponderosa pines have been dying due to its effects, while in the
Los Angeles basin itself a range of commercially produced

436

Figure 12.12 Los Angeles: Nitrogen dioxide 1984. Annual average values, pphm

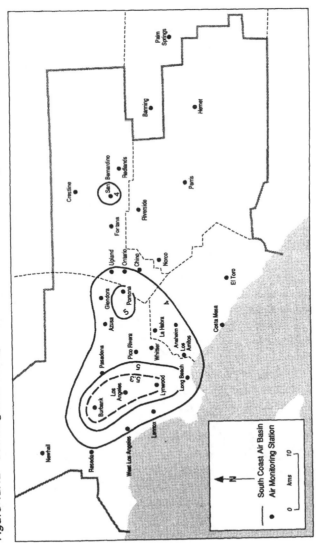

Source: Redrawn from South Coast Air Quality Management District, 1985, *Air quality trends in the South Coast Air Basin 1975–1984.* Reproduced by permission of the South Coast Air Quality Management District, El Monte, California.

Figure 12.13 Los Angeles: Ozone 1984. Number of days on which Federal Standard was exceeded (1-HR O_3 > 12 pphm)

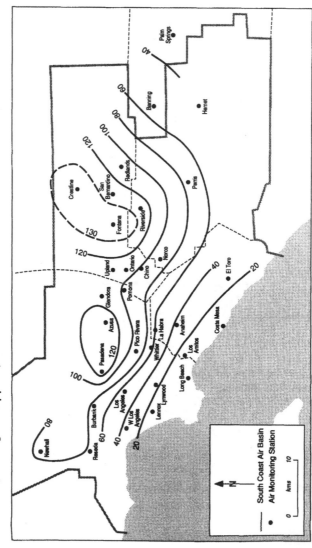

Source: Redrawn from South Coast Air Quality Management District, 1985, *Air quality trends in the South Coast Air Basin 1975–1984*. Reproduced by permission of the South Coast Air Quality Management District, El Monte, California.

vegetables and flowers can no longer be grown profitably, and many other crops have had their yields reduced.

12.6.3 Management of air pollution effects

The harmful effects of smog have been recognised since the mid-1950s, and from that time a number of control and warning systems have been brought into operation. The management of air pollution over the Los Angeles metropolitan area is currently the responsibility of the South Coast Air Quality Management District (SCAQMD). This was created by an act of the California legislature in February 1977 and covers the non-desert portions of Los Angeles, Orange, Riverside and San Bernardino Counties. The District succeeded the Southern California Air Pollution Control District (1975–7), which in turn had taken over responsibility from the individual county air pollution control districts. Initially, the SCAQMD was funded through local property taxes. However, by increasing permit fees and fines, and by charging the cost of air pollution control to those who caused it, the District now raises three-quarters of its funds from these sources.

The current emergency programme to react to air pollution events was adopted by SCAQMD in 1977. This procedure, known as Regulation VII, sets out the actions to be taken by industry, commerce, government and the public to prevent air pollution concentrations attaining levels which would be harmful to the public, or to alleviate such conditions if they were to occur (SCAQMD, 1985a). Three episode levels are recognised, based largely on concentrations of ozone, carbon monoxide and sulphur dioxide in the atmosphere (Table 12.2). Episodes are officially declared whenever any of the specified contaminant levels are reached. Every day the district provides an air pollution forecast for the following day. If episode conditions are predicted procedures which reduce emissions are introduced.

If an episode condition is reached, the information is disseminated to the news media and to schools by radio and telephone. When an alert is called at the lowest level (stage one episode) the mass media warn people with respiratory and heart conditions to limit their activities, while in schools strenuous exercising is reduced. Such alerts occur quite often. Thus, for example, a stage one ozone episode was called on a hundred days during 1980, though carbon monoxide stage one episodes only occurred on four

Table 12.2: Air Pollution Episode criteria — Los Angeles (values in ppm)

Contaminant	Averaging time	Stage 1	Stage 2	Stage 3
Ozone	1 hour	0·20	0·35	0·50
Ozone in combination with sulphur dioxide	1 hour	0·20[a]	0·35[a]	0·50[a]
Carbon monoxide	1 hour	40·0	75·0	100·0
	12 hours	20·0	35·0	50·0
Sulphur dioxide	1 hour	0·5	1·0	2·0
	24 hours	0·2	0·7	0·9
Ozone in combination with sulphate	1 hour (ozone)		0·20	
	24 hours		25 micro g/m³	

Note: [a] These levels shall apply when the ozone concentration and the sulphur dioxide concentration each exceeds 0·10 ppm one-hour average, and shall be determined by adding the ozone and sulphur dioxide concentration.
Source: SCAQMD, 1985a.

days in the same year. At the next level, known as a stage two episode, the major industrial polluters are required to reduce their emissions of hydrocarbons and oxides of nitrogen by 20 per cent. Large employers are also asked to reduce car use by encouraging car sharing for commuters. Stage two episodes are estimated to cost the region about US$2 million a day. A stage three episode requires a state of emergency which can only be proclaimed by the state governor. This would effectively result in a public holiday being declared, with industry and commerce closing down. As yet no such episode has occurred, though estimates suggest that it would cost the area as much as $1 billion each day.

To facilitate its work the District has divided the region into 40 source/receptor areas which form the basis of its planning action. These 40 areas are based on regions with similar meteorological and air pollution characteristics. In each area air pollution conditions are monitored and forecasts of likely conditions prepared. However, for the rapid dissemination of information these 40 districts are grouped in 11 general areas which are more easily intelligible to the general public via the news media (Figure 12.14).

In an attempt to co-ordinate the activities of the various bodies involved in air pollution control in southern California, an Air Quality Management Plan (AQMP) was drawn up and approved

Figure 12.14 Air management areas, Los Angeles

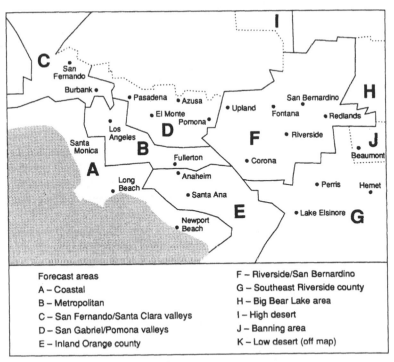

Forecast areas
A – Coastal
B – Metropolitan
C – San Fernando/Santa Clara valleys
D – San Gabriel/Pomona valleys
E – Inland Orange county

F – Riverside/San Bernardino
G – Southeast Riverside county
H – Big Bear Lake area
I – High desert
J – Banning area
K – Low desert (off map)

Source: Redrawn from South Coast Air Quality Management District, 1985, *Air quality trends in the South Coast Air Basin 1975–1984*. Reproduced by permission of the South Coast Air Quality Management District, El Monte, California.

by the boards of the Southern California Association of Governments and the SCAQMD in 1979. Later in the same year the California Air Resources Board adopted a revised plan and submitted it to the US Environmental Protection Agency (EPA), which in 1981 approved parts of the plan with conditions, but disapproved of the ozone and carbon monoxide sections. An application was made to the EPA for relaxation of the ozone and carbon monoxide standards until at least 1987, and this was granted.

The AQMP was designed to meet the requirements of both state and federal air pollution regulations. The California State–Lewis Air Quality Management Act (1976), which established the SCAQMD, had also required the development of a plan for the

441

South Coast Air Basin (SCAB), which would meet both the state and federal air quality standards as soon as possible. The Clean Air Act Amendments (1977) also demanded that a plan should be established, as it was projected that most Federal air pollution standards would continue to be violated by the early 1980s.

Federal air quality standards are divided into primary and secondary categories. Primary standards are set to protect public health, while the secondary standards are devised to safeguard public welfare, including plant and animal life, visibility and constructional materials. Primary and secondary standards have been set for sulphur dioxide, total suspended particulates, nitrogen dioxide, carbon monoxide, ozone and lead. The state air quality regulations are similar in concept to the Federal standards, although for certain pollutants (for example, sulphur dioxide and ozone) the state regulations are more stringent.

The South Coast Air Quality Management Plan was revised by the Southern California Association of Governments and the SCAQMD in 1982. The chief objective was the attainment of air quality standards at the earliest feasible date, with interim targets to promote continued improvement. It was also envisaged that the plan should be updated at 5-year intervals.

12.6.4 Temporal trends

One of the good points which has emerged from Los Angeles in recent years is that many pollutants have shown a decline since the worst years of the late 1960s. In more detailed studies between 1975 and 1984 air pollution over southern California has shown a marked overall improvement (SCAQMD, 1985b). Currently, ozone is the pollutant which it is most difficult to control and which exceeds air quality standards most frequently. However, it is difficult to analyse trends in ozone pollution, as ozone formation is greatly influenced by metereological conditions. Research has shown that the warmer the 850 millibar temperature — that is, the atmospheric temperature at about 1,463 m — the greater the ozone-forming potential of the air mass. The temperature values at 850 millibars, averaged over the period May to October, have been found to be a good predictor of likely ozone levels. These values show considerable fluctuations from year to year, with the 1982 level being the lowest during the period 1956–84 (SCAQMD, 1985b, p. 4). This meteorological indicator is therefore now used

Figure 12.15 Trends in ozone levels over Los Angeles 1975–85

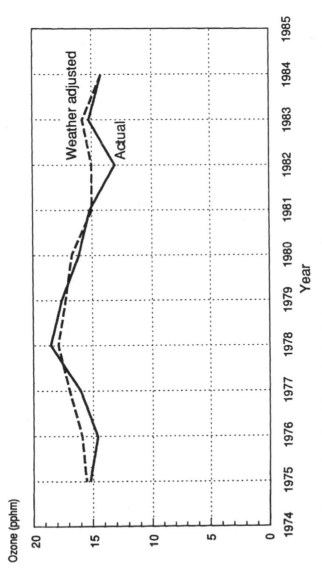

Source: Redrawn from South Coast Air Quality Management District, 1985, *Air quality trends in the South Coast Air Basin 1975–1984*. Reproduced by permission of the South Coast Air Quality Management District, El Monte, California.

Figure 12.16 Trends in carbon monoxide levels over Los Angeles 1975–84

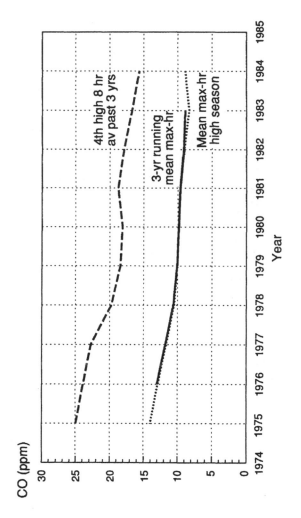

Source: Redrawn from South Coast Air Quality Management District, 1985, *Air quality trends in the South Coast Air Basin 1975–1984.* Reproduced by permission of the South Coast Air Quality Management District, El Monte, California.

Figure 12.17 Trends in lead levels over Los Angeles 1975–85

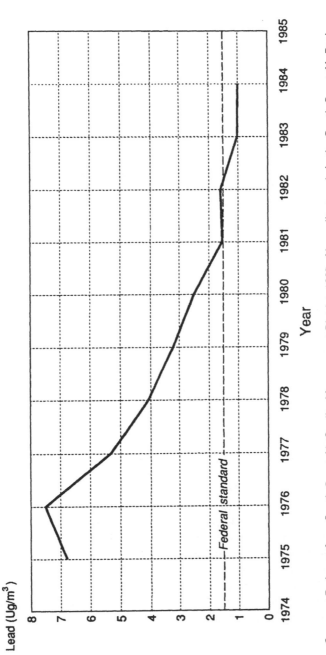

Source: Redrawn from South Coast Air Quality Management District, 1985, *Air quality trends in the South Coast Air Basin 1975–1984.* Reproduced by permission of the South Coast Air Quality Management District, El Monte, California.

to adjust the actual observed ozone values to provide more representative trend values (Figure 12.15). Over the observation period a slight fall in the weather-adjusted figures can be seen. However, in 1984 almost 10 per cent of the recording stations had more than 100 days on which the Federal ozone standard was exceeded.

Carbon monoxide levels have shown a substantial decline between 1975 and 1984 (Figure 12.16). Although carbon monoxide levels, like ozone, are influenced by meteorological conditions, the observed reduction in concentrations appears to be a real one. Using six stations as a representative sample it was found that the number of days on which the Federal carbon monoxide standard was exceeded decreased from 467 station days in 1975 to 147 days in 1984 (ibid., p. 8). This marked improvement in carbon monoxide levels is due to the growing efficiency of emission control devices fitted to motor vehicles and has occurred despite an increase in the number of vehicle-kilometres travelled (Austin *et al.*, 1982).

Mean annual nitrogen dioxide levels have also fallen by about 19 per cent over the observation period (1975–84), expressed in terms of three-year running means. Even more impressive, though, are the three-year running mean statistics for the number of days on which the state standard was exceeded at six representative stations. This revealed a reduction from 126 days in 1975–77 to only 21 days in 1982–4. Total suspended particulates have shown relatively little change over the observation period, with the three-year running means for eight representative stations only registering a fall of some 7 per cent. Control of this pollutant is — perhaps somewhat surprisingly — falling behind that of other pollutants. In contrast, the trend for lead content in the atmosphere (almost all of which originates from lead additives in gasoline) is a spectacular success story: for as the lead content of gasoline has been reduced and as more cars have been built to run on lead-free petrol, the atmospheric lead levels have declined about sevenfold between 1975 and 1984 at many stations (Figure 12.17).

By the mid-1980s the Los Angeles area is already complying with Federal standards for sulphur dioxide, and it is believed that with further controls it can continue to do so. It is also felt that the basin will be able to meet the Federal nitrogen dioxide and particulate standards by the late 1980s, and perhaps also the Federal eight-hour carbon monoxide standard by the year 2000. However, it would appear that ozone levels will continue to exceed both Federal and state standards, even at the end of the present century. In the longer term it is believed that the ozone standards can be attained, but that this will require far-reaching changes to the local economy, including perhaps the use of alternative fuels in place of petroleum products.

Numerous schemes have been suggested to restrict the use of the automobile, which is the main cause of the ozone problem, but few have had much success. Workers have been encouraged to share cars whenever possible in commuting to work, but as it would seem that only one-third of total vehicle-kilometres are the result of commuting, the impact, even if successful, would not be very great. Other research has suggested that the greatest pollution occurs during the first few kilometres over which an engine is operated. If correct this suggests that the number of trips is the crucial factor rather than the total kilometres travelled.

12.6.5 Conclusions

The management of air pollution in Los Angeles can be claimed as a success story in environmental management. In the 40 years since photochemical smog first began to have an effect on the inhabitants of the city, its true nature has been recognised by research and these findings have been widely accepted throughout the population. By the 1950s pollution control organisations had been established which, through guidelines and regulations, were extremely successful in reducing pollution from stationary point sources such as factories, power plants and even domestic dwellings. Less success occurred at this time with the control of emissions from vehicles, partly owing to the resistance of powerful organisations such as the car manufacturers and the petroleum industry. However, even here the victory was eventually won and by the late 1970s exhaust emissions from automobiles were only a fraction of those of vehicles in the late 1950s and 1960s.

To some extent, though, these clear successes have been diluted by continued population growth in the conurbation, which has meant still more cars and more industry. Fortunately current growth rates are well below those of the late 1950s and early 1960s when the Metropolitan area was growing at an annual rate of 300,000 people. The other area in which it has not proved possible to bring about changes which might further limit pollution is in terms of life-styles. In southern California the automobile provides a degree of mobility which cannot be obtained by other methods of transport. Car sharing schemes have been tried over the years, but with singularly little success. The only likely breakthrough here seems to be the advent of the electric car, but its widespread use still seems to be decades away.

Nevertheless, despite these shortcomings, Los Angeles appears to have weathered the worst of its air pollution and even if it has not been eliminated, it has at least been controlled and even reduced without any significant reduction in the pace of economic growth. Interestingly, it seems that it was the perception of other people in the USA in the early 1970s that Los Angeles was a smog-laden city which reduced its growth at this time, rather than strict economic factors. Management policies which quickly identified the nature of the problem, and which then set about to minimise the effects through regulatory control, have undoubtedly produced a present-day environment within the metropolitan area which is considerably better than would have been the case if no action had been taken. What, perhaps, is even more surprising is just how rapidly the problem has been got under control — a fact which surely reflects the dynamic character of the Los Angeles voters who had to pay for most of the changes which took place.

Part Three

Prospect

13

The Future of Drylands

As the twenty-first century is approached it has become increasingly clear that the arid lands of the world will be subjected to ever greater land use pressures as a result of the continued growth of · the population. Yet the way in which these pressures should be tackled is not always obvious, for what one is dealing with is a series of different approaches to drylands dependent upon the country being studied. Crucial factors are the wealth of the nation, its political system, and the relative importance of drylands in the overall economy of the country. For some countries drylands only account for a minor percentage of the total area, whereas in others they represent the only types of land present. The attitudes of 'outsiders', including organisations such as the United Nations and other international agencies, should not be overlooked either (Johnston, 1992; Kunugi, 1992). With these an abstract and academic approach is often followed in which a problem is identified and an ideal solution sketched out. Unfortunately, in many cases such an approach is unworkable as it bears little relation to the problems on the ground; chief amongst which is a growing population which must be fed and housed.

In recent years much has been written on 'sustainable development' as an objective for the developing countries, particularly with regards to agriculture (Doyle, 1991; O'Riordan, 1991; Turner, 1988). However, it is important to realise that all agricultural activity carries an environmental cost. All too often in the literature traditional or 'organic' agriculture is regarded as in some way superior to the modern, high fossil-fuel input, Western agriculture. Unfortunately, such views are oversimplified. All forms of agriculture begin by replacing the natural vegetation with

451

a crop which can be consumed by humans. This process inevitably involves disturbance of the top soil. Whilst it is true that modern Western methods affect the soil to greater depths and usually over larger areas, there is no doubt that traditional agricultural methods have caused considerable environmental damage. Nowhere is this better seen than in the Middle East and the Mediterranean region where subsistence agriculture practised over the last 10,000 years has degraded the natural vegetation and caused extensive soil erosion.

The differences between traditional and modern agricultural practices are, therefore, more to be assessed in the speed of environmental degradation, rather than in terms of good and bad practice. What is beyond doubt is that modern agricultural methods, with the greater energy inputs involved, are capable of causing much more disturbance to the environment and hence a more rapid rate of environmental degradation than traditional methods (Barrow, 1991). What must be considered is how the effects of agriculture can be minimised in terms of deleterious environmental effects. Following the 'dust bowl' conditions which prevailed on the southern Great Plains of the USA in the 1930s the Soil Conservation Service, after long periods of field studies, devised a series of conservation measures aimed at reducing soil erosion to a minimum (Steiner, 1990).

It would be unfair, though, to imply that the traditional agriculturalists were not aware of soil erosion problems (Thapa and Weber, 1991). Throughout the Mediterranean and Middle East check dams were constructed along stream channels to cause sediment deposition and to provide extra fertile land for cultivation. Such structures had the added advantage of conserving moisture in the soil and so permitted a more rapid vegetative growth to take place. On slopes terraces were constructed. These were highly labour intensive, but they did minimise soil movement and provided optimum environments for crop growth under what were often extremely difficult conditions. There is no doubt, therefore, that soil conservation methods are available which could be applied throughout the world's drylands. The problem is the cost of implementing them (Bojo, 1991). Today the agriculture practised in almost all countries is influenced by product prices on world markets. This implies that areas which can produce a crop cheaply to a large extent determine the world price of a commodity. In effect, this means that the cost of conservation measures in drylands can often increase the price of a crop so that it is no longer competitive even in the country in which it is produced.

An example of the above situation is provided in the Mediterranean region. Here the basic infrastructure of the terraces is already in place. However, the cost of labour is so high today that it is impossible even to maintain the terraces, let alone build new ones, and still grow crops competitively. As a result throughout the Mediterranean the conservation structures built by traditional societies have begun to collapse over the last few decades. As more and more of these structures deteriorate vast amounts of sediment and soil, which have been locked behind them for decades, begin to pour into the river systems.

These examples suggest that conservation measures are unlikely to be introduced into the dryland areas of developing countries without substantial aid from the developed world. This might seem a rather dispiriting conclusion to reach but it is probably better to take a realistic approach rather than to envisage any self-help solution to the soil erosion problem. The message which must be learnt is that soil conservation techniques do have a real cost which many developing countries will not be able to afford. While it is true that certain of these measures improve crop growth by increasing moisture content or providing some similar advantage, most do not supply sufficient benefit to warrant implementing them on a purely commercial basis. Certain techniques, such as contour ploughing, have no significant extra cost implication and could easily be implemented where appropriate (Vogel, 1992). Regrettably, though, these simple yet effective modifications to current practice are not commonly introduced. Many developing countries are, therefore, being placed in difficult positions. Their growing populations need extra food and as a result the cultivated area is often being extended into ever more marginal lands with steeper slopes, lower rainfalls or thinner soils (Bowonder, 1987). Under such conditions the rates of soil erosion and land degradation are bound to increase even if conservation measures are utilised. Although numerous works exist describing the physical nature of the land degradation process, Bojo (1991) has pointed out that remarkably little information exists as to the economic costs which this land degradation actually causes (Dregne, 1992). Until such information is available it will be difficult to assess the relative benefits of different solutions to the problem.

An extensive literature now exists covering the topic of desertification in terms of its causes and the impacts which it has produced

453

on the environment (Grainger, 1990; Mainguet, 1991; Poulsen and Lawesson, 1991). However, in recent years a number of workers have begun to question whether the concept of desertification has been exaggerated (Binns, 1990; Forse, 1989; Hellden, 1988, 1991). Such workers claim that there is little evidence to indicate a sustained degradation of the land. The reality is perhaps a little more complex. During the 1960s and especially the 1970s, too much emphasis was undoubtedly placed on the impact which society was having on the desertification process. The great mistake which these workers made was to underestimate the importance of natural climatic variation and change and the impact that these have had on delicate ecosystems.

Our knowledge of climatic change in the Quaternary era is now so detailed that it is possible to identify more than 20 cold phases which have led to ice advances of varying magnitude. The impact of these climatic changes in the world's drylands remains much less well known, though research on lake levels and other climatic indicators suggests that considerable changes in hydrological conditions have taken place (Le Houerou, 1992). The significance of these changes means that it is rather naive to speak of 'normal' climatic or hydrological conditions for a region. It is easy to appreciate that similar 'average' figures can be obtained for either a 'rising' or 'falling' trend situation. However, the overall impact on the environment of either increasing dryness or increasing wetness is vastly different. What environmental managers must learn to do is to interpret what is actually occurring in time intervals from as short as two years to in excess of a hundred years (Thompson, 1989). What must be avoided are attempts to massage the available data to fit sequences of various periodicities. This is not to deny the possible existence of such fluctuations, but to date it would seem correct to conclude that they have not been established in many of the world's arid regions with any degree of certainty (Tyson, 1986).

What is also not sufficiently appreciated is that indications of increasing aridity or wetness may not be simply the result of lower or higher precipitation respectively. All too often the complexity of the hydrological cycle is overlooked. Precipitation is the major water input to the system, but losses can be by both evapotranspiration and by streamflow. Given this it is perfectly possible for increasing precipitation, if accompanied by rising temperatures, to produce a situation where there is less available moisture for plant growth or river flow than existed previously. Similarly, lower precipitation, together with falling temperatures could, theoretically, produce increased runoff.

Research on the Colorado River in the USA suggests that climatic changes may produce annual runoff fluctuations of up to 30 per cent (Nash and Gleick, 1991). From the hydrologic model which was used it was concluded that precipitation changes appeared to be more significant than temperature changes in affecting runoff patterns. In particular, it was felt that seasonal changes in runoff were likely to be greater than annual variations and so might be more useful indicators of climatic change.

What seems obvious is that environmental managers in drylands must be prepared to accept changing physical environmental conditions as the norm. Practices which rely over much on so-called 'average' or 'normal' conditions must be discarded. It is environmental trends which must be reacted to, with the recognition that some of these 'trends' might only go on for two or three years before a reversal sets in.

Throughout the drylands it is water shortages which pose the greatest problems to development policies (Agnew and Anderson, 1992; Biswas, 1992; Falkenmark, 1989b; Shata, 1992). Currently irrigation is by far the greatest user of water, but the plant biomass it produces is of low value in monetary terms. With the growth of urban areas there is much more competition for the available resources, and the new uses, for domestic or industrial supply, are able to pay much higher prices for each unit of water. In capitalist or market-orientated societies it seems inevitable that increasing quantities of water will be diverted to the new higher-value uses.

There is little doubt that many of the world's drylands provide almost ideal conditions for crop growth in terms of temperatures and levels of solar radiation. What is lacking, though, are adequate water supplies. However, as a result of engineering advances it is now feasible to deliver water to many arid zones from adjacent wetter areas. This delivery can be achieved by rivers, by groundwater systems or by aqueducts or pipelines. The more a system is 'engineered' the more costly it is likely to become and, theoretically, the more expensive the delivered water will be. Ultimately, however, it will be the actual or charged costs of water which will determine the uses to which the water can be put. Major irrigation projects have been a feature of the twentieth century with perhaps the United States Bureau of Reclamation being most closely associated with this type of development. In the early years of the century the opening up of the south-west of the USA by irrigation was an objective of the United States government. This was achieved by subsidising many aspects of the construction process. The most important method was the very low interest rates charged on infrastructure provision. Not surprisingly many other countries followed this strategy in their own planning policies. Subsidies are not a problem if a specific aim is being targetted nor if the agency providing the subsidy is extremely affluent. However, difficulties do arise when subsidies are given to irrigation projects which are constructed with the objective of being financially solvent in terms of their construction and management costs.

The chief difficulty with subsidies is that they distort the operation of the market. This means that the water appears to be cheaper than it really is and so tends to be utilised for purposes for which it is, in reality, too expensive. For example, if true water costs are calculated it would be discovered in many countries that certain crops could not be grown with current water utilisation rates. When adequate water resources are available the subsidised use of water and, therefore, almost by definition the inefficient use of water is something which can be accepted. Problems occur, though, when water shortages develop and there are increased demands on the available water resources. In such circumstances the ideal solution is to allow market forces to operate and water will be shifted from less valuable to more valuable uses. Unfortunately when concealed subsidies are introduced for irrigation agriculture the market mechanism is distorted and so will not be able to function properly. Under such conditions water will continue to be used for irrigation even when it could be employed more profitably elsewhere. However, even with a subsidised system there does come a time when the advantages of using water for non-agricultural uses become so strong that at least some of the irrigated land is abandoned. In the south-west of the USA, in states like Arizona, water supply organisations have for many years been purchasing water rights used primarily for irrigation so that they can be utilised for public supply purposes (Chapter 3). On the High Plains of Texas a similar mechanism has also come into operation. Here, however, it is the increasing cost of pumping the groundwater which has convinced the farmers that they will have to survive in future by using less water for their crop production (Chapter 8).

This leads on to what the future of irrigated agriculture is likely to be in drylands. In the Western nations it seems unlikely that many new major irrigation projects will be embarked upon for economic reasons as the financial returns of existing schemes have been extremely disappointing in recent years. Some might still be constructed, however, as an element of land settlement policy when a particular strategy is pursued at whatever the cost, or as a symbol of material advancement. There still remains the question of what to do with existing projects, which are often developing ever-greater management problems as the result of over-watering and the build-up of soil salinity. Such problems are particularly severe in the Murray Basin of Australia (Collett, 1978; Chambers *et al.*, 1992; Simpson and Herczeg, 1991). Even though operational costs are sometimes greater than returns, many governments seem content to continue to subsidise such projects rather than to close them down completely.

In developing countries a more complex situation prevails. Here large irrigation networks have been constructed as prestige projects with subsidies at many different levels. All too often political considerations make it impossible to remove these subsidies and wasteful water use continues unchecked. Israel provides an interesting example (Chapter 10) of a country which has heavily subsidised its irrigated agriculture for political and strategic reasons, but which has eventually had to accept that such a policy could not be continued in view of demands from other sectors of the economy. Although to date relatively little water has been switched out of agriculture to industrial/urban use there seems little doubt that this will occur on a growing scale in the future. It is now finally recognised by the Israeli government that irrigated agriculture is capable of producing only low economic returns for each cubic metre of water used. In contrast, industrial use of water can produce very high returns. Bearing the above in mind it would seem that the twentieth century trend of ever-increasing areas of irrigated agriculture in drylands must be coming to an end.

A factor which will help to prolong the use of irrigated agriculture is the more efficient use of water employing sophisticated water application techniques such as sprinkler, trickle and drip methods (Nakayama and Bucks, 1986; Rydzewski, 1987). Israel itself has already reduced its water consumption per hectare by about 30 per cent since the state was first established in 1948. This has undoubtedly delayed the decline of irrigated agriculture which in terms of total water use was still growing in Israel in the late 1980s. What many of the dryland nations of the world must now begin to plan for is, at best, a maintenance of irrigated areas at current levels or more likely a significant contraction as demands for water elsewhere in the economy increase. Water will become too valuable to be used for what is essentially low value food or fibre production. However, with some of the poorest dryland countries there is still little opportunity for industrial development and so here the competition for water from other users will be less.

A problem currently being addressed by many dryland countries, and especially those from the Third World, concerns the question of food provision. Most of the world's cereal varieties originated in dryland areas and so production conditions for these crops are near ideal in such regions. A feature of subsistence agriculture is self-sufficiency in staple products. This has meant that until the current century and, indeed in many cases until the time of the Second World War, most developing countries have been able to provide for their basic food needs. Since about 1900 the populations of these countries have begun to expand rapidly largely as a result of increased hygiene and elementary medical care for young children. Associated with this population growth have been advances in agriculture which have meant that crop yields have increased. These increases, together with often considerable expansion of the actual crop areas, have ensured that growing food production has for many years been able to keep up with the demands of an expanding population.

In the Middle East, the ability of indigenous agricultural production to feed local populations seems to have ended in many countries during the late 1960s and 1970s (Beaumont and McLachlan, 1985; Richards, 1986). By this time there was little new land of satisfactory quality which could be opened up to agriculture without enormous investment and the easy productivity gains of extra fertiliser use and seed selection had already been obtained. What happened, therefore, was that many countries developed rapidly increasing food deficits in cereals. This led to the importation of crops like wheat, particularly in years when domestic production was curtailed by drought or other circumstances. As populations continued to grow so did wheat imports so that by the late 1980s the vast majority of the countries of the Middle East were now totally dependent on these external food sources (Beaumont, 1989). In many countries imports have now grown to exceed domestic production. It seems inevitable that this trend will continue in the future. However, an extra factor which needs considering here is the effect which climatic change may have on agricultural production (Parry, 1990). In some cases this is likely to reduce local food availability still further.

When discussing wheat production and imports in the Middle East it is necessary to comment on the situation in Saudi Arabia. In the late 1970s Saudi Arabia had a total wheat production of only 100,000 tonnes and was importing substantial quantities of wheat for domestic consumption. Then for strategic reasons the government of Saudi Arabia decided that it wished to become self-sufficient in wheat supply. To achieve this end it decided to give large subsidies to farmers for wheat production. As a result the farmers were receiving during the 1980s prices for their wheat which were anything from four to six times higher than the prevailing price for wheat on the world market. The net effect of these subsidies was that production of wheat expanded tremendously.

In Saudi Arabia, owing to the very dry climate, this wheat had to be grown using irrigation with water supplied from deep aquifer systems. Therefore, wheat was really being classed as an expensive crop, because of the subsidies, and was using very valuable water resources for its production. In reality, though, wheat is a cheap bulk crop. The problem was made worse because the groundwater being used is 'fossil' water, which once consumed will not be replaced (Al-Saleh, 1992). In effect the government was 'mining' the water to meet a strategic objective which could have been achieved much more cheaply using alternative methods (Al-Ibrahim, 1991). For example, purchase of wheat on the world market and the construction of large local wheat storage facilities would probably have been a far more sensible solution to the problem perceived by the government. Despite these obvious shortcomings the government persisted with its policy and by the late 1980s wheat production was over two and a half million tonnes. Indeed, so great was the production that home demands were satisfied and wheat exports soon reached a value in excess of one million tonnes. The Saudi Arabian example is instructive in that it shows that crops can be produced in harsh dryland environments if sufficiently high subsidies are made available to farmers to overcome these difficulties. What there is no doubt at all about is that the Saudi Arabian scenario is not one which can be applied to poorer dryland nations. Inevitably, therefore, growing food dependency seems likely to become an increasingly common feature of dryland countries.

It is difficult to generalise as to what is the best food policy strategy for a country to follow as everything depends on individual circumstances. For example, with the oil rich countries of the Middle East various opportunities present themselves. Some could, like Saudi Arabia, pursue a policy of food self-sufficiency based on subsidised agriculture. Oil revenues do, however, allow a country to import food in large quantities and money can, therefore, be thought of as a substitute for water. Indeed, for many of the rich dryland countries of the world this is probably the most sensible approach, particularly as staple foods are quite cheap in world market terms. Other countries, like Jordan, which has no oil revenues, face difficult choices. Population numbers dictate that food imports have to be purchased as insufficient cereals can be produced locally. Lack of water resources means that this situation is unlikely to change significantly in the future. Therefore, increasing amounts of valuable foreign exchange will have to be expended to prevent people from starving.

Perhaps the greatest single factor affecting drylands is population growth (El-Badry, 1992). Even here, however, it is difficult to make general statements. For example, some countries, like Spain, have a current population of around 36 million and a projected population for 2020 AD of only 41 million. In contrast, other dryland states, such as Egypt, Iran, and Pakistan are likely to register population increases of in excess of 80 per cent between 1985 and 2020. It is population more than anything else which dictates the pressure being imposed on a dryland environment. When population rises the environment inevitably suffers (Myers, 1992). It is vital, therefore, that all dryland countries develop strategies to limit population growth. The Western countries have shown conclusively that the best way to discourage population growth is to increase the standards of living of their inhabitants. Unfortunately, this does not appear a feasible solution for many countries and, indeed, it would seem more likely that standards of living will actually fall. In turn this will probably increase the population still further and put an even greater strain on the available resources. Government induced population control policies have rarely succeeded and have often caused much resentment amongst the people who have been subjected to them. This was certainly the case in India during the 1980s. In China government policy is imposed through economic measures and these do seem to have had some effect, but how unpopular they have been is difficult to assess in a country with a totalitarian regime.

With most of the developing dryland countries populations do not seem likely to stabilise over at least the next two decades. Therefore, environmental managers and planners will have to accept that any idea of 'sustainable' development will have to be postponed (Falkenmark and Suprapto, 1992). Instead they will be forced to adopt what is effectively a 'damage limitation' scenario. Western nations will come to realise that in some dryland countries continued degradation of the environment is an inevitable consequence of current population numbers, irrespective of future increases. This is not a gospel of doom, but rather an acceptance of what is going to happen. Once it is acknowledged it does mean that a strategy can be developed to minimise the total impact and to ensure that maximum degradation is steered towards those areas of least economic value.

In Third World countries the rural areas are often suffering from depopulation as young people, who are no longer willing to tolerate the harsh conditions of subsistence agriculture, migrate to the urban areas. This migration is often greatest from the remote environments on the desert margins where the continual toil to keep the desert at bay is greatest. In the past these were some of the most conservative areas in a country, but with the advent of the radio and television the attractions of urban life have become apparent to everyone. In other rural areas opportunities for movement to urban areas may not exist. Under such conditions the population will continue to increase and marginal land which is unsuitable for cultivation will be brought into crop production.

Throughout the poorer countries of the arid zone land reform has also disrupted the traditional way of life. Most land reform programmes have been undertaken for political rather than for social reasons, and it is, therefore, not surprising that many of the poorest people have benefited little. All too often the chief reason for land reform has been the curtailing of the political power of the landlords rather than the improvement of the lot of the peasants (Hooghlund, 1982). Although some groups have undoubtedly benefited from certain of the programmes, for many there has been little change. Of particular concern is the growing indebtedness of rural families as they attempt to introduce Western farming technologies in the form of fertilisers and machinery. Under this type of farming the capital inputs are much greater than under the traditional systems, yet these countries lack the financial institutions which can lend money at reasonable rates of interest.

Equally repressive for the rural dweller is the strategy of rural despoilation employed by many governments. Although it is never portrayed as such, what in effect is happening is that governments strip assets from the countryside so that the urban areas can prosper. As the power base of most Third World governments is dependent on the urban populations, such governments go out of their way to ensure the maximum support of these people. One of the easiest ways of achieving this is through a cheap food policy. However, what started off as political expediency soon comes to have a major impact on the rural areas. In order to save money the prices of staple products, such as cereals, are held artificially low by government action and as a result it is the farmers who end up subsidising their urban brethren. The net effect is that the farmers do not have enough money to operate successfully and so are forced to borrow yet more. As a result rural indebtedness spirals upwards. More young people leave for the cities and the urban food problem gets worse. In addition, the rural dwellers all too often 'mine' available resources in an attempt to lower their costs. Under such conditions the pressure on the land seems set to increase to such an extent that famines will recur more and more frequently, until they are accepted by the people of the Western nations as almost inevitable consequences of life in such regions.

What has to be clearly accepted is that different parts of the environment have different 'recovery' times. For example, the overgrazing of a grassland ecosystem in one year may lead to large numbers of animal deaths, but is unlikely to have any significant effect on grass production in the long-term provided that seed has been formed and shed (Fuls, 1992; Livingstone, 1991). The point here is that with annual species each year is a new start. With destruction of shrub vegetation the situation is more serious and recovery times in such cases may be from five to in excess of thirty years (Polis, 1991). With fully grown tree ecosystems it is decades or perhaps even centuries before an equivalent vegetation pattern is restored. The monitoring of such changes will become increasingly important in the future and it is likely that satellite systems will play a vital role in this respect (Hanan *et al.*, 1991; Justice, 1986; Kaushalya, 1992; Verstraete and Pinty, 1991).

With soils the situation is more serious still, but even here generalisations are often difficult to make (Dregne, 1992; Higgitt, 1992; Loughran, 1989). With some soils, such as Terra Rossa developed on hard crystalline limestone, the soil profile is usually thin, with depths of less than 25 cm. Once the soil is removed down to bedrock it would probably take thousands of years for a comparable new profile to form. However, with soils developed on soft limestones, like rendzinas, different conditions prevail. With these soil depths can be considerable, perhaps as much as 100 cm. In such conditions the soil erosion problem is less severe as the removal of the top 25 cm will still leave weathered material which, with the addition of organic matter, can produce an acceptable growing medium. It is also possible to come across deeply weathered material or alluvial deposits which in certain areas may be many metres thick. Here it is possible for considerable erosion to occur without any significant deterioration in the quality of the growing medium as far as the plants are concerned. The only reservation which must be mentioned is that where thick unconsolidated material occurs the main type of soil erosion is often of the gully variety. This can quickly result in large areas being made uncultivable by gully formation and the production of a 'badlands' topography with steep slopes being the dominant land unit.

Too often in the past deterioration of vegetation through drought has been considered by some workers to be evidence of serious environmental degradation. In fact, although the visual situation looks bleak it is often the case that almost complete recovery of the vegetation cover can occur in the next wet period. With drylands research much more attention has to be focused on the state of the soil itself which is the growing medium for all vegetation. What needs to be observed are indicators of degradation including sheet and gully erosion. If such processes are rapidly occurring then the productivity of that particular environment is likely to deteriorate rapidly unless control measures are introduced. All too often though the process is one of increasing sheet erosion which it is very difficult to identify conclusively in the absence of detailed monitoring.

Throughout the developed world there is increasing concern being expressed about the long-term survival of arid land ecosystems (Epps and Crittendun, 1992; Evanari et al., 1986; Looney, 1991). Such environments are fragile and easily damaged by the ever-increasing pressures to which they are being subjected (Cole, 1987; Schmida et al., 1987; Steiner, 1990). In the USA an approach which is widely used is the establishment of a National Park, which sterilises land use and permits the so-called 'natural' environment to be preserved. Any tourist activity which is generated is steered to 'honey-pot' sites, such as particularly spectacular visual landscapes, where the overall impact of the activity can be contained (Figure 13.1). The effective destruction of the natural ecosystem in these small areas is permitted so that larger areas can be preserved. Many countries, particularly those with large areas of arid lands, often regard these areas as of little intrinsic value, and believe they can be used for anything. In the USA, USSR and Australia in the first half of the twentieth century, arid lands were used for the testing of atomic weapons in the atmosphere. Even today underground nuclear weapons testing still occurs in Nevada, and large areas of desert are used for bombing ranges and military manoeuvres (Figure 13.1).

One of the greatest pressures in drylands is likely to be caused by urbanisation. It is a process which affects both developed and developing countries alike and its cause is population growth and internal migration (Qutub, 1992). Indeed, the process may even speed up in certain parts of the world, including the 'Sun Belt' of the USA and the more populous countries of the Third World. Urbanisation does have the beneficial aspect of generating new economic activity, but at the same time it necessitates a larger burden on the food supply industry. It should not be forgotten that dryland environments do hold considerable attraction for human settlement, with high temperatures, low humidities and lots of sunshine. Nowhere is this better seen than in the 'Sun Belt' of the USA where desert oases such as Albuquerque, Tucson, Phoenix and Salt Lake City have shown rapid growth rates since 1950. In particular many retirement communities have been established in the drylands of the south-west of the USA where people can enjoy the outdoor life-style pioneered by the inhabitants of Los Angeles in the early years of the twentieth century. It would seem, therefore, that urbanisation in drylands is likely to be an ever more important phenomenon in the late twentieth and early twenty-first centuries in both the developing and developed countries. In the latter it is possible that large areas of drylands may be completely turned over to urban leisure and recreational activities.

As a phenomenon urbanisation dates mainly from the post-Second World War period. One of its most obvious effects concerns the sheer scale of the area covered by the urban system, which seems to grow outwards at ever-increasing rates. Los Angeles is perhaps the classic example, yet it is by no means unique. Mexico City, Ankara, Tehran, Cairo and Lima all show this similar pattern of rapid spatial growth in the developing world. This is a growth which continues for the most part in random fashion annexing high class agricultural land as easily as desert waste. As a result local agricultural production can often reveal

Figure 13.1 Recreational and military lands in the arid south-west
of the USA

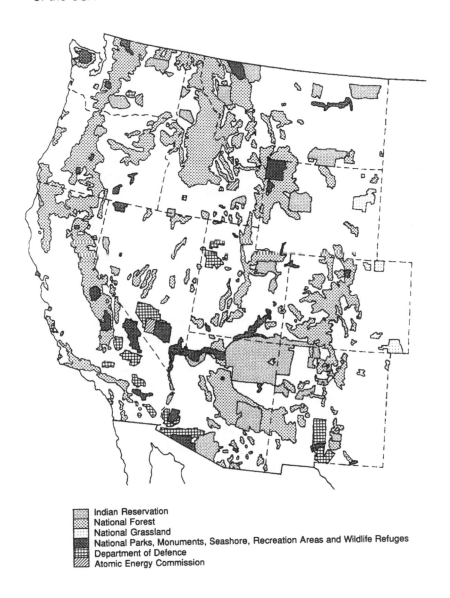

Indian Reservation
National Forest
National Grassland
National Parks, Monuments, Seashore, Recreation Areas and Wildlife Refuges
Department of Defence
Atomic Energy Commission

Source: Redrawn from *The National Atlas of the United States of America*, 1970, US Department of the Interior, Geological Survey, Washington D.C.

marked declines as high quality land disappears beneath luxury bungalow or squatter settlements (Ferguson-Bisson, 1992).

Our knowledge of the impact of dryland urban systems on the atmosphere above them has stemmed initially from detailed studies of the Los Angeles smog. There is now no doubt that this photochemical pollution is caused largely by the exhaust emissions from internal combustion engines, though undoubtedly exacerbated by local environmental conditions. What is encouraging though is that the citizens of Los Angeles have shown that if people care enough this form of pollution can be reduced by stringent legislation on motor vehicle emissions (Chapter 12). Smog is not, however, confined to the developed countries. In Mexico city smog is now much worse than in Los Angeles and the situation is likely to deteriorate further owing to the lack of proper controls on emissions from motor vehicles. However, smog is not the only atmospheric pollution in dryland urban areas. In Ankara, which is surrounded by low hills, air pollution is greatest in winter. Here it is a particulate and gaseous pollution caused by the burning of lignite. Its effects on people with lung problems can be considerable.

Of growing significance as standards of living rise is the disposal of solid waste which for a modern Western city of one million people can be around 2,000 tonnes per day. By far the commonest means of disposal is the landfill site, but it has become increasingly recognised that this is ever more likely to cause groundwater pollution as the range of chemicals appearing in the wastes increases. Difficulties are already being experienced in many cities in so far as most obvious holes in the ground are already filled in. This necessitates movement of waste over ever greater distances to new landfill sites. Certain Western cities have tried to overcome some of these difficulties by incineration of combustible solid waste. Unfortunately this can often add considerably to local air pollution and energy costs are high.

As cities grow in dryland environments the continued supply of water for domestic and industrial use becomes increasingly difficult (Bell, 1988; McGarry, 1991; Shata, 1992). Detailed examples of this have already been cited for Tucson, Arizona (Chapter 3) and Los Angeles (Chapter 12). Around 600,000 cubic metres per day are needed for every one million inhabitants in an advanced Western city. As these cities grow the meagre local water resources soon become inadequate to supply the needs. This means that water has to be brought from elsewhere, often over long distances, by pipeline or aqueduct system. Istanbul and Tehran already show similar trends obtaining water from sources many tens of kilometres away. Such systems are expensive and also mean that the water cannot be used for other purposes. With Tucson and Los Angeles irrigated agriculture has already declined in the surrounding regions. In the case of Los Angeles all of the water from the Owens Valley is now sent to be used for urban supply. In Israel a transition phase is just being entered upon as cities like Tel Aviv grow and demand more water. In these cities it seems likely that some form of demand management, perhaps based on the price mechanism, will have to be introduced in an effort to slow the rate of increase in demand (Wichelns and Cone, 1992).

Certain dryland cities, such as those in the Gulf, including Kuwait, Bahrain and Doha, have been forced to opt for a policy of water supply based on desalinated seawater (Chapter 9). For them access to other land based water sources is either restricted or impracticable. Today the technology of desalination is well tried and reliable. Flash distillation continues to be the dominant method, but reverse osmosis is becoming more important as membrane technology improves (Khan, 1986). The basic problem with the desalination process is the high cost of energy used to process the water. Although it is possible to recover waste heat from electricity generating plant the overall costs of water production, including maintenance and plant replacement, are very high. Inevitably this means that the water produced can only be utilised for high cost purposes such as drinking water supply. Desalination of sea water is certainly not a solution to all the problems of water supply in dryland cities and even the oil rich states are beginning to question the long-term viability of such operations.

Waste water disposal is itself a difficult issue (Biswas and Arar, 1988; McGarry, 1991; Shelef, 1991). Until the 1960s many dryland cities in developing countries did not have efficient sewage systems and wastes were dumped into the nearest water course or allowed to percolate into the ground. Apart from the very real disease potential pollution was largely of the organic variety which could be broken down by naturally occurring organisms. Oxygen depletion was locally a problem, but little lasting harm appears to have been done to the environment by these methods of disposal. However, as cities have grown the range of pollutants has increased as well (Akhter, 1990). Lead from petrol, oil and rubber

467

from cars, as well as a wide range of new chemicals used by homes and industries as standards of living have increased. This has meant that the waste water has changed from a potential asset to a real danger. Many countries, including the USA and Israel, have spent considerable sums in trying to devise systems by which urban waste waters can be recovered for either domestic supply or more likely for irrigation usage (Sheikh, 1991; Shuval, 1987). To date all of these attempts have met with considerable problems.

Attitudes to land in arid areas in the more developed countries reveal tremendous variations, dependent largely on government policies. In the past it has not been unusual for governments to use arid land settlement as an arm of policy and to heavily subsidise such operations. In Israel the settlement of dry rural areas was for strategic reasons and the government, through its agricultural policies, provided substantial financial incentives to ensure success. The aim in Australia has always been to settle the arid interior, and this has been partially achieved by subsidies for the provision and operation of railways and by special tariffs for the movement of wheat. In the driest areas the pastoralists have enjoyed cheap land prices and generous disaster relief programmes when environmental conditions have become particularly severe. Finally, in the USA the irrigation movement of the early decades of the twentieth century, undertaken by the Bureau of Reclamation, has had all its projects subsidised by low interest rates.

With all these examples it is the policy objective, in nearly all cases the settlement of arid lands, which is much more important than any economic returns which can be generated. The cost of achieving a particular objective is often very high. Such policies can indeed only be pursued when a country's dependence on its arid lands for survival is limited. In other words the higher productivity of the more humid lands is used to subsidise the arid lands in what is in effect a transfer of resources.

In many arid lands a 'mining' approach is applied to the resources available. Such a policy can be generated by indigenous governments or, more likely, by a multinational corporation. Oil in the Middle East, iron ore and uranium in Australia, and gold and diamonds in South Africa, provide examples of this type of activity. They all lead to 'boom and bust' economies, whereby prosperity is generated for a few years or a few decades, before resource exploitation ceases, usually for economic reasons. It is possible during this period of boom, however, that sufficient capital is generated to provide self-sustaining economic activity within the country as a whole. Certainly, in the case of South Africa in the nineteenth century the exploitation of gold and diamonds proved an invaluable spur to economic growth. Similarly in the Middle East today the revenues from oil are so large that sufficient economic activity may be generated to sustain the countries when the oil begins to run out (Chapter 9).

Concern for specific issues about the arid zone is often expressed by international organisations. One such is UNEP (United Nations Environment Programme) which is involved with a programme to combat desertification. The net result of this approach by international agencies is to focus attention on very poor countries, such as those of the Sahel, which are unable to do much about the droughts they experience owing to the very severe population pressures they are facing (Glantz and Katz, 1985) (Figure 13.2). Aid assistance may be made available to remedy specific problems, but it stands little chance of success as the basic nature of the problem remains (Warren and Khogali, 1992). The result is to produce a growing number of people who can only survive by the provision of international food aid.

In 1992 the national leaders of the world came together for the Conference on the Environment in Rio de Janeiro, Brazil. It was a Conference on which high hopes were placed, but like so many international gatherings the achievements were rather modest. Certainly as far as the world's drylands are concerned the impact of the Conference would appear to be minimal. The agreement to attempt to hold carbon dioxide emissions by the year 2000 to 1990 levels might, if actually implemented, help to slow down the global warming process (Jones and Henderson-Sellars, 1990; Woodwell and Ramakrishna, 1989). However, little was achieved with respect to the problems of land degradation and population growth.

A key feature of the twentieth century, and particularly of the period since 1945, is the way in which individual decision-makers are influenced or even controlled by government or international agency policies. Previously, individual farmers following a subsistence life-style reacted to environmental forces and little else. They were free to make any decision within the framework of the accumulated wisdom of the area in which they operated, and this affected such things as the crops grown and the times of planting and harvesting. All this has changed with the growth of market economies and the penetration of their influence into even the

Figure 13.2 Rainfall for Gao, Mali

April–October rainfall (mm)

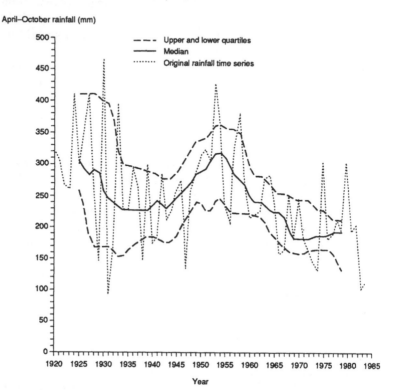

Note: Time series of 11 year moving quartiles of wet season (April – October) rainfall. The top line is the upper quartile (i.e. the 75th percentile) for the given 11 year period. The middle line is the median (50th percentile) and the bottom line is the lower quartile (i.e. 25th percentile). The dotted line indicates the original rainfall time series for the period 1920–1984 from which the moving quartiles were derived.

Source: Redrawn from Glantz M.H. & Katz R.W., 1985, 'Drought as a constraint to development in Sub-Saharan Africa', *Ambio*, v. 14, pp. 334–9. Reproduced by permission of the Royal Swedish Academy of Sciences.

remotest dryland regions (Yesilada *et al.*, 1987). The result is that today farmers' decisions about the crops to be grown can be affected by happenings many hundreds and even thousands of kilometres away from where they live and work. Government food policies influence the prices which can be obtained for cereals and even determine which crops can be successfully cultivated. Similarly, subsidies provided by governments for water provision or for electricity for irrigation pumping can give rise to a type of land use which may not be feasible under different conditions.

Given these changes the governments of today have a greater responsibility for the successful management of their drylands than was the case in the past. As a result it is crucial that ministers are advised by managers with a broad knowledge of environmental issues and a holistic approach to problem-solving. All too often in the past managers have been specialists in one discipline and have lacked the broad overview needed for successful environmental planning. In such circumstances it is perhaps not surprising that so many development projects initiated in drylands have not lived up to their expectations. During the 1950s and 1960s projects which stressed technological development were pre-eminent. Many of these, such as large dam projects, were constructed with an inadequate knowledge of the local environment. As a result the dams were often rapidly filled with silt or else did not provide as much water as had been predicted (Cummings, 1990). During the 1970s there was a movement away from technological approaches to those which stressed the importance of the social aspects in development. Unfortunately these were no more successful than their earlier counterparts, as they tended to ignore environmental effects completely. What is needed for the future is an integrated approach which considers social effects within an environmental context. As yet, though, there are few managers with the necessary skills to practise such a programme.

What is quite obvious from the foregoing survey is that there is no single agreed strategy for how best to develop drylands. Everything depends on what is trying to be achieved. Here there are so many different objectives, many of which may be incompatible with each other, including preserving the arid zone ecosystem; maintaining optimum biological productivity through agriculture; the settlement of arid lands; obtaining maximum economic returns (i.e. 'mining' resources); urbanisation; and recreational activity. Only governments can decide what they wish to achieve.

Much has been written on the Nile valley and the significance of this region as an area which has been peopled continuously for over five thousand years. While this is true it does not emphasise the unusual nature of the Nile environment with a river whose waters and silt used to provide new inputs of nutrients on a yearly basis. In effect what one had here was a self-sustaining system of which there are very few in nature. However, it is equally important to remember that at the beginning of the nineteenth century, when Egypt was still self-sufficient in staple foodstuffs produced by subsistence agriculture using basin irrigation, the population of the country was only about two million.

In Turkey, Iran and Syria, all countries with large areas of semi-arid lands in which rain-fed agriculture was possible, only very low populations were able to be supported prior to the end of the nineteenth century. For example, in Turkey the population was only 14 million in 1900, compared with approximately 57 million today (1992). The point being made here is that traditional societies, living at a subsistence level, whether by arable farming, pastoralism or hunter gathering, had over thousands of years stabilised at very low population numbers. Values such as these quoted above represent, therefore, the true 'sustainable' levels which can be achieved in drylands without serious deterioration of the environment. Even this, though, could be challenged, as the subsistence agriculturalists of the Middle East and the Mediter-ranean have, over some eight thousand years of occupancy, caused serious environmental degradation of vegetation and soil resources. This does not mean, of course, that drylands are unable to carry greater population numbers. They obviously can. How-ever, this can only be achieved by two conditions being met. The first is by making considerable fossil-fuel subsidies into the production system. The second is by 'mining' the environmental resources, such as soils, vegetation and water, over time spans of a few years to many decades.

What too many environmental managers seem to assume is that large population numbers can be maintained in drylands without any extra energy costs or environmental degradation occurring. It is forgotten that nature usually makes adjustments to excessive population numbers in animals and humans by causing the deaths of those which the environment cannot provide food for. In our modern society the death of human beings through starvation is quite rightly regarded as unacceptable. However, if these people are to be kept alive and to remain in those same dryland areas, then certain 'costs' will have to be paid. In the Mediterranean and the Middle East the 'cost' has been a severely degraded natural vegetation and the removal of soils over large areas. In Egypt a current population in excess of 50 million is supported by massive fertiliser inputs, huge investments in capital projects like the

Aswan High Dam and annual grain imports of more than 7 million tonnes (1989).

The future of the Sahel region, perhaps the largest area of continuous semi-arid grasslands in the world, looks bleak (Watts, 1989; Warren and Khogali, 1992). In recent decades drought has devastated the region. Rarely does the drought extend over the whole area, but what happens instead is that the drought affects different parts at different times (Binns and Mortimore, 1991; Tucker *et al.*, 1991). Over a long time scale, though, all areas are eventually affected. Given the high population numbers which now exist local agriculture is incapable of supplying the food needs. People, therefore, become dependent on food distributed by their governments or on food aid given directly by international agencies such as the United Nations or the Red Cross. This dependency culture which is being established does not bode well for the future. The longer it continues, the greater the population growth, and the more the local population numbers become out of touch with the carrying capacity of the local environment which has already been put under stress with drought and overuse of the biological resources.

It is always difficult to learn from the past, and yet it is essential if we are to avoid making the same mistakes as our forebears. History shows the crucial lesson that most drylands have only low and limited carrying capacities if long-term usage is planned. Theories of 'sustainable' development are often put forward with little understanding of the operation of the environment in which the development is to take place. As a result expectations are raised which cannot be attained in the long-term except by the mining of resources. Somehow the feeling has been generated that drylands are potentially rich regions with resources just waiting to be tapped. Experience, in fact, points in other directions. What would seem to be beyond dispute is that prior to the industrial revolution the carrying capacities of the world's drylands, as witnessed by human population numbers, were low.

Any future strategies which are adopted will have to recognise the naturally low carrying capacity of drylands; their susceptibility to human induced degradation and the ways in which they can be severely affected by relatively minor climatic peturbations. This is not to say, of course, that drylands cannot share in rapid economic development. California and Texas are examples of drylands with economic systems comparable with the most advanced in the world, though it must be admitted that they do form part of the giant economy of the USA. Their twentieth century development has been based largely on fossil-fuel subsidies until self-sustaining economic growth within an urban context took place. What is certain is that these spectacular growth patterns cannot be re-created in the dryland states of Africa, such as Chad, Niger and Mali where wealth creation is actually declining as the soil and vegetation resources continue to be degraded. Environmental managers, working closely with politicians, must decide the strate-gies for individual dryland regions and not be afraid to recognise that in certain cases accelerated land degradation may be a sacrifice which has to be paid in order that local populations can survive future drought or famine.

Bibliography

Abbas, J.A. and El-Oqlah, A.A. (1992) 'Distribution of communities of halophytic plants in Bahrain', *Journal of Arid Environments*, vol. 22, no. 3, pp. 205–18

Abbey, D. (1979) 'Energy production and water resources in the Colorado River basin', *Natural Resources Journal*, vol. 19, pp. 275–314

Abdullah, M.M. (1977) *A modern history of the United Arab Emirates*, Croom Helm, London

Aberbach, S.H. and Sellinger, A. (1967) 'Review of artificial groundwater recharge in the coastal plain of Israel', *Bulletin of the International Association of Scientific Hydrology*, vol. 12, pp. 65–77

Abraham, C.M. and Rosencranz A. (1986) 'An evaluation of pollution control legislation in India', *Columbia Journal of Environmental Law*, vol. 11, pp. 101–18

Adam, H. and Giliomee, H. (1979) *Ethnic power mobilized: can South Africa change?*, Yale University Press, New Haven

Adams, M.E. and Holt, J.M. (1985) 'The use of land and water in modern agriculture', in P. Beaumont and K.S. McLachlan (eds), *Agricultural development in the Middle East*, John Wiley, Chichester, pp. 63–83

Adams, R.M. (1965) *Land behind Baghdad*, University of Chicago Press, Chicago

—— (1972) 'Patterns of urbanisation in early Southern Mesopotamia', in P.J. Ucko, R. Tringham and G.W. Dimbleby (eds), *Man, settlement and urbanism*, Schenkman, Cambridge, Mass., pp. 735–49

—— (1978) 'Strategies of maximization, stability and resilience in Mesopotamian society, settlement and agriculture', *Proceedings of the American Philosophical Society*, vol. 122, pp. 329–35

Adams, W.M. and Carter, R.C. (1987) 'Small-scale irrigation in sub-Saharan Africa', *Progress in Physical Geography*, vol. 11, no. 1, pp. 1–27

Adar, E.M., Rosenthal, E., Issar, A.S. and Batelaan, O. (1992) 'Quantitative assessment of the flow pattern in the southern Arava valley (Israel) by environmental tracers and a mixing cell model', *Journal of Hydrology*, vol. 136, pp. 333–52

Addison, (1959), *Sun and shadow at Aswan*, Chapman and Hall, London

Adelman, I., Taft Morris, C., Fetini, H. and Golan-Hardy, E. (1992) 'Institutional change, economic development and the environment', *Ambio*, vol. 21 , no. 1, pp. 106–11

Adhana, A.H. (1991) 'Peasant response to famine in Ethiopia, 1975–1985', *Ambio*, vol. 20, no. 5, pp. 186–8

Agnew, C. and Anderson, E. (1992) *Water resources in the arid realm*, Routledge, London

Akhter, M.S. (1990) 'Trace metal analysis of sewage sludge and soils in Bahrain', *Water, Air and Soil Pollution*, vol. 51, nos 1/2, pp. 147–52

Al-Barawy, R. (1972) *Economic development in the United Arab Republic (Egypt)*, Anglo-Egyptian Bookshop, Cairo

Al-Feel, M.R. (1985) 'Kuwait: small-scale agriculture in an oil economy', in P. Beaumont and K.S. McLachlan (eds), *Agricultural development in the Middle East*, John Wiley, Chichester, pp. 279–88

Al-Ibrahim, A.A. (1991) 'Excessive use of groundwater resources in Saudi Arabia: impacts and policy options', *Ambio*, vol. 20, no. 1, pp. 34–7

Al-Mallakh, R. (1980) *Saudi Arabia: rush to development*, Croom Helm, London

Al-Saleh, M.A. (1992) 'Declining groundwater level of the Minjur Aquifer, Tebrak area, Saudi Arabia', *Geographical Journal*, vol. 158, no. 2, pp. 215–22

Alexander, G. and Williams, O.B. (eds) (1973) *The pastoral industries of Australia: practice and technology of sheep and cattle production*, Sydney University, Sydney

Allan, J.A. (1981) *The Sahara: ecological change and early economic history*, Menas Press, London

—— (1984) 'Oases', in J.L. Cloudsley-Thompson (ed.), *Sahara Desert*, Pergamon Press, Oxford, pp. 325–33

Arab Report and Record (1970a) 'Dam devastates Mediterranean fishing', *Arab Report and Record*, issue 23, 1–15 December, p. 677

—— (1970b) 'Aswan Dam increases Mediterranean salinity', *Arab Report and Record*, issue 11, 1–15 June, p. 346

—— (1970c) 'High Dam blamed for fishing decline', *Arab Report and Record*, issue 3, 1–14 February, p. 103

—— (1975) 'Iraq-political-foreign, and Syria-political-foreign', *Arab Report and Record*, issue 9, 1–15 May, pp. 276 and 289

Arab Republic of Egypt (1972) *The High Dam*, Ministry of Culture and Information, State Information Office, Cairo

Arnon, T. (1972) *Crop production in dry regions*, Vol. II — *Systematic treatment of the principal crops*, Leonard Hill, London

Attia, F.A.R. (1992) 'Economic evaluation of goundwater development plans in upper Egypt', *International Journal of Water Resources Development*, vol. 8, no. 2, pp. 135–44

Austin, T., Cross, R. and Heinen, P. (1982) *The California vehicle emission control program — past, present and future*, California Air Resources Board, Society of Automotive Engineers, Technical Paper Services, no. 811232, Los Angeles

Australia (1962–73) *Atlas of Australian resources*, Department of National Development, Canberra

Australian Info International (1989) 'Australian aboriginal culture', Australian Government Publishing Service, Canberra

Avakyan, A.B. and Polyushkin, A.A. (1989) 'Flood control experience in the USA', *Hydrotechnical Construction*, vol. 23, no. 1, pp. 53–8

Avigour, A., Magaritz, M. and Issar, A. (1992) 'Pleistocene paleoclimate of the arid region of Israel as recorded in calcite deposits along regional transverse faults and veins', *Quaternary Research*, vol. 37, no. 3, pp. 304–14

Avnimelech, Y., Dasberg, S., Harpaz, A. and Levin, I. (1978) 'Prevention of nitrate leakage from the Hula basin, Israel: a case study in watershed management', *Soil Science*, vol. 125, no. 4, pp. 233–9

476

Ayers, R.S. and Westcot, D.W. (1985) *Water quality for agriculture*, FAO Irrigation and Drainage Paper, no. 29, Rev. 1, Rome

Babikir, A.A.A. (1986) 'Some aspects of climate and economic activities in the Arab Gulf States', *GeoJournal*, vol. 13, no. 3, pp. 211–22

Bach, W. and Jain, A.K. (1991) 'Towards climate conventions scenario analysis for a climatic protection policy', *Ambio,* vol. 20, no. 7, pp. 322–9

Badeaym, R.A. (1974) *The water seekers*, Peregrine Smith, Santa Barbara

Bailey, H.P. (1966) *The climates of Southern California*, Natural History Guide, no. 17, University of California Press, Berkeley

Baker, B.W. and Wright, G.L. (1978) 'The Murray valley: its hydrologic regime and the effects of water development on the river', *Proceedings of the Royal Society of Victoria*, vol. 90, pp. 103–10

Baker, R.G., Maher, L.J., Chumbley, C.A. and Van Zant, K.L. (1992) 'Patterns of Holocene environmental change in the Midwestern United States', *Quaternary Research*, vol. 37, no. 3, pp. 379–89

Bakiewicz, W., Milne, D.M. and Noori, M. (1982) 'Hydrogeology of the Umm Er Radhuma aquifer, Saudi Arabia, with reference to fossil gradients', *Quarterly Journal of Engineering Geology (London)*, vol. 15, pp. 105–26

Barker, P. (1982) *Saudi Arabia: the development dilemma*, Economist Intelligence Unit, Special Report no. 16, London

Barnard, A. (1958) *The Australian wool market 1840–1900*, Melbourne University, Carlton

Barnard, A. (ed.) (1962) *The simple fleece: studies in the Australian wool industry*, Melbourne University, Partsville

Barnea, J. (1983) 'Small-scale mining in semi-arid areas', in E. Campos-Lopez and R.J. Anderson (eds) *Natural resources and development in arid regions*, Westview Press, Boulder, Colorado, pp. 159–63

Barnes, C.J., Jacobsen, G. and Smith, G.D. (1992) 'The origin of high-nitrate ground waters in the Australian arid zone', *Journal of Hydrology*, vol. 137, pp. 181–97

Barrow, C.J. (1991) *Land degradation*, Cambridge University Press

Barth, M. and Titus, J. (eds) (1984) *Greenhouse effect and sea level rise: a challenge for this generation*, Van Nostrand Reinhold, New York

Barth, H.K. and Quiel, F. (1986) 'Development and changes in the Eastern province of Saudi Arabia', *GeoJournal*, vol. 13, no. 3, pp. 251–9

Bastin, G., Hyde, K.W. and Foran, B.D. (1983) 'Range assessment' in G. Crook (ed.), *Man in the centre — Proceedings of a Symposium held at CSIRO, Alice Springs, 3–5 April 1979*, CSIRO, Melbourne, pp. 150–66

Bauer, F.H. (1983) 'The coming of European man', in G. Crook (ed.), *Man in the centre — Proceedings of a Symposium held at CSIRO, Alice Springs, 3–5 April 1979*, CSIRO, Melbourne, pp. 26–45

Baur, J.E. (1959) *Health seekers of southern California*, Huntingdon Library, San Marino

Beaumont, P. (1968) 'Qanats on the Varamin Plain, Iran', *Transactions of the Institute of British Geographers*, vol. 45, pp. 169–79

477

—— (1971) 'Qanat systems in Iran' *Bulletin of the International Association of Scientific Hydrology*, vol. 16, pp. 39–50

—— (1973) *River regimes in Iran*, Occasional Publications (New Series) no. 1, Department of Geography, University of Durham

—— (1977a) 'Water in Kuwait', *Geography*, vol. 62, no. 3, pp. 187–97

—— (1977b) Water and development in Saudi Arabia, *Geographical Journal*, vol. 143, pp. 42–60

—— (1978a) 'The Euphrates River — an international problem of water resources development', *Environmental Conservation*, vol. 5, pp. 35–43

—— (1978b) 'Man's impact on river systems — a world wide view', *Area*, vol. 10, pp. 38–41

—— (1981) 'Changing patterns of land and water use in the Isfahan oasis, Iran', in W. Mechelein (ed.), *Desertification in extremely arid environments*, Stuttgarter Geographische Studien, Band 95, pp. 29–63

—— (1985a) 'Irrigated agriculture and groundwater mining on the High Plains of Texas, USA', *Environmental Conservation*, vol. 12, no. 2, pp. 119–30

—— (1985b) 'Trends in Middle Eastern agriculture', in P. Beaumont and K.S. McLachlan (eds), *Agricultural development in the Middle East*, John Wiley, Chichester, pp. 305–22

—— (1988) 'Arid land management — a physical geographer's perspective', in *Proceedings of the Conference on Arid Lands — today and tomorrow*, Office of Arid Lands Studies, Tucson, Arizona, Westview, Belhaven, pp. 1089–102

—— (1989) 'Wheat production and the growing food crisis in the Middle East', *Food Policy*, vol. 14, no. 4, pp. 378–84

—— 1990) 'Water scarcity as a limiting factor to development in the Middle East', in *Obstaculos al desarrollo en el Mediterraneo Oriental (Constraints on development in the Eastern Mediterranean)*, A.V. Lorca and R. de Andrés (eds), I Seminario International sobre los paises de Oriente Medio, Instituto de Cooperacion con el Mundo Arabe, Madrid, pp. 91–9

—— (1992) 'Water — a resource under pressure', in *The Middle East and Europe — an integrated communities approach*, (ed.), G. Nonneman, Federal Trust for Education and Research, London, 1992, pp. 179–84

Beaumont, P. and Atkinson, K. (1969) 'Soil erosion and conservation in northern Jordan', *Journal of Soil and Water Conservation*, vol. 24, pp. 144–7

Beaumont, P. and McLachlan, K.S. (eds) (1985) *Agricultural development in the Middle East*, Wiley, Chichester

Beaumont, P., Bonnie, M. and McLacklan, K. (eds) (1989) *Qanat and Khattara: traditional water systems in the Middle East and North Africa*, Middle East and North African Studies Press, Wisbech

Beaumont, P. Blake, G.H. and Wagstaff, J.M. (1988) *The Middle East — a geographical study*, Fulton, London

Bedoian, W.H. (1978) 'Human use of the Pre-Saharan ecosystem and its impact on desertization', in N.L. Gonzalez (ed.), *Social and technological management in drylands*, Westview, Boulder, pp. 61–109

Bell, A. (1985–6) 'Fire v. shrubs in the semi-arid rangelands', *Ecos*, vol. 46, pp. 3–8

—— (1987) 'A satellite eye on soil erosion', *Ecos*, vol. 51, pp. 14–17

Bell, B. (1970) 'The oldest records of the Nile floods', *Geographical Journal*, vol. 136, pp. 569–73

Bell, F.C. (1988) 'The sharing of scarce water resources', *Geoforum*, vol. 19, no. 3, pp. 353–66

Bender, B. (1975) *Farming in Prehistory*, J. Baker, London

Bernstein, L. and Francois, L.E. (1973) 'Comparison of drip, furrow and sprinkler irrigation', *Soil Science*, vol. 115, no. 1, pp. 73–86

Bigger, R. (1959) *Flood control in Metropolitan Los Angeles*, University of California Publications in Political Science, vol. VI, University of California Press, Berkeley

Binns, J.A. (1990) 'Is desertification a myth?', *Geography*, vol. 75, pp. 106–13

Binns, J.A. and Mortimore, M.J. (1991) 'Ecology, time and development in Kano State, Nigeria', in Swindell, K., Baba, J.M. and Mortimore, J.M. (eds), *Inequality and development: case studies from the Third World*, Macmillan, London, pp. 359–80

Birks, J.S. (1981) 'The impact of economic development on pastoral nomadism in the Middle East: an inevitable decline?' in J.I. Clarke and H. Bowen-Jones (eds), *Change and development in the Middle East*, Methuen, London, pp. 82–93

Biswas, A.K. (1991) 'Water resources in the 21st century', *Water International*, vol. 16, no. 3, pp. 142–44

—— (1992) 'Water for Third World development: a perspective from the South', *International Journal of Water Resources Development*, vol. 8, no. 1, pp. 3–9

Biswas, A.K. and Arar, A. (eds) (1988) *Treatment and reuse of wastewater*, Butterworth, London

Biswas, A.K., Masakhalia, Y.F.O., Odego-Ogwal, L.A. and Palnagyo, E.P. (1987) 'Land use and farming systems in the Horn of Africa', *Land Use Policy*, vol. 4, pp. 419–43

Biswas, M.R. and Biswas, A.R. (eds) (1980) *Desertification*, Environmental Sciences and Applications, vol. 12, Pergamon Press, Oxford

Bittinger, M.W. and Green, E.B. (1980) *You never miss the water till . . . (The Ogallala story)*, Water Resources Publication, Littleton, Colorado

Blackburn, G. and McLeod, S. (1983) 'Salinity of atmospheric precipitation in the Murray-Darling drainage division, Australia', *Australian Journal of Soil Research*, vol. 21, pp. 411–34

Blackman, Jr, W.C., Rouse, J.V., Schillinger, G.R. and Schafer, Jr, W.H. (1973) 'Mineral pollution in the Colorado River basin', *Journal of the Water Pollution Control Federation*, vol. 45, pp. 1517–57

Blainey, G. (1981) *The rush that never ended — a history of Australian mining*, Melbourne University Press, Melbourne

Blake, G.H. (1972) 'Israel: immigration and dispersal of population', in J.J. Clarke and W.B. Fisher (eds), *Populations of the Middle East and North Africa — a geographical approach*, University of London Press, London, pp. 182–201

Blaney, H.F. and Criddle, W.D. (1950) *Determining water requirements in irrigated areas from climatological and irrigation data*, U.S. Department of Agriculture, Soil Conservation Technical Paper, 96

Blouet, B.W. and Lawson, M.P. (1975) *Images of the Plains: the role of human nature in settlement*, University of Nebraska Press, Lincoln, Nebraska

Blouet, B.W. and Luebke, F. (1979) *The Great Plains: Environment and culture*, University of Nebraska Press, Lincoln, Nebraska

Bocco, G. (1991) 'Traditional knowledge for soil conservation in central Mexico', *Journal of Soil and Water Conservation*, vol. 46, pp. 346–8.

Bojo, J.P. (1991) 'Economics of land degradation', *Ambio*, vol. 20, no. 2, pp. 75–9

Borchert, J.R. (1971) 'The Dust Bowl in the 1970s', *Annals of the Association of American Geographers*, vol. 61, pp. 1–22

Bowen-Jones, H. (1978) 'Oman's First Plan', in *Middle East Yearbook 1978*, IC Magazines Ltd, London, pp. 85–9

Bowonder, B. (1987) 'Environmental problems in developing countries', *Progress in Physical Geography*, vol. 11, no. 2, pp. 246–59

Braidwood, R.J. *et al.* (1960) *Prehistoric investigations in Iraqi Kurdistan*, Chicago

Brawer, M. (1968) 'The geographical background of the Jordan water dispute' in C.A. Fisher (ed.), *Essays in political geography*, Methuen, London, pp. 225–42

Bredehoeft, J. (1984) 'Physical limitations of water resources', in E.A. Englebert and A.F. Scheuring (eds), *Water scarcity: impacts on Western agriculture*, University of California Press, Berkeley, pp. 17–50

Brittain, R. (1986) 'Casa del Aqua: an experimental and demonstration urban residence', *Arid Lands Newsletter*, no. 24, pp. 2–14

Brookes, A. (1989) *Channelized rivers: perspectives for environmental management*, John Wiley, Chichester

Brunsden, D., Doornkamp, J.C. and Jones, D.K.C. (1979) 'The Bahrain Surface Materials Resources Survey and its application to planning', *Geographical Journal*, vol. 145, pp. 1–35

Budd, W.W. (1992) 'What capacity the land?', *Journal of Soil and Water Conservation*, vol. 47, no. 1, pp. 28–31

Bull, W.B. (1977) 'The alluvial fan environment', *Progress in Physcial Geography*, vol. 1, pp. 222–70

Burdon, D.J. (1973) *Groundwater resources of Saudi Arabia*, Groundwater resources in Arab Countries, ALESCO Science Monographs, no. 2

—— (1982) Hydrogeological conditions in the Middle East, *Quarterly Journal of Engineering Geology (London)*, vol. 15, pp. 71–82

Bureau of Agricultural Economics (1970) *The Australian beef cattle industry: an economic survey 1964–65 and 1966–67*, Bureau of Agricultural Economics, Canberra

Buru, M.M., Ghanem, S.M. and McLachlan, K.S. (1985) *Planning and development in modern Libya*, MENAS Press, Wisbech

Butzer, K.W. (1959) 'Environment and human ecology in Egypt during pre-dynastic and early dynastic times', *Bull. Soc. Geogr. Egypte.*, vol. 32

—— (1976) *Early hydaulic civilization in Egypt: a study in cultural ecology*, University of Chicago Press, Chicago

—— (1984) 'Long-term Nile flood variation and political discontinuities in Pharaonic Egypt', in J.D. Clark and S.A. Brandt (eds), *From Hunters to Farmers — the causes and consequences of food production in Africa*, University of California Press, Berkeley, pp. 102–12

Butzer, K.W. and Hansen, C.L. (1968) *Desert and river in Nubia*, University of Wisconsin Press, Madison

Caceres, L., Gruttner, V.E. and Contreras, R.N. (1992) 'Water recycling in arid regions: Chilean case', *Ambio*, vol. 21, no. 2, pp. 138–44

Calf, G.E., McDonald, P.S. and Jacobson, G. (1991) 'Recharge mechanism and groundwater age in the Ti-Tree Basin, Northern Territory', *Australian Journal of Earth Science*, vol. 38, pp. 299–306

Carneiro, R. and Hilse, D.F. (1966) 'On determining the probable rate of population growth during the Neolithic', *American Anthropologist*, vol. 68, pp. 177–81

Carter, H.B. (1964) *His Majesty's Spanish flocks: Sir Joseph Banks and the merinos of George III of England*, Angus and Robertson, Sydney

Cauvin, J. (1972) *Nouvelles fouilles à Tell Mureybet (Syrie): 1971–1972*, Rapport préliminaire AAAS XXII, pp. 105–15

—— (1973) 'Découverte sur l'Euphrate d'un village natoufien du IXe millénaire av. J.-C. à Mureybet (Syrie)', *Annales archéologiques de Syrie*, 276, pp. 1985–7

Cereceda, P., Schemenauer, R.S. and Suit, M. (1992) 'An alternative water supply for Chilean coastal desert villages', *International Journal of Water Resources Development*, vol. 8, no. 1, pp. 53–9

Chambers, L.A., Williams, B.G., Barnes, C.J. and Wasson, R.J. (1992) 'The effects of irrigation waste-water disposal in a former discharge zone of the Murray Basin, Australia', *Journal of Hydrology*, vol. 136, pp. 303–32

Chatters, J.C. and Hoover, K. A. (1992) 'Response of the Columbia River fluvial system to Holocene climatic change', *Quaternary Research*, vol. 37, no. 1, pp. 42–59

Chisholm, D.A. (1983) 'Rural European man as a resource manager', in G. Crook (ed.), *Man in the Centre — Proceedings of a Symposium held at CSIRO, Alice Springs, 3–5 April 1979*, CSIRO, Melbourne, pp. 189–98

Chitale, M.A. (1992) 'Development of India's river basins', *International Journal of Water Resources Development*, vol. 8, no. 1, pp. 30–84

Christian Science Monitor (1970) 'Sea of Galilee — in five years a pool of undrinkable water', *Christian Science Monitor*, 12 December

Christopher, A.J. (1976) 'The emergence of livestock regions in the Cape Colony 1855–1911', *South African Geographer*, vol. 5, pp. 310–20

—— (1982) *South Africa*, Longman, London

Clarke, J.I. (1981) 'Contemporary urban growth in the Middle East', in J.I. Clarke and H. Bowen-Jones (eds), *Change and development in the Middle East*, Methuen, London, pp. 154–70

—— (1985) 'Islamic populations: limited demographic transition', *Geography*, vol. 70, pp. 118–28

Cloudsley-Thompson, J.L. (1977) *Man and the biology of arid zones*, Arnold, London

Cloudsley-Thompson, J.L. (ed.) (1984) *Sahara Desert*, Pergamon Press, Oxford

Coffey, M. (1978) 'The dust storms on the Great Plains of the USA', *Natural History*, vol. 87, no. 2

Cohen, M.N. (1975) 'Population pressure and the origins of agriculture', in S. Folgar (ed.), *Population, ecology and social evolution*, Mouton, The Hague, pp. 79–121

Cole, J.P. (1984) *Geography of the Soviet Union*, Butterworths, London

Cole, K.L. and Webb, R.H. (1985) 'Late Holocene vegetation changes in Greenwater Valley, Mojave Desert, California', *Quaternary Research*, vol. 23, no. 2, pp. 227–35

Cole, M.M. (1987) 'The Savannas', *Progress in Physical Geography*, vol. 11, no. 3, pp. 334–55

Collett, K.O. (1978) 'The present salinity position in the River Murray basin', *Proceedings of the Royal Society of Victoria*, vol. 90, pp. 111–23

Collins, R.A. (1990) *The waters of the Nile: hydropolitics of the Jonglei canal*, Oxford University Press, Oxford

Condon, R.W., Newman, J.C. and Cunningham, G.M. (1969a) 'Soil erosion and pasture degeneration in central Australia. Part I. Soil erosion and degeneration of pastures and top feeds', *Journal of the Soil Conservation Service, NSW*, vol. 25, pp. 47–92

—— (1969b) 'Soil erosion and pasture degeneration in central Australia. Part II. Prevention and control of soil erosion and pasture degeneration', *Journal of the Soil Conservation Service, NSW*, vol. 25, pp. 161–82

—— (1969c) 'Soil erosion and pasture degeneration in central Australia. Part III. The assessment of grazing capacity', *Journal of the Soil Conservation Service, NSW*, vol. 25, pp. 225–50

—— (1969c) 'Soil erosion and pasture degeneration in central Australia. Part IV. Management of livestock', *Journal of the Soil Conservation Service, NSW*, vol. 25, pp. 295–321

Considine, M.L. (1984) 'The search for improved salt tolerance', *Ecos*, vol. 46, pp. 12–18

Cooke, R.U. (1984) *Geomorphological hazards in Los Angeles*, London Research Series in Geography, 7, Allen and Unwin, London

Cooke, R.U. and Reeves, R.W. (1976) *Arroyos and environmental change in the American south-west*, Clarendon Press, Oxford

Cooke, R.U. and Warren, A. (1973) *Geomorphology in deserts*, Batsford, London

Cooke, R.U., Brunsden, D., Doornkamp, J.C. and Jones, D.K.C. (1982) *Urban geomorphology in drylands*, Oxford University Press, Oxford

Cooke, R.U., Goudie, A.S. and Doornkamp, J.C. (1978) 'Middle East — review and bibliography of geomorphological contributions', *Quarterly Journal of Engineering Geology*, vol. 11, pp. 9–18

Cordes, R. and Scholz, F. (1980) *Bedouins, wealth and change*, United Nations University, Tokyo

Cottrell, A.G. (1980) *The Persian Gulf States: a general survey*, Johns Hopkins University Press, Baltimore

Crawford, C.S. and Gosz, J.R. (1982) 'Desert ecosystems: their resources in space and time', *Environmental Conservation*, vol. 9, no. 3, pp. 181–95

Crisp, M.D. and Lange, R.T. (1976) 'Age structure, distribution and survival under grazing of the arid zone shrub *Acacia burkittii*, *Oikos*, vol. 27, pp. 86–92

Crook, G. (ed.) (1983) *Man in the centre — Proceedings of a Symposium held at CSIRO, Alice Springs, 3–5 April 1979*, CSIRO, Melbourne

Crouchley, A.E. (1938) *Economic development of modern Egypt*, Longman, London

CSIRO (1960) *The Australian environment*, Melbourne University, Parkville

Cummings, B.J. (1990) *Dam the rivers, damn the people*, Earthscan, London

Cunningham, R.B. and Morton, R. (1983) 'A statistical method for the estimation of trend in salinity in the River Murray', *Australian Journal of Soil Research*, vol. 21, pp. 123–32

Dahl, G. (1991) 'The Beja of Sudan and the famine of 1984–1986', *Ambio*, vol. 20, no. 5, pp. 189–91

Dan, J. (1973) 'Arid zone soils', in B. Yaron, E. Danors and Y. Vaadia (eds), *Arid zone irrigation*, Chapman and Hall, London, pp. 11–28

Darrah, W.C. (1969) *Powell of the Colorado*, Princeton University Press, Princeton

Davis, O.K. (1992) 'Rapid climatic change in coastal Southern California inferred from pollen analysis of San Joaquin marsh', *Quaternary Research*, vol. 37, no. 1, pp. 89–100

Davison, B.R. (1966) *The Northern Myth: a study of the physical and economic limits to agricultural and pastoral development in tropical Australia*, Melbourne University, Carlton

Department of Science, Bureau of Meteorology (1975a) *Climatic atlas of Australia, Map set 1 — temperature*, Australian Government Publishing Service, Canberra

—— (1975b) *Climatic atlas of Australia, Map set 3 — evaporation*, Australian Government Publishing Service, Canberra

—— (1977) *Climatic atlas of Australia, Map set 5 — rainfall*, Australian Government Publishing Service, Canberra

—— (1978) *Climatic atlas of Australia, Map set 6 — relative humidity*, Australian Government Publishing Service, Canberra

Department of Water Resources, State of California (1980) *California flood management: an evaluation of flood damage prevention programs*, Department of Water Resources, State of California, Bulletin, 199

Dincer, T., Al-Mugrin, A. and Zimmerman, U. (1974) 'Study of the infiltration and recharge through sand dunes in arid zones with special reference to the stable isotopes and thermonuclear tritium', *Journal of Hydrology*, vol. 23, pp. 79–109

Dixon, C. (1990) *Rural development in the Third World*, Routledge, London and New York

Dixon, J.A., James, D.E. and Sherman, P.B. (1990) *Dryland management: economic case studies*, Earthscan Publications, London

Doornkamp, J.C., Brunsden, D. and Jones, D.K.C. (eds) (1980) *Geology, geomorphology and pedology of Bahrain*, Geobooks, Norwich

Dorn, R.I., Clarkson, P.B., Nobbs, M.F., Loendorf, L.L. and Whitley, D.S. (1992) 'New approach to the radio-carbon dating of rock varnish, with examples from drylands', *Annals of the Association of American Geographers*, vol. 82, no. 1, pp. 136–51

Dorten, W.S., Elbaz-Poulichet, F., Mart, L.R. and Martin, J-M. (1991) 'Reassessment of the river input of trace metals into the Mediterranean Sea', *Ambio*, vol. 20, no. 1, pp. 2–6

Doyle, D. (1991) 'Sustainable development: growth without losing ground', *Journal of Soil and Water Conservation*, vol. 46, pp. 8–13

Drakakis-Smith, D. (1987) *The Third World city*, Methuen, London

Dregne, H.E. (1976) *Soils of arid regions*, Developments in Soil Science, 6, Elsevier Scientific Publishing Company, Amsterdam

—— (1983) *Desertification of arid lands*, Harwood Academic Publishers, London

—— (1985) 'Aridity and land degradation', *Environment*, vol. 27, no. 8, pp. 16–33

—— (1992) 'Erosion and soil productivity in Asia', *Journal of Soil and Water Conservation*, vol. 47, no. 1, pp. 8–13

Dregne, H.H. and Tucker C.J. (1988) 'Desert encroachment', *Desertification Control Bulletin (UNEP, Nairobi)*, vol. 16, pp.16–19

Duncan, R. (1967) *The Northern Territory pastoral industry 1863–1910*, Melbourne University, Carlton

Dutton, R. (1985) 'Agricultural policy and development: Oman, Bahrain, Qatar, and the United Arab Emirates', in P. Beaumont and K.S. McLachlan (eds), *Agricultural development in the Middle East*, John Wiley, Chichester, pp. 227–40

Dworkin, J. (1974) *Global trends in natural disasters, 1947–1973*, Working Paper No. 26, Natural Hazards Research and Applications Information Center, Boulder, Colorado

Dzurik, A.A. (1990) *Water resources planning*, Rowman and Littlefield Publishers, Inc, Savage, Maryland

Ek, R. and Karadawi, A. (1991) 'Implications of refugee flows on political stability in the Sudan', *Ambio*, vol. 20, no. 5, pp. 196–203

El-Arifi, S.A. (1986) 'The nature of urbanisation in the Gulf countries', *GeoJournal*, vol. 13, no. 3, pp. 223–35

El-Badry, M.A. (1992) 'World population change: a long-range perspective', *Ambio*, vol. 21, no. 1, pp. 18–23

El-Sharief, A.S. (1986) 'Climatic constraints and potential corn production in Saudi Arabia', *GeoJournal*, vol. 13, no. 2, pp. 119–27

Elphick, R. and Giliomee, H. (eds) (1979) *The shaping of South African society 1652–1820*, Longman, London

Emiliani, C. and Shackleton, N.J. (1974) The Brunhes epoch: isotopic palaeotemperatures and geohydrology, *Science, NY*, vol. 183, pp. 511–14

Engelbert, E.A. and Scheuring, A.F. (eds) (1982) *Competition for California water: alternative resolutions*, University of California Press, Berkeley

—— (1984) *Water scarcity: impacts on Western agriculture*, University of California Press, Berkeley

English, P.W. (1968) 'The origin and spread of qanats in the Old World', *Proceedings of the American Philosophical Society*, vol. 112, pp. 170–81

Enzel, Y., Brown, W.J., Anderson, R.Y., McFadden, L.D. and Wells, S.G. (1992) 'Short-duration Holocene lakes in the Mojave River drainage basin, southern California', *Quaternary Research*, vol. 38, no. 1, pp. 60–73

Epps, W.R. and Crittenden, R. (1992) 'Appraisal of land degradation in Australia: rectification of agricultural land management', *Land Use Policy*, vol. 9, no. 3, pp. 199–208

Eriksson, E. and Khunakasem, V. (1969) 'Chloride concentration in groundwater, recharge rate and the rate of deposition of chloride in the Israel coastal plain', *Journal of Hydrology*, vol. 7, pp. 179–97

Evanari, M., Noy-Meir, I. and Goodall, D.W. (eds) (1986) *Ecosystems of the world: hot deserts and arid shrublands*, Elsevier, Oxford

Evanari, M., Shanan, L. and Tadmor, N. (1971) *The Negev — the challenge of a desert*, Harvard University Press, Cambridge, Mass

Evans, R.G., Walker, W.R. and Skogerboe, G.V. (1982) 'Optimal salinity control program for the upper Colorado River basin', *Water Supply and Management*, vol. 6, nos. 1–2, pp. 169–97

Evensen, W.E. (1985) 'Water for Salt Lake City', *Public Works*, vol. 116, pp. 102–4

Falkenmark, M. (1989a) 'The massive water scarcity now threatening Africa — why isn't it being addressed', *Ambio*, vol. 18, no. 2, pp. 112–18

—— (1989b) 'Middle East hydropolitics: water scarcity and conflicts in the Middle East', *Ambio*, vol. 18, no. 6, pp. 350–52

—— (1989c) 'Water scarcity and food production in Africa', in D. Pimentel and C.W. Hall (eds), *Food and natural resources*, Academic Press, pp. 163–89

Falkenmark, M. and Suprapto, R.A. (1992) 'Population-landscape interactions in development: a water perspective to environmental sustainability', *Ambio*, vol. 21, no. 1, pp. 31–6

FAO (1966) *Production Yearbook 1965*, FAO, Rome

—— (1986) *Production Yearbook 1985*, FAO, Rome

Ferguson-Bisson, D. (1992) 'Rational land management in the face of demographic pressure: obstacles and opportunities for rural men and women', *Ambio*, vol. 21, no. 2, pp. 90–4

Field, M. (1973) 'Developing the Nile', *World Crops*, vol. 25, no. 1, pp. 11–15

Flannery, K. (1973) 'The origins of agriculture', in B.J. Siegal *et al.* (eds), *Annual review of anthropology*, vol. 2, Annual Reviews Press, Palo Alto, California, pp. 270–310

Fleming, P.M. (1983) 'Hydrology of the Centre', in G. Crook (ed.), *Man in the centre — Proceedings of a Symposium held at CSIRO, Alice Springs, 3–5 April 1979*, CSIRO, Melbourne, pp. 70–84

Flug, M., Walker, W.R. and Skogerboe, G.V. (1979) 'Optimal water use and salinity control for energy — upper Colorado River basin', *Water Resources Bulletin*, vol. 15, pp. 964–73

—— (1982) 'Impact of energy development upon water and salinity in the upper Colorado River basin', *Water Supply and Management*, vol. 6, nos 1–2, pp. 199–220

Fogelson, R.M. (1967) *The fragmented metropolis, Los Angeles, 1850–1930*, Harvard University Press, Cambridge, Mass.

Foley, J.C. (1957) *Droughts in Australia: review of records from earliest years of settlement to 1955*, Bulletin, no. 43, Commonwealth Bureau of Meteorology, Melbourne

Fookes, P.G. and Collis, L. (1975) 'Problems in the Middle East', *Concrete*, vol. 9, no. 3, pp. 12–17

Fookes, P.G. and Higginbottom, I.E. (1980a) 'Some problems of construction aggregates in desert areas, with particular reference to the Arabian peninsula. 1. Occurrence and special characteristics', *Proceedings, Institution of Civil Engineers*, vol. 68, part 1, pp. 39–67

—— (1980b) 'Some problems of construction aggregates in desert areas, with particular reference to the Arabian peninsula. 2. Investigation, production and quality control', *Proceedings, Institution of Civil Engineers*, vol. 68, part 1, pp. 69–90

Forse, W. (1989), 'The myth of the marching desert', *New Scientist*, vol. 1650, pp. 31–2

Fradkin, P.L. (1981) *A river no more: the Colorado River and the West*, University of Arizona Press, Tucson

Frederick, K.D. and Hanson, J.C. (1982) *Water for Western agriculture*, Resources for the Future, Washington, DC

Fricke, C.A.P. and Pedersen, D.T. (1979) 'Groundwater resources management in Nebraska', *Groundwater Journal*, vol. 17, no. 6, pp. 544–9

Frihy, O.E. (1988) 'Nile delta shoreline changes: aerial photographic study of a 28 year period', *Journal of Coastal Research*, vol. 4, pp. 597–606

Fritts, H.C. and Gordon, G.A. (1980) 'Reconstructed annual precipitation for California', in M.K. Hughes *et al.*, *Climate from tree rings*, Cambridge University Press, Cambridge, pp. 185–91

Frood, A.M. (1967) 'The Aswan High Dam and the Egyptian economy', in R.W. Steel and R. Lawton (eds), *Liverpool essays in geography — a jubilee collection*, Longman, London, pp. 363–79

Fuls, E.R. (1992) 'Semi-arid and arid rangelands: a resource under siege due to patch-selective grazing', *Journal of Arid Environments*, vol. 22, no. 2, pp. 191–3

Furley, P.A. and Newey, W.N. (1983) *Geography of the biosphere*, Butterworths, London

Gale, S.J. (1992) 'Long-term landscape evolution in Australia', *Earth Surface Processes and Landforms*, vol. 17, pp. 323–43

Galnoor, I. (1980) 'Water policymaking in Israel', in H.I. Shuval (ed.), *Water quality management under conditions of scarcity — Israel as a case study*, Academic Press, New York, pp. 287–314

Gamble, C. and Soffer, O. (1989) *The world at 18000 B.P. Vol. 2, Low latitudes*, Unwin Hyman, London

Garbell, M.A. (1965) 'The Jordan Valley Plan', *Scientific American*, vol. 212, no. 3, pp. 23–31

Gardner, D. (1982) 'High Plains Dust Bowl furor swirls round heads of conservers', *Irrigation Age*, vol. 17, no. 3, pp. 6,7 and 12

Gavaghan, H. (1986) 'A saline solution to Israel's drought', *New Scientist*, 10 July, pp. 26–7

Gentilli, J. (1969) 'Evaluation and valorization of Australian landscapes', *Proceedings Ecological Society, Australia*, vol. 4, pp. 101–14

Georgakakos, A. P., Sadaka, T. and Barr, T.W. (1992) 'Tradeoffs in the regulation of the Equatorial lakes', *International Journal of Water Resources Development*, vol. 8, no. 1, pp. 10–16

Gerasimov, I.P. *et al.*, (1976) 'Large-scale research and engineering programs for the transformation of nature in the Soviet Union and the role of geographers in their implementation', *Soviet Geography — Review and Translation*, vol. 17, no. 4, pp. 235–45

Ghosh, S and Lahiri, D. (1992) 'Social conflicts and their remedies in the distribution of canal water: case study of the Damodar Valley Irrigation Project, India', *International Journal of Water Resources Development*, vol. 8, no. 1, pp. 65–72

Glantz, M.H. and Ausubel, J.H. (1984) 'The Ogallala aquifer and carbon dioxide: comparison and convergence', *Environmental Conservation*, vol. 11, no. 2, pp. 123–31

Glantz, M.H. and Katz, R.W. (1985) 'Drought as a constraint to development in Sub-Saharan Africa', *Ambio*, vol. 14, pp. 334–9

Gleick, P.H. (1987) 'The development and testing of a water balance model for climate impact assessment: modelling the Sacramento Basin', *Water Resources Research*, vol. 23, pp. 1049–61

—— (1988) 'The effects of future climatic changes on international water resources: the Colorado River, the United States and Mexico', *Policy Sciences*, vol. 21, pp. 23–39

—— (1989) 'Climate change and international politics: problems facing developing countries', *Ambio*, vol. 18, no. 6, pp. 333–9

Goldberg, D. (1967) 'Modern irrigation for increased agricultural production', in *Water for peace*, vol. 3. *Water supply technology*, International Conference on Water for Peace, US Government Printing Office, Washington, DC, pp. 395–406

Goldsmith, E. and Hilyard, N. (1984) *The social and environmental effects of large dams*, vol. 1, Wadebridge Ecological Centre, Camelford, Cornwall

Gondwe, Z.S. (1986) 'Agricultural policy in Tanzanian at the crossroads', *Land Use Policy*, vol. 3, pp. 31–6

Goodie, A.S., Warren, A., Jones, D.K.C. and Cooke, R.U. (1987) 'The character and possible origins of the aeolian sediments of the Wahiba Sand Sea, Oman', *Geographical Journal*, vol. 153, no. 2, pp. 231–56

Gordon, A.H. and Lockwood, J.G. (1970) 'Maximum one day falls of precipitation in Tehran', *Weather, London*, vol. 25, pp. 2–8

Gould, C.N. (1906) *The geology and water resources of the eastern portion of the Panhandle of Texas*, United States Geological Survey, Water Supply and Irrigation Paper, no. 154

Government of Niger (1980) 'Desertification in the Eghazer and Azawak region', in J.A. Mabbutt and C. Floret (eds), *Case studies on desertification*, UNESCO, Paris, pp. 115–46

Graf, W.L. (1988) *Fluvial processes in dryland rivers*, Springer-Verlag, London

Grainger, A. (1990) *The threatening desert: controlling desertification*, Earthscan Publications, London

Green, D.E. (1973) *Land of the underground rain*, University of Texas Press, Austin

Gregory, S. (1986) 'The climatology of drought', *Geography*, vol. 71, pp. 97–104

Gregory, S. (ed.) (1988) *Recent climatic change*, Bellhaven Press, London

Griffin, G.F. and Freidel, M.H. (1985) 'Discontinuous change in central Australia: some implications of major ecological events for land management', *Journal of Arid Environments*, vol. 9, pp. 63–80

Griffiths, I.L. and Binns, J.A. (1988) 'Hunger, help and hypocrisy: crisis and response to crisis in Africa', *Geography*, vol. 73, no.1, pp. 48–54

Grove, A.T. (1973) 'Desertification in the African environment', in D. Dalby and R.J. Harrison-Church (eds), *Report on the 1973 Symposium, Drought in Africa*, School of Oriental and African Studies, University of London

Guelke, L. (1976) 'Frontier settlement in early Dutch South Africa', *Annals of the Association of American Geographers*, vol. 66, pp. 25–42

Hadley, N.F. and Szarek, S.R. (1981) 'Productivity in desert ecosystems', *BioScience*, vol. 31, pp. 747–52

Haktanir, T. (1992) 'Comparison of various flood frequency distributions using annual flood peaks data of rivers in Anatolia', *Journal of Hydrology*, vol. 136, pp. 1–31

Halim, Y., Guergues, S.K. and Saleh, H.H. (1967) 'Hydrographic conditions and plankton in the south east Mediterranean during the last normal Nile flood (1964)', *Internationale Revue der Gesamten Hydrobiologie*, Band 52, Heft 3, Berlin, pp. 401–25

Halliday, F. (1979) *Iran: dictatorship and development*, Penguin, Harmondsworth

Hamdan, G. (1961) 'Evolution of irrigation agriculture in Egypt', in *History of land use in arid regions*, Arid Zone Research, vol. 17, UNESCO, New York, pp. 119–42

Hammad, H.Y. (1970) *Ground water potentialities in the African Sahara and the Nile valley*, Beirut Arab University, Beirut

Hammerton, D. (1972) 'The Nile River — a case history' in R.T. Oglesby, C.A. Carson and J.A. McCann (eds), *River ecology and man*, Academic Press, New York, pp. 171–214

Hanan, N.P., Prevost, Y., Diouf, A., and Diallo, O. (1991) 'Assessment of desertification around deep wells in the Sahel using satellite imagery', *Journal of Applied Ecology*, vol. 28, pp. 173–86

Harlan, J. (1978) 'Origins of cereal agriculture in the Old World', in C. Reed (ed.), *Origins of Agriculture*, Mouton, The Hague, pp. 357–83

Harpaz, Y. and Schwarz, J. (1967) 'Operating a limestone aquifer as a reservoir for a water supply system', *Bulletin of the International Association of Scientific Hydrology*, vol. 12, no. 1, pp. 78–90

Harris, C.D. (1940) *Salt Lake City — a regional capital*, University of Chicago Libraries, Chicago

Harris, M. (1980) *Culture, people, nature: an introduction to general anthropology*, Harper and Row, New York

Harris, R.C. and Guelke, L. (1977) 'Land and society in early Canada and South Africa', *Journal of Historical Geography*, vol. 3, pp. 135–53

Hassan, F. (1973) 'On the mechanisms of population growth during the Neolithic', *Current Anthropology*, vol. 14, pp. 535–40

Haws, F.W. (1973) *A study of water institutions in Utah and their influence on the planning, developing and managing of water resources*, Utah Water Research Laboratory, College of Engineering, Utah State University, Logan

Hays, J.R. (1948) *TVA on the Jordan*, Washington, DC

Heady, E.O. and Nicol, K.J. (1976) 'Models of agricultural water, land use and the environment', in R.M. Thrall *et al.* (eds), *Economic modeling for water policy evaluation*, North-Holland/TIMS Studies in the Management Sciences, vol. 3, North-Holland Publishing Company, Amsterdam, pp. 29–56

Heathcote, R.L. (1966) *Back of Bourke: a study of land appraisal and settlement in semi-arid Australia*, Melbourne University, Carlton

—— (1975) *Australia*, Longman, London

—— (1983) *The arid lands: their use and abuse*, Longman, London

Hefny, K., Farid, M.S. and Hussein, M. (1992) 'Groundwater assessment in Egypt', *International Journal of Water Resources Development*, vol. 8, no. 2, pp. 126–33

Heier, K.W. (1990) 'Drinking water for Riyadh', *Pipes and Pipelines International*, vol. 35, no. 3, pp. 13–16

Hellden, U. (1988) 'Desertification monitoring: Is the desert encroaching?', *Desertification Control Bulletin (UNEP, Nairobi)*, vol. 17, pp. 8–12

—— (1991) 'Desertification — time for an assessment', *Ambio*, vol. 20, no. 8, pp. 372–83

Helms, C.M. (1981) *The cohesion of Saudi Arabia*, Croom Helm, London

Hewes, L. (1965) 'Causes of wheat failure in the dry farming region, central Great Plains, 1939–1957', *Economic Geography*, vol. 41, pp. 313–50

Hidore, J.J. and Albokhair, Y. (1982) 'Sand encroachment in Al-Hasa Oasis, Saudi Arabia', *Geographical Review*, vol. 71, no. 1, pp. 350–6

Higgitt, D. (1992) 'Soil erosion and soil problems', *Progress in Physical Geography*, vol. 16, no. 2, pp. 230–8

Higgs, E.S. and Jarman, M.R. (1972) 'The origins of animal and plant husbandry', in E.S. Higgs (ed.), *Papers in economic prehistory*, Cambridge University Press, Cambridge, pp. 3–13

High Plain Associates (1982) *Six state High Plains Ogallala Aquifer Regional Resources Study* (A Report to the US Department of Commerce and the High Plains Study Council), Department of Commerce, Washington, DC

Hill, A.G. (1972) 'Population growth in the middle East since 1945 with special reference to the Arab countries of west Asia', in J.I. Clarke and W.B. Fisher (eds), *Change and development in the Middle East*, Methuen, London, pp. 130–53

Hill, A.R. (1987) 'Ecosystem stability', *Progress in Physical Geography*, vol. 11, no. 3, pp. 315–33

Hillel, D.I. (1982) *Negev: land, water and life in a desert environment*, Praeger, London

Himida, I.H. (1970) 'The Nubian Artesian Basin, its regional hydrogeological aspects and palaeohydrological reconstruction', *Journal of Hydrology, New Zealand*, vol. 9, pp. 89–116

Hjort, A.F., Ornas, and Salim, M. (1991) 'Research and development issues for African drylands', *Ambio*, vol. 20, no. 5, pp. 388–94

Hoffman, A. (1981) *Vision or villainy: origins of the Owens Valley — Los Angeles water controversy*, Texas A and M University Press, College Station, Texas

Holburt, M.B. (1982a) 'Colorado river water allocation', *Water Supply and Management*, vol. 6, nos. 1–2, pp. 63–73

—— (1982b) 'International problems on the Colorado River', *Water Supply and Management*, vol. 6, nos. 1–2, pp. 105–14

Hollingworth, C. (1971) 'Egypt's Aswan balance-sheet', *The Times*, London, 15 January

Hollis, G.E. (1978) 'The falling levels of the Caspian and Aral Seas', *Geographical Journal*, vol. 144, pp. 62–80

Holz, R.K. (1968) 'The Aswan High Dam', *Professional Geographer*, vol. 20, no. 4, pp. 230–7

Hooghlund, E.J. (1982) *Land and revolution in Iran 1960–1980*, Texas University Press, Austin, Texas

Hooper, A.D.L. (1983) 'The pastoral industry in central Australia', in G. Crook (ed.), *Man in the centre — Proceedings of a Symposium held at CSIRO, Alice Springs, 3–5 April 1979*, CSIRO, Melbourne, pp. 167–73

Hopkinson, N. (1990) *Desertification, debt and structural adjustment in sub-Saharan Africa*, Wilton Park Papers 18, London, HMSO

Houghton, R.A. and Skole, D. L. (1990) 'Changes in the global carbon cycle between 1700 and 1985', in B.L. Turner (ed.), *The earth transformed by Human Action*, Cambridge University Press, Cambridge

Howell, P.P. (ed.) (1953) 'The Equatorial Nile Project and its effects in the Sudan', *Geographical Journal*, vol. 119, pp. 33–48

Hulme, M. (1989) 'Is environmental degradation causing drought in the Sahel?', *Geography*, vol. 74, pp. 38–46

Hurst, H.E. (1952) *The Nile: a general account of the river and the utilisation of its waters*, Constable, London

Idris, K.H. (1986) 'The Gulf Co-operation Council: industrial co-operation in "A Gulf of Difference"', *GeoJournal*, vol. 13, no. 3, pp. 245–50

International Soil Reference and Information Centre (ISRIC) (1990) *World map of the status of human-induced soil degradation*, ISRIC, Wageningen, Netherlands

Iorns, W.V., Hembree, C.H. and Oakland, G.L. (1965) *Water resources of the upper Colorado River basis — technical report*, US Geological Survey, Professional Paper no. 441, Washington, DC

Isaac, E. (1970) *Geography of domestication*, Prentice-Hall, Englewood Cliffs, NJ

Ismael, J.S. (1982) *Kuwait*, Syracuse University Press, Syracuse

Israelsen, O.W., Maughan, J.H. and South, G.P. (1946) *Irrigation companies in Utah — their activities and needs*, Bulletin 322, Agricultural Experiment Station, Utah State Agricultural College, Logan, Utah

Issar, A., Bein, A. and Michaeli, A. (1972) 'On the ancient water of the

Upper Nubian Sandstone aquifer in central Sinai and southern Israel', *Journal of Hydrology*, vol. 17, pp. 353–74

Jackson, R.H. (1978) 'Mormon perception and settlement', *Annals of the Association of American Geographers*, vol. 68, no. 3, pp. 317–34

Jacobsen, G. (1988) 'The hydrology of Lake Amadeus, a groundwater-discharge playa in central Australia', *Bureau of Mineral Resources, Journal of Australian Geography and Geophysics*, vol. 10, pp. 301–8

Jacobsen, T. and Adams, R.M. (1958) 'Salt and silt in ancient Mesopotamian agriculture', *Science, NY*, vol. 128, pp. 1251–8

Jahns, R.E. (ed.) (1954) *Geology of southern California*, Californian Division of Mines, Sacramento, Bulletin, 70

Jakeman, A.J. and Dietrich, C.R. (1989) 'Solute transport in a stream-aquifer system. 2 — Application of model identification to the River Murray', *Water Resources Research*, vol. 25, pp. 2177–85

Johnson, D.E., Bormann, M.M. and Ben Ali, M.N. (1992) 'Evaluation of plant species for land restoration in central Tunisia', *Journal of Arid Environments*, vol. 22, pp. 305–22.

Johnson, D.L. (1969) *The nature of nomadism: a comparative study of pastoral migrations in southwestern Asia and northern Africa*, Department of Geography, University of Chicago Research Paper no. 118

Johnston, R.J. (1992) 'Laws, states and super states: international law and the environment', *Applied Geography*, vol. 12, no. 3, pp. 211–28

Jones, M.D.H. and Henderson-Sellars, A. (1990) 'History of the greenhouse effect', *Progress in Physical Geography*, vol. 14, no. 1, pp. 1–18

Jonglei Investigation Team (1954) *The Equatorial Nile Project and its effects on the Anglo-Egyptian Sudan*, Sudan Government, London

Justice, C.O. (ed.) (1986) 'Monitoring the grasslands of semi-arid Africa using NOAA-AVHRR data', *International Journal of Remote Sensing*, vol. 7, pp. 1383–622

Kahrl, W.L. *et al.*, (1979) *The California water atlas*, State of California, Sacramento

Karmon, Y. (1960) 'The drainage of the Huleh swamps', *Geographical Review*, vol. 50, pp. 169–93

Kashef, A.I. (1983) 'Salt water intrusion in the Nile delta', *Groundwater*, vol. 21, pp. 160–7

Kaushalya, R. (1992) 'Monitoring the impact of desertification in western Rajasthan using remote sensing', *Journal of Arid Environments*, vol. 22, no. 3, pp. 293–304

Keith, R.W. (1980) *A climatological air quality profile, California South Coast Air Basin*, South Coast Air Quality Management District, El Monte

Kellner, K. and Bosch, O.J.H. (1992) 'Influence of patch formation in determining the stocking rates for southern African grassland', *Journal of Arid Environments*, vol. 22, no. 1, pp. 99–105

Kelly, P.M. and Gribbin, J. (1979) 'Climatic impact of Soviet river diversions', *New Scientist*, no. 84, pp. 762–5

Kemp, D. (1990) *Global environmental issues: a climatological approach*, Routledge, London and New York

Kennedy, P. (1989) 'Monitoring the vegetation of Tunisian grazing lands using the normalized difference vegetation index', *Ambio*, vol. 18, no. 2, pp. 119–23

Kenyon, K.M. (1969–70) 'The origins of the Neolithic', *The Advancement of Science*, vol. 26, pp. 1–17

Kerley, G.I. and Erasmus, T. (1992) 'Small mammals in the semi-arid Karroo, South Africa: biomass and energy requirements', *Journal of Arid Environments*, vol. 22, no. 3, pp. 251–60

Khan, A.H. (1986) *Desalination processes and mulitstage flash distillation practice*, Elsevier, Oxford

Khogali, M.M. (1991) 'Famine, desertification and vulnerable populations: the case of Umm Ruwaba district, Kordofan region, Sudan', *Ambio*, vol. 20, no. 5, pp. 204–6

Khuri, F. (1982) *Tribe and state in Bahrain*, University of Chicago Press, Chicago

Kilani, A. (1988) 'Recharge of aquifers by wastewater in Jordan', in A.K. Biswas and A. Arar (eds), *Treatment and use of wastewater*, Butterworths, London, pp. 177–8

Kingdom of Saudi Arabia, Ministry of Planning (1981) *Third Development Plan 1400–1405 AH (1980–1985 AD)*, Riyadh

—— (1985) *Achievements of the Development Plans, 1390–1404 (1970–1984)*, Riyadh

Kirkbride, D. (1968) 'Beidha: early Neolithic village life south of the Dead Sea', *Antiquity*, vol. 42, pp. 263–74

Kiros, F.G. (1991) 'Economic consequences of drought, crop failure and famine in Ethiopia, 1973–1986', *Ambio*, vol. 20, no. 5, pp. 183–5

Kishk, M.A. (1986) 'Land degradation in the Nile valley', *Ambio*, vol. 15, pp. 226–30

Klein, C. (1965) *On the fluctuations of the Dead Sea since the beginning of the nineteenth century*, Hydrological Paper no. 7, Ministry of Agriculture, Water Commission, Hydrological Service, State of Israel, Jerusalem

Klemt, W.B. (1981) *Evaluating the ground-water resources of the High Plains in Texas — neutron-probe measurements of deep soil moisture as an indication of aquifer recharge rates*, Texas Department of Water Resources, Report LP-142, Austin, Texas

Klige, R.K. and Myagkov, M.S. (1992) 'Changes in the water regime of the Caspian Sea', *GeoJournal*, vol. 27, no. 3, pp. 299–307

Knapp, P.A. (1991) 'Long term soil and vegetation recovery in five semiarid Montana ghost towns', *Professional Geographer*, vol. 43, pp. 486–99

Knott, J.M. (1973) *Effects of urbanisation on sedimentation and floodflow in Colma Creek basin, California*, US Department of Interior, Geological Survey, Water Resources Division, Menlo Park, California, Open File Report

Koppen, W. (1931) *Die Klimate der Erde*, Water de Gruyter, Berlin

Kovda, V.A., Samoilova, E.M., Chorley, J.L. and Skujins, J.J. (1979) 'Soil processes in arid lands', in D.W. Goddall and R.A. Perry (eds), *Arid land ecosystems*, vol. 1, Cambridge University Press, Cambridge, pp. 439–70

Kraenzel, C.F. (1955) *The Great Plains in transition*, University of Oklahoma Press, Tulsa

Kunugi, T. (1992) 'The role of international institutions in promoting sustainable development', *Ambio*, vol. 21, no. 1, pp. 112–15

Lado, C. (1992) 'Problems of wildlife management and land use in Kenya', *Land Use Policy*, vol. 9, no. 3, pp. 169–84

Lambton, A.K.S. (1953) *Landlord and peasant in Persia*, Oxford University Press, London

Lancaster, N. (1989) *The Namib sand and sea dune forms, processes and sediments*, Balkema, Rotterdam and Brookfield

Laster, R.E. (1980) 'Legal aspects of water quality management in Israel', in H.I. Shuval (ed.), *Water quality management under conditions of scarcity — Israel as a case study*, Academic Press, New York, pp. 263–86

Law, Jr, J.R. and Hornsby, A.G. (1982) The Colorado River salinity problem, *Water Supply and Management*, vol. 6, nos. 1–2, pp. 87–103

Lawless, R.I. (1985) 'The agricultural sector in development policy', in P. Beaumont and K.S. McLachlan (eds), *Agricultural development in the Middle East*, John Wiley, London, pp. 107–22

Le Houerou, H.N. (1989) *The grazing land ecosystems of the African Sahel*, Springer-Verlag, Berlin

—— (1992) 'Outline of the biological history of the Sahara', *Journal of Arid Environments*, vol. 22, no. 1, pp. 3–30

Lea, J. (1988) *Tourism and development in the Third World*, Routledge, London

Levintanus, A. (1992) 'Saving the Aral Sea', *International Journal of Water Resources Development*, vol. 8, no. 1, pp. 60–4

Liphschitz, N. and Biger, G. (1992) 'Building in Israel throughout the ages — one cause for the destruction of the Cedar forests of the Near East', *GeoJournal*, vol. 27, no. 4, pp. 345–52

Little, T. (1965) *High Dam at Aswan*, Methuen, London

—— (1971) 'Why cry havoc at Aswan?', *Middle East International*, June, no. 3, pp. 5–6

Livingstone, I. (1991) 'Livestock management and "overgrazing" among pastoralists', *Ambio*, vol. 20, no. 2, pp. 80–5

Lockwood, J.G. (1980) 'Milankovitch theory and ice ages', *Progress in Physical Geography*, vol. 4, pp. 79–87

Looney, J.W. (1991) 'Land degradation in Australia: the search for a legal remedy', *Journal of Soil and Water Conservation*, vol. 46, pp. 256–9

Lorius, C., Jouzel, J., Ritz, C., Merlivat, L., Barkov, N.I., Korotkevich, Y.S. and Kotlyakov, V.M. (1985) 'A 150,000 year climatic record from Antarctic ice', *Nature*, vol. 316, pp. 591–6

Los Angeles County Flood Control District (1983) *1983 storm report*, Los Angeles County Flood Control District, Los Angeles

Los Angeles, Department of Public Works, Bureau of Engineering (1982) *Wastewater Facilities Plan — Final Report*, City of Los Angeles, Los Angeles

Loughran, R.J. (1989) 'The measurement of soil erosion', *Progress in Physical Geography*, vol. 13, no. 2, pp. 216–33

Lovett, J.V. (ed.) (1973) *The environmental, economic and social significance of drought*, Angus and Robertson, Sydney

Low, B.S. and Low, W.A. (1975) 'Feeding interactions of red kangeroos and cattle in an arid ecosystem', in R.L. Reed (ed.), *Proceedings of III World Conference on Animal Production*, Sydney University Press, Sydney, pp. 87-94

Lowdermilk, W.C. (1944) *Palestine — land of promise*, Victor Gollancz, London

Lvovitch, M.I. (1973) 'The global water balance', *United States, International Hydrological Decade Bulletin*, vol. 23, pp. 28-42

Maadhah, A.G., Hamid, S.H. and Amin, M.B. (1986) 'Overview of the petrochemical industry in Saudi Arabia', *Arab Journal for Science and Engineering*, vol. 10, pp. 327-38

Mabbutt, J.A. (1984) 'A new global assessment of the status and trends of desertification', *Environmental Conservation*, vol. 11, no. 2, pp. 100-13

Mabbutt, J.A. and Floret, C. (eds) (1980) *Case studies on desertification*, UNESCO, Paris

Mageed, Y.A. (1984) 'The Jonglei Canal: a conservation project of the Nile', *International Journal of Water Resources Development*, vol. 2, nos. 2-3, pp. 85-101

Main, C.T. (1952) *The unified development of the water resources of the Jordan valley basin*, Boston, Mass

Mainguet, M. (1991) *Desertification — natural background and human mismanagement*, Springer-Verlag, Berlin

Mann, D.E. (1975) 'Conflict and coalition: political variables underlying water resources development in the upper Colorado River basin', *Natural Resources Journal*, vol. 15, pp. 141-69

Marinov, U. and Ragen, N. (1974) 'Environmental management of the Sea of Galillee', *Environmental Management*, vol. 2, no. 6, pp. 487-90

Martin, C.W. (1992) 'Late Holocene alluvial chronology and climate change in the central Great Plains', *Quaternary Research*, vol. 37, no. 3, pp. 315-22

Martin, W.E., Ingram, H.M., Laney, N.K. and Griffin, A.H. (1984) *Saving water in a desert city*, Resources for the Future, Washington, DC

Martinez-Cob, A. and Cuenca, R.H. (1992) 'Influence of elevation on regional evapotranspiration using multivariate geostatistics for various climatic regimes in Oregon', *Journal of Hydrology*, vol. 136, pp. 353-80

Martonne, E. (1926) 'L'indice d'aridité', *Bull. Ass. Geogr. de France*, vol. 9

Mason, R.J. and Mattson, M.T. (1990) *Atlas of United States environmental issues*, Macmillan Publishing Company, New York

Mather, J.R. (1974) 'Climate, clothing and human comfort', in J.R. Mather (ed.), *Climatology — fundamentals and applications*, McGraw-Hill, New York, pp. 219-61

Mattson, J.O. and Rapp, A. (1991) 'The recent droughts in western Ethiopia and Sudan in a climatic context', *Ambio*, vol. 20, no. 5, pp. 172-5

Matz, R. (1967) 'Application of electrodialysis in Israel', in *Water for peace*, vol. 12. *Water supply technology*, International Conference on

Water for Peace, US Government Printing Office, Washington, DC pp. 43–57

McCabe, G.J. and Wolock, D.M. (1991) 'Detectability of the effects of a hypothetical temperature increase on the Thornthwaite Moisture Index', *Journal of Hydrology*, vol. 125, pp. 25–35

McCarthy, F.D. (1959) 'Habitat, economy and equipment of the Australian Aborigine', *Australian Journal of Science*, vol. 19,4a, pp. 88–96

McCaull, J. (1969) 'Conference on the ecological aspects of economic development', *Nature and Resources, UNESCO*, vol. 5, no. 2, pp. 5–12

McGarry, M.G. (1991) 'Water supply and sanitation in the 1990s', *Water International*, vol. 16, no. 3, pp. 153–60

McHale, J. (1972) *World facts and trends*, Macmillan, New York

McKee, R. (1974) *The last West: a history of the Great Plains of North America*, Crowell, New York

McWilliams, C. (1973) *Southern California County, an island on the land*, Peregrine Smith, Santa Barbara

Meggitt, M.J. (1962) *Desert people*, Angus and Robertson, Sydney

—— (1971) *Desert people — a study of the Walbiri aborigines of central Australia*, University of Chicago Press, Chicago

Meigs, P. (1953) 'World distribution of arid and semi-arid homoclimates', in *Arid Zone Hydrology*, UNESCO Arid Zone Research Series no. 1, pp. 203–9

Meinig, D.W. (1965) 'The Mormon culture region: strategies and patterns in the geography of the American West, 1847–1964', *Annals of the Association of American Geographers*, vol. 55, no. 2, pp. 191–220

Mellaart, J. (1967) *Catal Hüyük, a Neolithic town in Anatolia*, McGraw-Hill, New York

—— (1975) *The Neolithic of the Near East*, Thames and Hudson, London

Mercado, A. (1980) 'The coastal aquifer of Israel: some quality aspects of groundwater management', in H.I. Shuval (ed.), *Water quality management under conditions of scarcity — Israel as a case study*, Academic Press, New York, pp. 93–146

Messer, J. and Mosley, G. (eds) (1983) *'What future for Australia's arid lands?'*, Australian Conservation Foundation, Melbourne

Metcalf, S.E. (1987) 'Historical data and climatic change in Mexico — a review', *Geographical Journal*, vol. 153, no. 2, pp. 211–22

Meyer, M.C. (1984) *Water in the Hispanic Southwest — a social and legal history 1550–1850*, University of Arizona Press, Tucson

Micklin, P.P. (1977) 'NAWAPA and the two Siberian water-diversion proposals', *Soviet Geography — Review and Translation*, vol. 18, no. 2, pp. 81–99

—— (1978a) 'Irrigation development in the USSR during the 10th Five Year Plan 1976–1980', *Soviet Geography — Review and Translation*, vol. 15, no. 1, pp. 1–24

—— (1978b) 'Environmental factors in Soviet water transfer policy', *Environmental Management*, vol. 6, no. 6, pp. 567–80

—— (1978c) 'Large-scale interbasin river diversions in the USSR: implications for the future', in W.A. Douglas Jackson (ed.), *Soviet resource management and the environment*, AAASS, Columbus, Ohio, pp. 71–2

—— (1981) 'A preliminary systems analysis of impacts of proposed Soviet river diversions on Arctic sea ice', *EOS Transactions, American Geophysical Union*, vol. 62, pp. 489–93

—— (1983) 'Water diversion proposals for the European USSR: status and trends', *Soviet Geography — Review and Translation*, vol. 24, no. 7, pp. 479–502

—— (1984) 'Recent developments in large-scale water transfers in the USSR', *Soviet Geography — Review and Translation*, vol. 25, no. 4, pp. 261–3

—— (1985) 'The vast diversion of Soviet rivers', *Environment*, vol. 27, no. 2, pp. 12–20 and pp. 40–5

—— (1986) 'Soviet river diversion plans: their possible environmental impact', in E. Goldsmith and N. Hilyard (eds), *The social and environmental effects of large dams*, Wadebridge Ecological Centre, pp. 91–106

—— (1988) 'Desiccation of the Aral Sea: a water management disaster in the Soviet Union', *Science*, vol. 241, pp. 1170–6

Middle East Economic Digest (1970) 'Fish production', *Middle East Economic Digest*, vol. XIV. no. 8, 20 February, p. 238

—— (1987a) 'Special Report — Kuwait', *Middle East Economic Digest*, vol. 31. no. 28, pp. 21–42

—— (1987b) 'Special Report — Saudi Arabia', *Middle East Economic Digest*, June

Middleton, N.J. (1985) 'Effect of drought on dust production in the Sahel', *Nature*, vol. 316, pp. 431–4

Mikhaylov, N.I., Nikolayev, V.A. and Timashev, I.Y. (1977) 'Environmental protection problems in relation to the diversion of the flow of Siberian rivers', *Soviet Hydrology — selected papers*, vol. 16, no. 3, pp. 232–5

Mikhiel, G.S., Meyer, S.E. and Pendleton, R.L. (1992) 'Variation in germination response to temperature and salinity in shrubby Atriplex species', *Journal of Arid Environments*, vol. 22, no. 1, pp. 39–49

Milankovitch, M. (1980) 'Matematische Klimalehr und astronomische theorie der Klimaschwankungen', in W. Koppen and R. Geiger (eds), *Handbuch der Klimatologie 1*, Teil A, Berlin

Milburn, M. (1984) 'Archaeology and prehistory', in J.L. Cloudsley-Thompason (ed.) *Sahara Desert*, Pergamon Press, Oxford, pp. 291–310

Milliman, J.D. and Meade, R.H. (1983) 'World-wide delivery of river sediment to the oceans', *Journal of Geology*, vol. 91, pp. 1–21

Ministry of National Development (1970) *Atlas of Australian resources*, Department of National Development, Canberra

Mirghani, M.A. (1986) 'Industrial development in the Kingdom of Saudi Arabia: achievements and potential critical success factors', *GeoJournal*, vol. 13, no. 3, pp. 261–8

Moghraby, A.I. El, and Sammani, M.O. El, (1985) 'On the environmental and socio-economic impact of the Jonglei Canal Project, Southern Sudan', *Environmental Conservation*, vol. 12, no. 1, pp. 41–8

Moiseev, N.N., Berezner, A.S., Ereshiko, F.I. and Lotov, A.V. (1980) 'A systems approach to the study of interbasin water transfers: the partial diversion of the USSR's northern rivers flow into the Volga', *Water Supply and Management*, vol. 4, pp. 323–37

Morgan, R. (1985) 'Development of early warning drought systems', *Disasters*, vol. 9, no. 1, pp. 44–50

Morrell, W.P. (1968) *The gold rushes*, Adam and Charles Black, London

Morrice, H.A.W. and Allan, W.N. (1959) 'Planning of the ultimate development of the Nile', *Proceedings of the Institution of Civil Engineers*, vol. 14, pp. 101–56

Mortimore, M.J. (1988) 'Desertification and resilience in semi-arid West Africa', *Geography*, vol. 73, no. 1, pp. 61–4

—— (1989) *Adapting to drought: farmers' famine and desertification in West Africa*, Cambridge University Press, Cambridge

Morton, R. and Cunningham, R.B. (1985) 'Longitudinal profile of trends in salinity in the River Murray', *Journal of Soil Research*, vol. 23, pp. 1–13

Myers, N. (1992) 'Population/environment linkages: discontinuities ahead', *Ambio*, vol. 21, no. 1, pp. 116–18

Nachtnebel, H.P and Kovar, K. (eds) (1991) *Hydrological basis of ecologically sound management of soil and groundwater*, International Association of Hydrological Sciences (IAHS) Press, Institute of Hydrology, Wallingford, UK.

Nadeau, R.A. (1974) *The water seekers*, Peregrine Smith, Santa Barbara

Naff, T. and Matson, R.C. (eds) (1984) *Water needs in the Middle East*, Westview Press, Boulder

Nakayama, F.S. and Bucks, D.A. (1986) *Trickle irrigation for crop production*, Elsevier, Oxford

Napier, T.L. (1990) 'The evolution of US soil-conservation policy: from voluntary adoption to coercion', in J. Boardman, I.D.L. Foster, and J.A. Dearing (eds) *Soil erosion on agricultural land*, John Wiley, Chichester, pp. 627–44

Narayana, V.V.D. and Rambabu, (1983) 'Estimation of soil erosion in India', *Journal of Irrigation and Drainage Engineering*, vol. 109, pp. 409–34

Nash, L.L. and Gleick, P.H. (1991) 'Sensitivity of streamflow in the Colorado basin to climatic changes', *Journal of Hydrology*, vol. 125, pp. 221–41

National Academy of Sciences (1968) *Water and choice in the Colorado basin: an example of alternatives in water management*, Publication 1689, Washington, DC

—— (1974) *More water for arid lands — promising technologies and research opportunities*, National Academy of Sciences, Washington, DC

National Water Commission (1973) *Water policies for the future*, US Government Printing Office, Washington, DC

National Water Resources Association (undated) *A National Investment . . .* , National Water Resources Association, Information leaflet, Washington, DC

Natural Research Council (1992) *Managing water resources in the West under conditions of climate uncertainty*, National Academy Press, USA

Nelson, H.J. (1983) *The Los Angeles metropolis*, Kendall/Hunt Publishing Company, Dubuque, Iowa

Nelson, L. (1952) *The Mormon Village — a pattern and technique of land settlement*, University of Utah Press, Salt Lake City

497

Nelson, R. (1988) 'Dryland management: the desertification problem', *Environment Department Working Paper, No. 8*, World Bank, Washington, DC

Nemec, J. (1985) 'Water resource systems and climatic change', in J.C. Rodda (ed.), *Facets of hydrology*, Wiley, Chichester, vol. 2, pp. 131–51

Neuman, J. (1953) 'On the water balance of Lake Tiberias 1935/36–1946/47', *Israel Exploration Journal*, vol. 3, no. 4, pp. 246–9

—— (1955) 'On the water balance of Lake Huleh and the Huleh Swamps 1942/43–1946/47', *Israel Exploration Journal*, vol. 5, pp. 49–58

Neumann, J. and Parpola, S. (1987) 'Climatic change and the eleventh-tenth century collapse of Assyria and Babylon', *Journal of Near Eastern Studies*, vol. 46, no. 3, pp. 161–82

Newman, D. (1991) *Population, settlement and conflict: Israel and the West Bank*, Cambridge University Press, Cambridge

Newman, J.C. and Condon, R.W. (1969) 'Land use and present condition', in R.O. Slayter and R.A. Perry (eds), *Arid lands of Australia*, Australian National University, Canberra, pp. 105–33

Newsome, A.E. (1975) 'An ecological comparison of two arid-zone kangaroos of Australia, and their anomalous prosperity since the introduction of ruminant stock to their environment', *Quarterly Review of Biology*, vol. 50, pp. 389–424

—— (1977) 'The red kangaroo — an example of biological indicators of environmental change', in H. Messel and J.T. Butler (eds), *Australian animals and their environment*, Shakespeare Head Press, Sydney, pp. 23–4

—— (1983) 'Native fauna as indicators of range condition', in G. Crook (ed.), *Man in the centre — Proceedings of a Symposium held at CSIRO, Alice Springs, 3–5 April 1979*, CSIRO, Melbourne, pp. 124–49

Norris, H.T. (1975) *The Tuaregs, their Islamic legacy and its diffusion in the Sahel*, Aris and Phillips, Warminster

—— (1984) 'Indigenous peoples of the Sahara', in J.L. Cloudsley-Thompson (ed.), *Sahara Desert*, Pergamon Press, Oxford, pp. 311–24

Noy-Meir, S. (1973) 'Desert ecosystems — environment and producers', *Annual Review of Ecology and Systematics*, vol. 4, pp. 25–51

O'Dea, T.F. (1957) *The Mormons*, University of Chicago Press, Chicago

O'Mara, G.T. (ed.) (1988) *Efficiency in irrigation: the conjunctive use of surface and groundwater resources*, World Bank Symposium, Washington, DC

O'Riordan, T. (1991) 'The new environmentalism and sustainable development', *The Science of the Total Environment*, vol. 108, pp. 5–15

Odum, H.Y. (1971) *Environment, power and society*, John Wiley, New York

Okazaki, S. (1986) 'The great Persian famine of 1870–71', *Bulletin of the School of Oriental and African Studies*, University of London, vol. XLIX, part 1, pp. 183–92

Olsson, K. and Rapp, A. (1991) 'Dryland degradation in central Sudan and conservation for survival', *Ambio*, Vol. 20, no. 5, pp. 192–5

Orange and Los Angeles County Water Reuse Study (1982) *Facilities Plan*, Vols 1 and 2, OLAC Water Reuse Study, Los Angeles

Orni, E. and Efrat, E. (1971) *Geography of Israel*, Israel Universities Press, Jerusalem

Owens, S. and Owens, P.L. (1991) *Environment, resources and conservation*, Cambridge University Press

Pandey, A.N. and Rokad, M.V. (1992) 'Sand dune stabilisation: an investigation in the Thar Desert of India', *Journal of Arid Environments*, vol. 22, no. 3, pp. 287–92

Parkes, D.N., Burnley, I.H. and Walker, S.R. (1985) *Arid zone settlement in Australia: a focus on Alice Springs*, The United Nations University, Tokyo

Parry, M. (1990) *Climatic change and world agriculture*, Earthscan Publications, London

Parsons, A.J., Abrahams, A.D. and Simanton, J.R. (1992) 'Microtopography and soil-surface materials on semi-arid piedmont hillslopes, southern Arizona', *Journal of Arid Environments*, vol. 22, no. 2, pp. 107–15

Paterson, J.H. (1971) *North America — a geography of Canada and the United States*, Oxford University Press, Oxford

Paylore, P. and Greenwell, J.R. (1979) 'Fools rush in: pinpointing the arid zone', *Arid Lands Newsletter*, vol. 10, pp. 17–18

Payne, B.R., Quijano, L. and Latorre, D.C. (1979) 'Environmental isotopes in a study of the origin of salinity of ground water in the Mexicali Valley', *Journal of Hydrology*, vol. 41, p. 201

Pearce, D. and Maler, K-G. (1991) 'Environmental economics and the Developing World', *Ambio*, vol. 20, no. 2, pp. 52–4

Pedersen, F.B. and Moller, J.S. (1981) 'Diversion of the River Neva: how it will influence the Baltic Sea, the Belts and Cattegat', *Nordic Hydrology*, vol. 12, no. 1, pp. 1–20

Penman, H.L. (1948) 'Natural evaporation from open water, bare soil and grass', *Proceedings of the Royal Society Sect. A*, vol. 193, pp. 120–45

—— (1963) *Vegetation and hydrology*, Commonwealth Agricultural Bureau, Farnham, Royal (England), Technical Communication, 53

Perry, T.M. (1966) 'Climate and settlement in Australia 1700–1930: some theoretical considerations', in J. Andrews (ed.), *Frontiers and men*, F.W. Cheshire, Melbourne, pp. 138–55

Petersen, D.F. and Crawford, A.B. (eds) (1978) *Values and choices in the development of the Colorado River basis*, University of Arizona Press, Tucson

Petersen, J.E. (1977) 'Tribes and politics in eastern Arabia', *Middle East Journal*, vol. 31, no. 3, pp. 297–312

—— (1978) *Oman in the twentieth century*, Croom Helm, London

—— (1986) 'The emergence of nation-states in the Arabian Peninsula', *GeoJournal*, vol. 13, no. 3, pp. 197–200

Pickford, J. (ed.) (1987) *Developing world water*, Grosvenor Press International, Hong Kong

Pike, J.G. (1970) 'Evaporation of groundwater from coastal playas (sabkhah) in the Arabian Gulf, *Journal of Hydrology*, vol. 11, pp. 79–88

—— (1983) 'Groundwater resources development and the environment in the central region of the Arabian Gulf', *Water Resources Development*, vol. 1, no. 2, pp. 115–32

Polis, G.A. (ed.) (1991) *The ecology of desert communities*, University of Arizona Press, Tucson, Arizona

Population Reference Bureau (1973) *World Population Data Sheet*, Population Reference Bureau, Washington, DC

—— (1985) *World Population Data Sheet (1985)*, Population Reference Bureau, Washington, DC

Poulsen, E. and Lawesson, J.E. (1991) *Dryland degradation: causes and consequences*, Aarhus, Denmark

Powell, J.W. (1878) *Report on the lands of the arid region of the United States*, 45th Congress, 2nd Session, H.R. Ecev. Doc 73, Washington, DC

Precoda, M. (1991) 'Requiem for the Aral Sea,' *Ambio*, vol. 20, no. 3–4, pp. 109–14

Price, D. (1985) *Ground water in Utah's densely populated Wasatch Front area — the challenge and the choices*, US Geological Survey Water Supply Paper, 2232, Washington, DC

Prushansky, Y. (1967) Water development, *Israel Digest, Israel Today*, no. 11, Jerusalem

Pryde, P.R. (1991) *Environmental management in the Soviet Union*, Cambridge University Press, Cambridge

Pye, K. (1992) 'Aeolian dust transport and deposition over Crete and adjacent parts of the Mediterranean Sea', *Earth Surface Processes and Landforms*, vol. 17, pp. 271–88

Qutub, I.Y. (1983) 'Urbanisation in contemporary Arab Gulf states', *Ekistics*, vol. 50, pp. 170–82

—— (1992) 'Rapid population growth and urban problems in Pakistan', *Ambio*, vol. 21, no. 1, pp. 46–9

Raina, P., Joshi, D.C. and Kolarkar, A.A. (1991) 'Land degradation mapping by remote sensing in the arid regions of India', *Soil Use and Management*, vol. 7, pp. 47–52

Rantz, S.E. (1970) *Urban sprawl and flooding in southern California*, United States Geological Survey Circular, no. 601–B, USGS, Washington, DC

Rawlins, S.L. and Raats, P.A.C. (1975) 'Prospects for high-frequency irrigation', *Science*, vol. 188, pp. 604–10

Reed, C.A. (ed.) (1977) *Origins of agriculture*, Mouton, The Hague

Riad, M. (1983) 'The Arabian Gulf and its peoples prior to 1930', *Journal of the Gulf and Arabian Peninsula Studies (Kuwait)*, vol. IX

Richards, A. (ed.) (1986) *Food, states and peasants: analyses of the agrarian question in the Middle East*, Westview Press, Boulder, Colorado

Rind, D. (1984) 'Global climate in the 21st century', *Ambio*, vol. 13, pp. 148–51

Roberts, S.H. (1968) *History of Australian land settlement 1799–1920*, Macmillan, South Melbourne

—— (1970) *The squatting age in Australia 1835–1847*, Melbourne University, Carlton

Robinson, J.W. (1977) *The San Gabriels*, Golden West Books, San Marino

Rodier, J.A. (1985) 'Aspects of arid zone hydrology', in J.C. Rodda (ed.), *Facets of hydrology*, vol. 2, Wiley, Chichester, pp. 205–47

Ron, Z.D. (1985) 'Development and management of irrigation systems in mountain regions of the Holy Land', *Transactions of the Institute of British Geographers*, vol. 10, pp. 149–69

Rubenson, S. (1991) 'Environmental stress and conflict in Ethiopian history: looking for correlations', *Ambio*, vol. 20, no. 5, pp. 179–82

Rydzewski, J.R. (ed.) (1987) *Irrigation development and planning*, John Wiley, Chichester

Saab, G. (1967) *The Egyptian agrarian reform*, Oxford University Press, Oxford

Saarinen, T.F. (1966) *Perceptions of the drought hazard on the Great Plains*, Dept. of Geography, Research Paper, no. 106, University of Chicago, Chicago

Sacks, L.A., Herman, J.S., Konikow, L.F. and Vela, A.L. (1992) 'Seasonal dynamics of groundwater-lake interactions at Doñana National Park, Spain', *Journal of Hydrology*, vol. 136, pp. 123–54

Saiki, M.K., Jennings, M.R. and Wiedmeyer, R.H. (1992) 'Toxicity of agricultural subsurface drainwater from the San Joaquin valley, California, to juvenile chinook salmon and striped bass', *Transactions of American Fisheries Society*, vol. 121, pp. 78–93

Sarris, A.H. (1987) *Agricultural stabilization and structural adjustment policies in developing countries*, FAO, Rome

Schemenauer, R. and Cereceda, P. (1991) 'Fog water collection in arid coastal locations', *Ambio*, vol. 20, no. 7, pp. 303–8

Schick, A.P. (1979) 'Fluvial processes and settlement in arid environments', *GeoJournal*, vol. 3, no. 4, pp. 351–60

Schick, A.P. and Inbar, M. (1972) 'Some effects of man on the hydrological cycle of the upper Jordan — Lake Kinneret watershed', in R. Keller (ed.), *Run-off regimen and water balance II*, Freiburger Geographische Hefte, Heft 12, 2nd Report of the IGU Commission of the International Hydrological Decade, Freiburg, pp. 27–43

Schilfgaarde, J.V. (1982) 'The Welton-Mohawk dilemma', *Water Supply and Management*, vol. 6, nos. 1–2, pp. 115–27

Schlesinger, W.H., Reynolds, J.R., Cunningham, G.L., Huenneke, L.F., Jarrel, W.M., Virginia, R.A. and Whitford, W.G. (1990) 'Biological feedbacks in global desertification', *Science*, vol. 247, pp. 1043–48

Schliephake, K. (1985) 'Industrial planning and new industrial towns in Saudi Arabian, Qatar and Oman', *Arabian Gulf studies 1: Nurnberger Wirtschafts-und Sozialgeographische Arbeiten*, vol. 37, pp. 85–134

Schmida, A., Evanari, M. and Noy-Meir, I. (1986) 'Hot desert ecosystems', in I. Noy-Meir and D.W. Goodall (eds), *Ecosystems of the world: hot deserts and arid shrublands*, Elsevier, Oxford, pp. 379–88

Scholte, P.T. (1992) 'Leaf litter and Acacia pods as feed for livestock during the dry season in Acacia-Commiphora bushland, Kenya', *Journal of Arid Environments*, vol. 22, no. 3, pp. 271–6

Schulze, F.E. and De Ridder, NA. (1974) 'The rising water table in the West Nubarya area of Egypt', *Nature and Resources*, vol. 10, pp. 12–18

Scoging, H. (1989) 'Runoff generation and sediment mobilization by water', in D. Thomas (ed.), *Arid zone geomorphology*, Bellhaven Press, London, pp. 87–116

Scoones, I. (1991) 'Wetlands in drylands: key resources for agricultural and pastoral production in Africa', *Ambio*, vol. 20, no. 8, pp. 366–71

Sengel, R. (1973) 'On the mechanism of population growth during the Neolithic', *Current Anthropology*, vol. 14, pp. 540–2

Sestini, G. (1989) 'Implications of climatic changes for the Nile delta', in L. Jeftie, J.D. Milliman and G. Sestini (eds), *Implications of climatic changes in the Mediterranean Sea*, Pergamon Press, Oxford.

Setia, P. and Piper, S. (1991) 'Ground water quality implications of soil conservation measures: an economic perspective', *Water Resources Bulletin*, vol. 27, pp. 201–8

Sharon, D. (1972) 'The spotiness of rainfall in a desert area', *Journal of Hydrology*, vol. 17, pp. 161–75

Shata, A.A. (1982) 'Hydrology of the Great Nubian Sandstone basin, Egypt', *Quarterly Journal of Engineering Geology (London)*, vol. 15, pp. 127–33

—— (1992) 'Water resources and the future of arid lands', *International Journal of Water Resources Development*, vol. 8, no. 2, pp. 87–97

Sheikh, B. (1991) 'Long-range planning for water reuse in the city of Los Angeles', *Water Science and Technology*, vol. 24, no. 9, pp. 11–17

Shelef, G. (1991) 'Wastewater reclamation and water resources management', *Water Science and Technology*, vol. 24, no. 9, pp. 251–65

Shumeli, E. (1967) 'The contribution of research to the efficient use of water in agriculture', in *Water for peace*, vol. 3. *Water supply technology*, International Conference on Water for Peace, US Government Printing Office, Washington, DC, pp. 407–22

Shuval, H.I. (1980a) 'Goals of water quality control', in H.I. Shuval (ed.), *Water quality management under conditions of scarcity — Israel as a case study*, Academic Press, New York, pp. 1–9

—— (1980b) 'Quality management aspects of wastewater reuse in Israel', in H.I. Shuval (ed.), *Water quality management under conditions of scarcity — Israel as a case study*, Academic Press, New York, pp. 211–42

—— (1980c) 'Conclusions: the impending water crisis', in H.I. Shuval (ed.), *Water quality management under conditions of scarcity — Israel as a case study*, Academic Press, New York, pp. 315–37

—— (1987) 'The development of water reuse in Israel', *Ambio*, vol. 16, pp. 186–90

Sigurdson, J. (1977) 'Water policies in India and China', *Ambio*, vol. 6, no. 1, pp. 70–6

Simaika, Y.M. (1967) 'Multipurpose development of the Nile basin', in *Water for peace*, Vol. 8 — *Planning and developing water programs*, Washington, DC, pp. 214–22

Simpson, H.J. and Herczeg, A.L. (1991) 'Salinity and evaporation in the River Murray basin, Australia', *Journal of Hydrology*, vol. 124, pp. 1–27

Sinden, J.A. (ed.) (1972) *The national resources of Australia: prosperity and problems of development*, Angus and Robertson, Sydney

Skarpe, C. (1991) 'Impact of grazing in Savanna ecosystems', *Ambio*, vol. 20, no. 8, pp. 351–6

Skogerboe, G.V. (ed.) (1982a) 'Water and energy development in an arid environment: the Colorado River basin', *Water Supply and Management*, vol. 6, nos. 1–2 (Special issue)

—— (1982b) 'Development issues in the Colorado River basin', *Water Supply and Management*, vol. 6, nos 1–2, pp. 3–10

—— (1982c) 'The physical environment of the Colorado River basin', *Water Supply and Management*, vol. 6, nos 1–2, pp. 11–27

—— (1982d) 'Colorado river development', *Water Supply and Management*, vol. 6, nos 1–2, pp. 29–61

Skogerboe, G.V. and Radosevich, G.E. (1982) 'Future water development policies', *Water Supply and Management*, vol. 6, nos 1–2, pp. 221–32

Skogerboe, G.V., Walker, W.R. and Evans, R.G. (1982) 'Salinity control measures for Grand Valley', *Water Supply and Management*, vol. 6, nos 1–2, pp. 129–67

Slayter, R.O. and Perry, R.A. (1969) *Arid lands of Australia*, Australian National University, Canberra

Sloggett, G. (1982) *Energy and US agriculture: irrigation pumping, 1974–1980*, National Resources Economics Division, Economic Research Service, US Dept of Agriculture, Economic Report, no. 495

Smith, C.G. (1966) 'The disputed waters of the Jordan', *Transactions of the Institute of British Geographers*, vol. 40, pp. 111–28

Smith, K.R. (1991) 'Allocating responsibility for global warming: The Natural Debt Index', *Ambio*, vol. 20, no. 2, pp. 95–6

Smith, S.E. (1986) 'Drought and water management: the Egyptian response', *Journal of Soil and Water Conservation*, vol. 41, pp. 297–300

Smith, S.E. and Abdel-Kader, A. (1988) 'Coastal erosion along the Egyptian delta', *Journal of Coastal Research*, vol. 4, pp. 245–55

Smith, S.K. (1982) 'American Indian water rights and the use of the Colorado River water', *Water Supply and Management*, vol. 6, nos 1–2, pp. 75–86

Soil Survey Staff (1960) *Soil classification: a comprehensive system (7th Approximation)*, Soil Conservation Service, US Dept of Agriculture, US Government Printing Office, Washington, DC

Solecki, R.L. (1964) 'Zawi Chemi Shanidar, a post-Pleistocene village site', *Report VIth INQUA Congress 1962*, vol. IV, pp. 405–12

Solecki, R.S. (1963) 'Prehistory in Shanidar Valley, North Iraq', *Science*, vol. 139, pp. 179–93

—— (1964) 'Shanidar cave, a late Pleistocene site in N. Iraq', *Report VIth INQUA Congress 1962*, vol. IV, pp. 413–23

—— (1971) *Shanidar: the first flower people*, Knopf, New York

Solley, W.B., Chase, E.G. and Mann IV, W.B. (1983) *Estimated use of water in the United States in 1980*, United States Geological Survey Circular 1001, Washington, DC

Sophocleous, M. (1992) 'Groundwater recharge estimation and regionalisation: the Great Bend Prairie of central Kansas and its recharge statistics', *Journal of Hydrology*, vol. 137, pp. 113–40

South Africa (1970) *Second Report of the Commission of Enquiry into agriculture*, R.P. 84/1970, Government Printer, Pretoria

South Coast Air Quality Management District (SCAQMD) (1981) *Air Quality Digest*, South Coast Air Quality Management District, El Monte, California, September

—— (1985a) *Air pollution episodes — what they are — what they do*, South Coast Air Quality Management District, El Monte, California

—— (1985b) *Air quality trends in the South Coast Air Basin 1975–1985*, South Coast Air Quality Management District, El Monte, California

Southern Arizona Water Resources Association (SAWARA) (undated) *Supply — Demand*, Southern Arizona Water Resources Association, Tucson, Publicity leaflet

Speidel, D.H., Ruedisili, L.C. and Agnew, A.F. (eds) (1988) *Perspectives on water use and abuses*, Oxford University Press, Oxford.

Speigler, K.S. and Laird, A.D.K. (eds) (1980) *Principles of desalination — Parts A and B*, Academic Press, London

Splinter, W.E. (1976) 'Centre-pivot irrigation', *Scientific American*, vol. 234, no. 6, pp. 90–9

Stadfeld, O.R. and Schlaweck, K.I. (1988) 'International comparison of water prices', *Aqua*, vol. 4, pp. 173–7

Stanley, D.J. (1988) 'Subsidence in the north-eastern Nile delta: rapid rates, possible cause and consequences', *Science*, vol. 240, pp. 497–500

Stebbing, E.P. (1935) 'The encroaching Sahara: the threat to the West African colonies', *Geographical Journal*, vol. 85, pp. 506–24

Steinbeck, J. (1974) *The grapes of wrath*, Heinemann, London

Steiner, F.R. (1990) *Soil conservation in the United States: policy and planning*, Johns Hopkins University Press, Baltimore

Steiner, R. (1981) *Los Angeles: the centrifugal city*, Kendall/Hunt, Dubuque, Iowa

Steinhart, J.S. and Steinhart, C.E. (1974) 'Energy use in the US food industry', in P. Abelson (ed.), *Energy: use, conservation and supply*, AAAS, Washington, DC

Stiles, D. (1984) 'Desertification: a question of linkage', *Desertification Control Bulletin (UNEP Nairobi)*, vol. 11, pp. 1–6

Street, F.A. and Grove, A.T. (1979) 'Global maps of lake-level fluctuations since 30,000 yr B.P.', *Quaternary Research*, vol. 12, pp. 83–118

Stuart, J.M. (1861) 'Journal of Australian exploration', *Journal of the Royal Geographical Society*, vol. 31, pp. 65–145

Suliman, M.M. (1988) 'Dynamics of range plants and desertification monitoring in the Sudan', *Desertification Control Bulletin (UNEP, Nairobi)*, vol. 16, pp. 27–31

Sutcliffe, J.V. (1974) 'A hydrological study of the southern Sudd region of the upper Nile', *Bulletin of the International Association of Hydrological Sciences*, vol. 19, pp. 237–55

Swanson, G.J. (1985) 'First in the nation', *Water Well Journal*, vol. 39, no. 9, pp. 53–61

Swift, J. (1975) 'Pastoral nomadism as a form of land-use: the Twareg of the Adrar n Iforas', in T. Monad (ed.), *Pastoralism in Tropical Africa*, IAI, Oxford, pp. 443–53

—— (1977) 'Desertification and man in the Sahel', in *Land Use and*

Development, African Environment Special Report, No. 5, London, pp. 171–8

—— (1982) 'The future of African hunter-gatherer and pastoral peoples', *Development and Change*, vol. 13, pp. 159–81

Tahal (1977) *The water economy — development and management*, Tahal, Tel Aviv

Talling, J.F. and Rzoska, J. (1967) 'The development of plankton in relation to the hydrological regime in the Blue Nile', *Journal of Ecology*, vol. 55, pp. 637–62

Texas Department of Water Resources (1979) *Groundwater availability in Texas*, Texas Department of Water Resources, Austin, Texas

—— (1983) *Water for Texas — Planning for the future*, Texas Department of Water Resources, Austin, Texas

Texas Water Development Board (1968) *The Texas Water Plan*, Texas Water Development Board, Austin, Texas

Thapa, G.B. and Weber, K.E. (1991) 'Soil erosion in developing countries: a politico-economic explanation', *Environmental Management*, vol. 15, pp. 461–73

Thatcher, L., Rubin, M. and Brown G. (1961) 'Dating desert groundwater', *Science, NY*, vol. 134, p. 105

Thomas, D.S. (ed.) (1989) *Arid zone geomorphology*, Bellhaven Press, London

Thomas, D.S.G. (1992) 'Desert dune activity: concepts and significance', *Journal of Arid Environments*, vol. 22, no. 1, pp. 31–8

Thomas, D.S.G. and Shaw, P.A. (1991) *The Kalahari environment*, Cambridge University Press

Thomas, G.A. and Jakeman, A.J. (1985) 'Management of salinity in the River Murray basin', *Land Use Policy*, vol. 2, no. 2, pp. 87–102

Thompson, R., Turner, G.M., Stiller, M. and Kaufman, A. (1992) 'Near East paleomagnetic secular variation recorded in sediments from the Sea of Galilee (Lake Kinneret)', *Quaternary Research*, vol. 23, no. 2, pp. 175–88

Thompson, B.W. (1965) *The climate of Africa*, Oxford University Press, Nairobi

Thompson, J.W. (1966) 'Governmental responses to the challenges of water resources in Texas', *Southwestern Historical Quarterly*, vol. 70, pp. 44–64

Thompson, R.D. (1989) 'Short-term climatic change: evidence, causes, environmental consequences and strategies for action', *Progress in Physical Geography*, vol. 13, no. 3, pp. 315–47

Thompson, R.S. (1992) 'Late Quaternary environments in Ruby Valley, Nevada', *Quaternary Research*, vol. 37, no. 1, pp. 1–15

Thornes, J.B. (1985) 'The ecology of erosion', *Geography*, vol. 70, pp. 222–35

Thornthwaite, C.W. (1948) 'An approach towards a rational classification of climate', *Geographical Review*, vol. 38, pp. 55–94

Thornthwaite, C.W. and Mather, J.W. (1957) *Instructions and tables for computing the potential evapotranspiration and the water balance*, Publications in Climatology, Drexel Institute of Technology. X, no. 3

Timberlake, L. (1985) *Africa in crisis*, Earthscan, London

Tindale, N.B. (1959) 'Ecology of primitive Aboriginal man in Australia', in R.L. Crocker (ed.), *Biogeography and ecology in Australia*, Junk, The Hague

Tolba, M.K. (1990) 'Climate change and water management', *Water International*, vol. 15, pp. 56–7

Tong, S.T.Y. (1992) 'The use of non-metric multidimensional scaling as an ordination technique in resource survey and evaluation: a case study from south-east Spain', *Applied Geography*, vol. 12, no. 3, pp. 243–60

Tout, D. (1990) 'The horticultural industry of Almeria Province, Spain', *Geographical Journal*, vol. 156, no. 3, pp. 304–12

Townsend, J. (1977) *Oman: the making of a modern state*, Croom Helm, London

Tucker, C.J., Dregne, H.E. and Newcomb, W.W. (1991) 'Expansion and contraction of the Sahara desert from 1980 to 1990', *Science*, vol. 253, pp. 299–301

Turner, R.K. (ed.) (1988) *Sustainable environmental management*, Bellhaven Press, London

Tyson, P.D. (1980) 'Atmospheric circulation changes and the occurrence of extended wet and dry spells over Southern Africa', in J.A. Mabbutt and S.M. Berkowicz (eds), *The threatened drylands — regional and systematic studies of desertification*, University of New South Wales, Australia, pp. 95–110

—— (1986) *Climate change and variability in southern Africa*, Oxford University Press, Oxford

Tyson, P.D., Dyer, T.G.J. and Mametse, M.N. (1975) 'Secular changes in South African rainfall: 1880–1972', *Quarterly Journal of the Royal Meteorological Society*, vol. 101, pp. 817–33

Ucko, P.J. and Dimbleby, G.W. (eds) (1969) *The domestication and exploitation of plants and animals*, Duckworth, London

Ukayli, M.A. and Husain, T. (1988) 'Comparative evaluation of surface water availability, waste water reuse and desalination in Saudi Arabia', *Water International* , vol. 13, pp. 215–25

UNESCO (1969) *Discharge of selected rivers of the world. Vol. 1 — General and regime characteristics of stations selected*, Paris

—— (1979) *Map of the world distribution of arid regions*, MAB Technical Notes, 7, UNESCO, Paris

UNESCO, FAO (1963) *Bioclimatic map of the Mediterranean zone*, Explanatory Notes, UNESCO, Paris, Arid Zone Research, vol. 21

United Arab Republic, Information Department (1963) *The High Dam — bulwark of our future*, Cairo

United Arab Republic, Ministry of the High Dam (1968) *The High Dam approaching the last year of construction January 1968*, Survey of Egypt

United Arab Republic, Ministry of the High Dam, Aswan High Dam Authority (1968) *Aswan High Dam — commissioning of the first units — transmission of power to Cairo*, Ministry of the High Dam, Aswan

United Nations Conference on Desertification (UNCOD) (1977) *Desertification: its causes and consequences*, Pergamon, Nairobi

US Department of Agriculture (1976) *Egypt — major constraints to increasing agricultural productivity*, USDA, Foreign Agricultural Economic Report no. 120, Washington, DC

US Department of Agriculture (USDA) *Land capability*, USDA Handbook no. 210, Washington, DC

US Department of Commerce, Bureau of the Census (1975) *Historical statistics of the United States — Colonial times to 1970*, vols. 1 and 2, Washington, DC

—— (1981) *Statistical abstract of the United States, 1981*, US Government Printing Office, Washington, DC

—— (1984) *Utah — state and county data*, 1982 Census of Agriculture, vol. 1, Geographic area series, Part 44 Utah — State and County data, US Government Printing Office, Washington, DC

US Department of the Interior, Bureau of Reclamation (1985) *1984 annual report*, Bureau of Reclamation, Washington DC

Vardi, Y. (1980) 'National water resources planning and development in Israel — the endangered resource', in H.I. Shuval (ed.), *Water quality management under conditions of scarcity — Israel as a case study*, Academic Press, New York, pp. 37–49

Verstraete, M.M. and Pinty, B. (1991) 'The potential contribution of satellite remote sensing to the understanding of arid lands processes', *Vegetatio*, vol. 91, pp. 50–72

Verstraete, M.M. and Schwartz, S.A. (1991) 'Desertification and global change', *Vegetatio*, vol. 91, pp. 3–13

Viessman, W. (1990) 'Water management: challenge and opportunity', *Journal of Water Resources Planning and Management*, vol. 116, no. 2, pp. 155–69

Vilentchuk, I. (1967) 'Desalted water for Israel's agriculture' in *Water for peace, Vol. 2 — Water supply and technology*, International Conference on Water for Peace, US Government Printing Office, Washington, DC, pp. 13–24

Visher, S.S. (1966) *Climatic atlas of the United States*, Harvard University Press, Cambridge, Mass.

Vogel, H. (1992) 'Effects of conservation tillage on sheet erosion from sandy soils at two experimental sites in Zimbabwe', *Applied Geography*, vol. 12, no. 3, pp. 229–42

Voropaev, G.V. and Velikanov, A.L. (1984) 'Partial southward diversion of northern and Siberian rivers', *International Journal of Water Resources Development*, vol. 2, nos 2–3, pp. 67–83

Voskresenskiy, K.P., Sokolov, A.A. and Shiklomanov, I.A. (1973) 'Surface water resources of the USSR and their change under the effect of industrial and agricultural activity', *Soviet Hydrology*, vol. 2, pp. 123–41

Wade, N. (1985) 'Choice of desalination and power plants', *Water and Sewage*, vol. 9, no. 5, pp. 3–6

Waggoner, P.E. (ed.) (1990) *Climatic change and US water resources*, Wiley, New York

Walker, A.S., Olsen, J.W., and Bagen, (1987) 'The Badain Jaran desert: remote sensing investigations', *Geographical Journal*, vol. 153, no. 2, pp. 205–10

Wallace, J.S. (1991) 'The measurement and modelling of evaporation from semiarid land', in M.V.K. Sivakumar, *et al.* (eds), *Proceedings of Workshop on soil water balance in the Sudano-Sahelian zone*, February 18–23, IAHS Publication 199, Wallingford, pp. 131–48

Wallace, J.S., Gash, J.H.C. and Sivakumar, M.V.K. (1990) 'Preliminary measurements of net radiation and evaporation over bare soil and fallow bushland in the Sahel', *International Journal of Climatology*, vol. 10, pp. 201–10

Wallen, C.C. (1967) 'Aridity defintions and their applicability', *Geografiska Annaler*, vol. 49A, nos. 2–4, pp. 367–84

Waller, D.H. (1989) 'Rain water — an alternative source in developing and developed countries', *Water International*, vol. 14, pp. 27–36

Walters, M.O. (1989) 'A method of estimating design flood hydrograph shape in an arid region', *Water International*, vol. 14, pp. 2–5

Warren, A. (1984) 'The problems of desertification', in J.L. Cloudsley-Thompson (ed.), *Sahara Desert*, Pergamon Press, Oxford, pp. 335–42

Warren, A. and Agnew, C. (1988) 'An assessment of desertification and land degradation in arid and semi-arid areas', *International Institute for Environment and Development, Drylands Programme, Paper 2*, IIED, London

Warren, A. and Khogali, M. (1992) *Assessment of desertification and drought in the Sudano-Sahelian region 1985–1991*, United Nations Sudano-Sahelian Office (UNSO), New York

Waterbury, J. (1979) *Hydropolitics of the Nile valley*, Syracuse University Press, Syracuse

—— (1982) *Riverains and Lacustriner: towards international co-operation in the Nile basin*, Discussion Paper no. 107, Research Program in Development, Princeton University

Watson, J.W. (1982) *The United States: habitation of hope*, Longman, London

Watts, M.J. (1989) 'The agrarian question in Africa: debating the crisis', *Progress in Human Geography*, vol. 13, no. 1, pp. 1–41

Weaver, J.D. (1980) *Los Angeles: the enormous village, 1781–1981*, Capra Press, Santa Barbara

Webb, W.P. (1931) *The Great Plains*, Gunn, New York

Weeks, W.F. and Campbell, W.J. (1973) 'Icebergs as a fresh-water source — an appraisal', *Journal of Glaciology*, vol. 12, no. 65, pp. 207–33

Weinbaum, M.G. (1982) *Food development and politics in the Middle East*, Westview Press, Croom Helm, London

Weiner, A. (1972) 'Comprehensive water resources development — case study — Israel', in A. Weiner, *The role of water in development*, McGraw-Hill, New York, pp. 401–11

Wells, S.G. and Haragan, D.R. (eds) (1983) *Origin and evolution of deserts*, University of New Mexico Press, Albuquerque

Wessel, T.R. (ed.) (1977) *Agriculture on the Great Plains, 1876–1930: a symposium*, University of California Press, Davis, California

West, N.E. (1981) 'Nutrient cycling in desert ecosystems', in D.W. Goodall and R.A. Perry (eds), *Arid land ecosystems: structure, functioning and management*, vol. 2, Cambridge University Press, Cambridge, pp. 301–24

Westcoat Jr, J.L. (1984) *Integrated water development — water use and conservation practice in western Colorado*, University of Chicago, Department of Geography, Research paper no. 210

Western, D. (1982) 'The environment and ecology of pastoralists in arid savannas', *Development and Change*, vol. 13, pp. 183–211

Westing, A.H. (ed.) (1986) *Global resources in international conflict*, Oxford University Press, Oxford

Westing, A.H. (ed.) (1991) 'Environment and security: the case of Africa', *Ambio*, vol. 20, no. 5, pp. 167–206

Wheater, H.S., Butler, A.P., Stewart, E.J. and Hamilton, G.S. (1991) 'A multivariate spatial-temporal model of rainfall in southwest Saudi Arabia. I. Spatial rainfall characteristics and model formulation', *Journal of Hydrology*, vol. 125, pp. 175–99

Wheater, H.S., Onof, C., Butler, A.P. and Hamilton, G.S. (1991) 'A multivariate spatial-temporal model of rainfall in southwest Saudi Arabia. II. Regional analysis and long-term performance', *Journal of Hydrology*, vol. 125, pp. 201–20

White, G.F. (ed.) (1978) *Environmental effects of arid land irrigation in developing countries*, Man and the Biosphere (MAB), Technical Note 8

Whitford, W.G., Ludwig, J.A. and Noble, J.C. (1992) 'The importance of subterranean termites in semi-arid ecosystems in south-east Australia', *Journal of Arid Environments*, vol. 22, no. 1, pp. 87–91

Whitlach, E.E. and De Velle, C.S. (1990) 'Regionalisation in water resource projects', *Water International*, vol. 15, pp. 70–9

Whittington, D. and Haynes, K.E. (1985) 'Nile water for whom? Emerging conflicts in water allocation for agricultural expansion in Egypt and Sudan', in P. Beaumont and K.S. McLachlan (eds), *Agricultural development in the Middle East*, John Wiley, Chichester, pp. 125–49

Wichelns, D. and Cone, D. (1992) 'Tiered pricing motivates Californians to conserve water', *Journal of Soil and Water Conservation*, vol. 47, no. 2, pp. 139–44

Wickens, G.E. (1975) 'Changes in the climate and vegetation of the Sudan since 20,000 B.P.', *Boissiera*, vol. 24, pp. 43–65

Widtsoe, J.A. (1947a) 'A century of irrigation', *The Reclamation Era*, vol. 33, pp. 99–102

—— (1947b) *How the desert was tamed — a lesson for today and tomorrow*, Deseret Book Company, Salt Lake City

Wigley, T.M.L. and Raper, S.G.B. (1987) 'Thermal expansion of sea water associated with global warming', *Nature*, vol. 330, pp. 127–31

Wijkman, A. and Timberlake, L. (1985) 'Is the African drought an act of God or Man?', *Ecologist*, vol. 15, nos 1/2, pp. 9–18, 34

Wilcox, D.G. and McKinnon, E.A. (1972) *A report on the condition of the Gascoyne catchment*, Department of Agriculture and Department of Lands and Surveys, Western Australia

Wilcox, L.V. (1955) *Classification and use of irrigation waters*, US Department of Agriculture, Circular no. 969, Washington, DC

Wilhite, D. and Glantz, M. (1985) 'Understanding the drought phenomenon: the role of definitions', *Water International*, vol. 10, pp. 111–20

Wilkinson, G. (1987) 'Electrifying the Gulf', *Middle East Economic Digest*, vol. 31, no. 19, pp. 6–7

Willcocks, W. and Craig, J.I. (1913) *Egyptian irrigation*, Spon, London

Williams, M. (1989) 'Deforestation: past and present', *Progress in Human Geography*, vol. 13, no. 2, pp. 176–208

Williams, O.B. (1977) *Ecosystems of arid Australia*, IBP Cambridge University Press, Cambridge

Williams, O.B. Suijdendorp, H. and Wilcox, D.G. (1980) 'Australia — the Gascoyne basin', in M.R. Biswas and A.K. Biswas (eds), *Desertification, environmental sciences and applications*, vol. 12, Pergamon Press, Oxford, pp. 3–106

Winstanley, D. (1983) 'Desertification: a climatological perspective', in S.G. Wells and D.R. Haragan (eds) *Origin and evolution of deserts*, University of New Mexico Press, Albuquerque

Wischmeier, W.H. and Mannering, J.V. (1969) 'Relation of soil properties to its erodibility', *Soil Science American Proceedings*, vol. 33, pp. 131–7

Wischmeier, W.H. and Smith, D.D. (1965) *Predicting rainfall-erosion losses from cropland east of the Rocky Mountains*, Agr. Handbook 282, US Dept of Agriculture, Agriculture Research Service, Washington, DC

Wittfogel, K.A. (1957) *Oriental despotism: a comprehensive study of total power*, Yale University Press, New Haven

—— (1965) 'The hydraulic civilisation', in W.L. Thomas Jr. (ed.), *Man's role in changing the face of the earth*, University of Chicago Press, Chicago, pp. 152–64

Wood, T.G. (1991) 'Termites in Ethiopia: the environmental impact of their damage and resultant control', *Ambio*, vol. 20, no. 3–4, pp. 136–8

Woodwell, G.M. and Ramakrishna, K. (1989) 'Will there be a global greenhouse warming?', *Environmental Conservation*, vol. 16, no. 4, pp. 289–91

World Bank (1981) *World development report 1981*, Oxford University Press, Oxford

World Health Organisation (1989) *DDT and its derivatives — environmental aspects*, Environmental Health Criteria, No. 83, World Health Organisation, Geneva

World Meteorological Organisation (1966) *Measurement and estimation of evaporation and evapotranspiration*, Technical Paper no. 121, T.P.105, World Meteorological Organisation

Worster, D. (1979) *Dust Bowl — the Southern Plains in the 1930s*, Oxford University Press, New York

—— (1984) 'The Hoover Dam: a study of domination', in E. Goldsmith and N. Hildyard (eds), *The social and environmental effects of large dams*, Wadebridge Ecological Centre, pp. 17–23

Yesilada, B.A., Brockett, C.D. and Drury, B. (1987) *Agrarian reform in reverse; the food crisis in the Third World*, Westview Press, Boulder.

Young, E.A. (1992) 'Aboriginal land rights in Australia: expectations, achievements and implications', *Applied Geography*, vol. 12, no. 2, pp. 146–61

Zeid, M.A. (1992) 'Water resources assessment for Egypt', *International Journal of Water Resources Development*. vol. 8, no. 2, pp. 76–86

Zeist, W.V. and Wright Jr, H.E. (1963) 'Preliminary pollen studies at

Lake Zeribar, Zagros Mountains, southwestern Iran', *Science, NY*, vol. 140, pp. 65–9

Zhang, Y., Wang, X. and Rylander, M.K. (1992) 'Food habits of birds in a modified desert ecosystem in Central China', *Journal of Arid Environments*, vol. 22, no. 3, pp. 245–50

Zierer, C.M. (1934) 'The citrus fruit industry of the Los Angeles basin', *Economic Geography*, vol. 10, pp. 53–73

Zonn, I.S. (1982) 'Irrigation developments of deserts and amelioration of arid lands', in *Combatting desertification in the USSR — problems and experiences*, UNEP/USSR Commission for UNEP, Moscow

Index

Abalak deep well, Niger 209
Ab-Char regulator, Zayandeh
  River, Isfahan 290
Abduliya, Kuwait, water well
  field 351
Abilene, USA, cattle drives 310
Aboriginal lands, mineral rights
  161
Aborigines, Australia, 72,
  74–82, 183; Australoid 74;
  Tasmanoid 74
Abu Dhabi, oil production 336
Acacia shrubland, Gascoyne
  Basin, Western Australia 222
acorns 66, 67
Adelaide, South Australia 175
Adrar of the Iforas, Mali 199
Africa 8, 474; lake levels 51
Agadez, Niger 202, 204, 206,
  209, 211; livestock losses 212;
  rainfall 210
aggressive ground conditions
  116; salt weathering 114
agribusinesses, Khuzestan, Iran
  95
agricultural activity,
  environmental cost 451
agricultural investment, Oman
  89
agricultural production 464
agricultural sectors, neglect of
  92, 93; planning 93
agricultural subsidies, Great
  Plains 314
agricultural systems, Gulf States
  344
agricultural water use, Israel 382
agriculture, mechanisation 94
Ahaggar Massif, Algeria 196,
  199, 200
aid, from developed world 453
air conditioning systems, Gulf
  States 340
Air Highlands, Niger 199, 200,
  202

air management areas, Los
  Angeles 441
air pollution, Los Angeles
  431–48
air pollution control, Los
  Angeles 432
air pollution episode criteria,
  Los Angeles 440
air pollution management, Los
  Angeles 439–42
air pollutants, Los Angeles
  433–9
Air Quality Management Plan
  (AQMP) 440
Akkad 26
Alaska 327
Albuquerque 464
Alfisols 37, 38
Al Hofuf, Saudi Arabia, oasis
  353
Algeria 46, 195, 201; cereal
  production 94
Al Hassa, Saudi Arabia, date
  groves 344; wells 344;
Alice Springs, Australia 82,
  227–35; foundation of 229
Al Jubail, Saudi Arabia,
  desalination plant 352
Al Khobar, Saudi Arabia 346;
  desalination plant 352
All-American Canal 267, 268
alluvial deposits 463; fan 34;
  floods 110; groundwater
  recharge 29; irrigation systems
  72; Los Angeles 403; qanats
  130
alluvial mining, erosion in
  California 155
Alluvial plain 35
almonds 67
Al Qatif, oasis 353
Al Uqayr, Saudi Arabia,
  desalination plant 352
Amadror, Ahaggar Uplands 200
Amudarya River 387, 389, 398,

399; irrigation water demand
392
Amur River 387
Anatolia 66
Anglo-Iranian Oil Company 86,
338
animal dung, fertilizer 164; fuel
164
Ankara 464; air pollution 466
annual crops, Wadi Ziqlab,
Jordan 170
annual discharge fluctuations,
Atbara River 246; Blue Nile
River 246; Nile River 245, 246
annual precipitation, Sahara
desert 196, 197
Antarctic ice sheet 49
antelopes 45
ants 44
apartment complexes, Los
Angeles 403
aqueducts 455
aquents 38
aquifers 28; alluvial 29; confined
29; deep rock 29; Israel 363;
Nubian Sandstone 29;
Ogallala 29; porosity 28;
recharge 29, 30; recharge by
runoff 30; recharge rates 30;
specific retention 28; specific
yield 28; transmissibility 28;
unconfined 29; water yield 29
aquifer storage capacity, Israel
380
aquifer systems, Saudi Arabia
352
Arabian peninsula, aquifer
systems 30
Arabian/Persian Gulf, economic
development 332–56
Aral Sea 387, 398, 399; decrease
in area 399; environmental
problems 392; falling water
level 393
Arctic Ocean 398; effects of
water transfers on 400, 401
arid areas, attitudes of
governments 468
arid homoclimates, Meigs 5, 6
arid land ecosystems 463

arid land settlement 468
arid lands, policy objective,
settlement 468
aridity, classical definition 4;
definitions 3; index definition
4; water balance approach 4
arid land ecosystems, survival of
456
arid landforms 33–6; basin
model 34
arid nations, classification 84, 85
aridosols 37, 38
arid zone soils 36–40
Arizona Groundwater
Management Act 1980 118
Arizona, purchase of water
rights 456
Arizona, USA 11, 411; irrigation
268
Arroyos, 35
artesian system 30
arthropods 44
artificial recharge of
groundwater, Israel 365, 370
Asia 8
asses 45
Assyut Barrage, Nile River 249
Aswan, Nile discharge 245, 246,
260
Aswan Dam 236, 249, 251, 259
Aswan, Egypt, precipitation 243
Aswan High Dam 28, 239, 249,
251–60, 473; environmental
effects 253
Atbara River 239, 244
Atlas Mountains 195, 200
atomic weapon testing, Australia
456; USA 456
atomic weapons, dryland testing
464
Atrek River 389
Aurochs 64, 66
Australia 8, 10, 35, 45, 74, 464;
cracking soils 38; European
settlers 215; evaporation 17,
18; gold mines 159; iron ore
458; lake levels 51; maximum
temperatures 15; minimum
temperatures 15; mining 161,
468; pastoralism 195, 215–35;

precipitation 9; relative humidities 16; settlement of the arid interior 468; soil erosion 39; summer monsoon 16
automobiles, Los Angeles 402
Avdat, Negev 70
Avra Valley, Arizona, irrigation 117, 123
Azawak, Niger 202, 203
Azawak Zebu, Niger 211; Tuareg 207
Azov Sea, environmental problems 392; falling water level 395

Bactrian camel 46
'badlands', soil erosion 463
Bahariya oasis, Egypt 30
Bahrain 467; humidities 15, 332; oil discoveries 336
Bahrain Island, causeway 355; salt weathering 116; springs 353
Bahr al Gezel River 255
Bahr el Jebel River 243, 254; evaporation reduction scheme 254
Bakhtiari tribe, nomads 181
Ballona Creek, California 36
bananas 58
Banias River 362
Bantu 183
barbed wire 184; Great Plains 310
barley 64, 65, 66, 67, 69; wild ancestors 59
barley production, Middle East 93
Baro River 255
basin irrigation 130, 472; longevity in Nile valley 248; Nile valley 127, 246–8; origins in Nile valley 246, 247
Bays and Estuaries Plan, California 423
beach pollution, Pacific Coast, California 420
Bear River, Utah 293
Bear Valley 295

Beersheba, Israel 358
bees 44
Beidha, Jordan 66
Beit Netofa, Israel 369
Benue River 196
Berkeley, California, water balance 20
Bilharzia, Nile valley 253
Bilma, Nigeria 200
bioclimatic aridity 6
biological productivity 47
birds 44
bison 46; Great Plains 310
bitter vetch 65, 67
Biyadh aquifer 30
Blackfoot Indians 310
Blaney and Criddle formula 5, 19; potential evapotranspiration 19
blizzards, Great Plains 311
Blue Mountains, Australia 215
Blue Nile 239, 243, 244, 250, 251, 255, 259, 262
Boers 183
boll-weevil, Egypt 250
Bonneville basin 270
'boom and bust' economies 468
boreholes, Niger 208
Bororo zebu, Foulani 207
Boulder Canyon Project 267, 268
Brahmin cattle, Australia 218
Brazil 469
Brazos River, Texas 318
breadfruit 58
Brigham Young 294, 296, 297, 298; Mormons 293; Nebraska 294
Brucellosis, Alice Springs 231
building regulations, Los Angeles 425
Bureau of Reclamation 141–5, 266, 267, 268, 455; arid land management 142; irrigated agriculture 141; irrigation 468; sale of water and power 144; subsidies 143
Bushehr, Iran 13, 14; temperatures 14

Bushmen, Kalahari 72
butterflies 44

Cadmium 156
Cairo, Egypt 98, 464; population
   pressures 109; precipitation
   243; sewage disposal 109; slum
   communities 109; under- and
   unemployment 109; water
   provision 109
California 95, 313, 314, 315,
   473; Colma Creek 113;
   evaporation 17; gold 155;
   irrigation 145, 268; Los
   Angeles 402; sediment loads
   111; Sun Belt 97; tree ring
   chronologies 51; trickle
   irrigation 132, 133
California Air Resources Board
   441
California Aqueduct, water for
   Los Angeles 412
California State-Lewis Air
   Quality Management Act 1976
   441
California State Water Project
   411, 412
California State Water
   Resources Control Board 423
camel caravans 199; Sahara 200
camel herding 199
camels 43, 45, 46, 198; Niger
   209; nomads 180; Tuareg 207
Canada 327
Cape Colony 186, 187
Cape Province, South Africa 159
capers 67
Cape Town, South Africa 186
caravan trade 180; Sahara 199
carbon dioxide, emissions 469;
   levels in atmosphere 53
carbon monoxide, levels in
   atmosphere 439, 444; Los
   Angeles 433, 434, 446
carnivores 45, 46
carrying capacities 473; world's
   drylands 473
carrying capacity, nomadic
   animals 182; rangeland in
   Australia 237

car sharing, Los Angeles 448
Carson sewage treatment plant,
   California 421
Casa del Agua, Tucson, water
   use 124, 125
Cash crops, cotton 94; tobacco
   94
Caspian Sea 387, 396, 399;
   environmental problems 392;
   falling water level 395
Catal Huyuk, Turkey 66, 67
caterpillars 44
cattle 46, 62, 67, 69; Australia
   217; Alice Springs, Australia
   229, 230; vegetation
   degradation 234
cattle diseases, Australia 218
cattle drives 184
cattle prices, Alice Springs
   district 230; Australia 231
cattle production, South Africa
   188
cattle ranching, Great Plains
   310; Texas, 310
cattle stations, Walbiri tribe,
   Australia 77
Centennial Valley, Montana 327
centipedes 44
Central Arizona Project 118,
   124, 271, 276, 411
Central Utah Project 270
Central Valley, California 411;
   California Aqueduct 412
centre-pivot irrigation, Colorado
   basin 272; Kansas 323;
   Nebraska 323
cereal domestication 62
cereal production, Australia 176;
   Middle East 93; rain-fed
   agriculture 164
cereals 72, 94, 201
Chad 30, 474; famine 196
cheese 179
check dams, Middle East,
   Mediterranean 452
Chemchemal Plain 66
Cheyenne Indians 310
Chicago, Illinois 310
China, irrigation 136; population
   control 460

chlorides, salt weathering 114
cities, dryland 97, 98; growth
    rates 98, 99
City Creek, Salt Lake City,
    irrigation 299
City of Tucson Water
    Department 118
Class A pan, US Weather
    Bureau 17
classification of dryland states
    83–5
Clear Air Act Amendments
    1977 442
Clean Water Act 1977 423
climatic change 47–56, 69, 454,
    473; impact on agricultural
    production 458; instrumental
    records 56; Sahara desert 196,
    197, 198
climatic classification, Koeppen
    4
climatic fluctuations 51
Coachella Valley 268
coal 152
coastal erosion, Nile delta 253
coconuts 58
Collgran Project, Colorado 268
Colma Creek, California,
    erosion rates 113; frequency
    of flooding 111; land use
    changes 113; sediment
    movement 111; urbanisation
    111, 112
colonial powers, Sahara 201
Colorado 10; irrigation 145;
    Ogallala aquifer 323
Colorado basin, energy
    resources 275, 276; future
    water supply and demand
    272–6; total reservoir capacity
    274
Colorado-Big Thompson Project
    268
Colorado River 11, 138, 238,
    263, 455; annual flows 25;
    consumptive losses 148; floods
    25; groundwater recharge 265;
    Hoover Dam 142; hydraulic
    facilities 264; irrigation water
    use 146; river discharge 265;

snowmelt 21; storage projects
    269
Colorado River Aqueduct, Los
    Angeles 411, 413
Colorado River Basin Project
    Act 270, 275
Colorado River Compact 1922
    266, 267, 270, 411
Colorado River Storage Project
    270
Colorado Water Conservancy
    Law 1938 269
Colt revolver 184, 310
Columbia River 327
Commercial crops, Nile valley
    249
Commercial pastoralism 182–90;
    Australia 215, 235–6;
    development of 183; pressure
    from arable farmers 185;
    South Africa 186–90
Common Market 177
communication improvements,
    effects on nomadism 181
Company town, mining 153
concrete, deterioration 114; salt
    weathering 114
Condon Scheme, Australia 232
Conference on the Environment,
    Rio de Janeiro 469
confined aquifer, Australia 219
conservation structures, collapse
    of 453
conserving moisture in the soil
    452
constructional problems, 114–16;
    dryland cities 114
contour ploughing 453
contour walling, Wadi Ziqlab,
    Jordan 173
controlled environment
    agriculture, Gulf States 345
conveyance loss, USA 148
cooking utensils 164
Coolgardie 158, 159; gold 158
Co-operative, rural institutions
    95
copper 156; trade 200
Corn Laws, repeal of 175
Corps of Engineers, water

import to Great Plains 327, 329, 330, 331
cost of labour, terraces 453
cost of pumping groundwater, High Plains of Texas 456
cotton, Nile valley 250
County Board of Engineers on Flood Control, California 426
coyotes 46
creosote bush 43
cretaceous aquifer system 30; South Africa 353
crop area, Egypt 250, 257; Israel 383
crop prices 452
crop residues, animal feed 164; fertilizers 164; fuel 164
crop yields 458
crops grown on irrigated land, Soviet Central Asia 389
CSIRO, Australian pastoral farming 220
cultivated area, expansion in Egypt 252; extension of 453
cultivation, northward movement in Niger 214
cultivators, Maghreb 201; Sahel 201
Curecanti Project, Colorado 270

Dahna sand dunes, Saudi Arabia, groundwater recharge 30
Dair Abu Said, Jordan 170
Dakhla oasis, Egypt 30
Daly City, California 111
dam projects 471
Damascus, Syria 277
Dammam Limestone, aquifer 351
Dammam, Saudi Arabia 340, 348, 349
dams, construction 24, 25
Dan Region Wastewater Reclamation Project 374, 375, 376
Dan River 362
dates 201; nomads of the Sahara 198

Dead Sea 360; water flow into 363
Dead Sea lowlands 126, 168, 358, 367
debris basins, Los Angeles county 429
deep aquifer systems, Saudi Arabia 459
deep aquifers, Gulf States 344
deep mining 152
deep sea cores 49
deep wells, Sahel 207
degradation, indicators 463
degraded natural vegetation, Middle East 452
Delta Barrages, Nile valley 249
demand management for water, Kuwait 352, 467; Los Angeles 415
demographic transition model 90
dependency culture 473
depopulation, rural 461
desalinated seawater, Gulf 467
desalinated water, costs in Kuwait 351
desalination, Gulf States 345; Israel 385; Saudi Arabia 352
desalination process, high cost of energy 467
desertification 165, 453, 454, 469; Niger 202, 212; nomads 180
Desert Land Act 1877 139
desert pavements 38
desert scrub 40
development, social aspects 471
development plans 85–9; impact of oil revenues 338; Iran 86, 87, 89; Oman 87, 89
development process, Gulf States 340
Devensian glacial period 69
Devil's Gate Dam, Los Angeles 427
Dez Dam, Iran 95
Dharan, Saudi Arabia 340
diamonds 152, 157
diamonds, South Africa 468, 469
Dinder River 244

dingo 46; fence 220; Gascoyne
Basin, Australia 226
disposal of solid waste 466
District Groundwater Law 1949,
Texas 320
Diyala Plains, irrigation 127
Djibouti 83
Dneiper River, USSR 391
Dodge City, cattle drives 310
Doha, Qatar 467; urban growth
340
domestication of animals 58, 69,
70
domestication of plants 58, 69,
70
donkeys 43, 46, 168
Don River, USSR 391, 392, 396
drainage, Nile valley 248
draught animals 72
drip irrigation 457; Israel 383;
water use 133
dromedary 46
drought 16, 469, 473, 474; Alice
Springs district 231; animal
response 219; Australia 218;
effects on Australian wheat
production 177; effects on
commercial pastoralism 185;
effects on nomads 182;
Gascoyne Basin, Australia
223; Great Plains 311–14;
High Plains of Texas 316;
Niger 213; Sahel 53, 205,
209–13; Walbiri tribe 77, 82
dry farming 72; High Plains of
Texas 318
dryland ecosystems 46–7
drylands, definitions 3
Dubai 334; oil production 336
dust 17
Dust Bowl 17; Great Plains 309,
313
Dutch East India Company 186

East African plateau 243
Eastern Province, Saudi Arabia
347–50
East Ghor Canal, Jordan 369
economic depression, effects on
Australian wheat production
177; Great Plains 313
economic development 92
efficient use of water, Israel 457
Eghazer, Niger 202, 203
Egypt 83, 93, 195, 238, 243, 254,
259, 260 472; cropped area
256; financial constraints 109;
food imports 255; land reform
94; modernisation of economy
254; perennial irrigation 249,
256; population growth 460;
rapid population growth 255;
urban and industrial water
demands 262; water use 261,
262
Egyptian Water Master Plan 261
Eilat, Israel 358, 360
Einkorn 66, 67, 68
Elburz Mountains, Iran,
snowmelt 21
electrical conductivity, irrigation
waters 135
electricity, consumption in Gulf
States 342, 343; subsidies in
Gulf States 340
electricity generating plant,
desalination 467
electricity grid, Gulf States 343,
344
Emmer wheat 66, 67, 68
energy prices, Arab–Israeli war
232
Entisols 37, 38
environmental cost, agricultural
activity 451
environmental degradation 452,
472; particulate material 155;
mining 154
environmental impact of rapid
development, Gulf States
345–7
environmental management,
decision making 157; waste
disposal 107
environmental managers 454,
461, 472, 474; dealing with
environmental change 455
environmental pressures, Jordan
173

Environmental Protection
Agency (EPA) 441
environment, productivity 463
environmental trends 455
Eocene aquifer system 30; Saudi
Arabia 353
ephemerals 41
episode levels for smog, Los
Angeles 439
Equalisation Fund, Israel 383
erosion problems, 453
erosion vulnerability, Gascoyne
Basin, Australia 223, 225;
Northern Highlands, Jordan
172
Escalante, Utah 301–6; drought
304; economic depression 304;
grazing policies 304; irrigation
303; livestock production 303;
mechanisation 305
Eshed Kinrot, Israel 369
Ethiopia 239, 243; famine 196;
future water demands 260;
summer monsoon 195
Ethiopian Highlands 243
Euphrates River 21, 26, 64, 65;
irrigation systems 26; mean
annual flow 26; snowmelt 21;
Tabqa dam 26; water demands
26; water surpluses 23
eutrophication, Jordan River
372; USSR 399
evaporation, Israel 360
evapotranspiration 17–21, 53;
actual 4, 19, 47; potential 5,
19; rates 53

Falaj 128
famine 453, 462, 474; Sahel 196
Farafra oasis, Egypt 30
Farkhad Dam, Soviet Central
Asia 391
farm size, South Africa 188
farm unit amalgamation, wheat
farming in Australia 178
fauna 43–6
Feather River, California 411
Federal air quality standards 442
Federal carbon monoxide
standard, Los Angeles 446

Federal Flood Control Act 1936
427
Federal nitrogen dioxide
standard, Los Angeles 446
Federal ozone standard, Los
Angeles 446
Federal Water Pollution Control
Act 423
feed lots, commercial
pastoralism 185
fencing, of ranges in Australia
218
Fergana valley, USSR, irrigation
391
fertilisers 461, 472
fertilizer use, Egypt 250
fig 62
firewood 164
fisheries, Caspian Sea 395;
Onega River, USSR 399
fish ponds, Jordan valley 366
Flaming Gorge Project, Utah
270
flash distillation 467
flash floods 10, 23
flood canals, Nile valley 249
flood channels, Los Angeles 404
Flood control structures, Los
Angeles 427, 428, 429
flood damage, Los Angeles 431;
urban environments 110
Flood Insurance Rate Map
(FIRM) 430
floodplain management, Los
Angeles 430
flood protection, Aswan High
Dam 252
floods, behavioural approach
111; Los Angeles 425; Nile
valley 239; structural approach
111
flood warning system, urban
areas 110
Florida, Sun Belt 97
fluvents 38
fodder plants, Alice Springs 233
food aid 473
food deficits, cereals 458
food, imports 93; imports,

Jordan 460; urban prices 96;
US cities 107
food policies, Government 471
food policy, cheap supplies 462
food provision 458
food shortages 150; Egypt 256;
Niger 214
foreign debts 165
fossil fuel subsidies 472, 474;
agriculture 165
fossil groundwater 33; Israel
363; South Africa 353
'fossil' water, Saudi Arabia 459
Foulani 202; cultivators 204;
Niger 207, 211, 212
foxes 46
Fraser River 327
Fruitgrowers' Dam Project,
Colorado 268
furrow irrigation 130; High
Plains of Texas 317; water use
133

Gao, Mali 457
Gascoyne Basin, Australia,
sheep 221
Gazelles 43, 45, 64, 66, 67
Geomorphological hazards,
drylands 110; urban areas
109–14
gerbils 44
Gezira irrigation project, Sudan
250, 251
Ghajar springs 362
Ghana, gold 200
Ghav Khuni salt flats 284, 289
Ghazvin, Iran 283
Gila Project, Arizona 268
Gillespie Dam, Arizona 241
Glen Canyon Dam 263, 270
Glen Canyon Project, Colorado
270
Glendale, California 442
Glendale Narrows, Los Angeles
river 408
global temperature changes 55
global warming process 469
goats 46, 62, 64, 65, 66, 67, 69,
72, 98
gold 152, 157

gold bearing lodes, Kalgoorlie
159
gold mining boom, Western
Australia 177
gold, South Africa 468, 469
Golodnaye Steppe Project,
USSR 391
government policies, low urban
food prices 96
Gradoni terracing, Wadi Ziqlab,
Jordan 173
Grand Canyon 138, 266
Grand Valley Project, Colorado
267
Granite Reef Aqueduct 272
grasshoppers 44
grassland 40; expansion of wheat
farming in Australia 177;
Sahel 206, 207
grasslands 473
grazing resources, variability 190
grazing strategies, Alice Springs
233
'Great American Desert' 312
Great Basin 295, 296
Great Boulder mine 160
Great Fergana Canal, USSR 391
Great Fingall mine 160
Great Kavir, Iran 130
Great Plains, USA 307–31;
bison 46, 183; climate 308;
Dust Bowl 17, 452; growing
season 309; location 308; wind
16
Great Salt Lake Valley 293, 298;
agricultural potential 296;
Mormons 295
Great Trek, South Africa 187
Greece 62
Green Revolution 150
groundwater 28–33, 113; Los
Angeles 408; mining of 124,
152; Sahara 199; saline 114;
subsidence 114; Tucson 117
groundwater law 325
Groundwater Management Act
1975, Nebraska 325
groundwater mounds, Egypt 258
groundwater pollution, from
landfill 466

groundwater quality, Israel 363
groundwater recharge, Los
  Angeles 422; USA 148
groundwater, Saudi Arabia 459
groundwater withdrawals,
  Colorado basin 272
Gulf 467
Gulf of Carpentaria, evaporation
  17
Gulf War, pollution effects 347
gully erosion 39
gully formation 463
gully plugs, Wadi Ziqlab, Jordan
  166, 173

halomorphic soils 42
halophytes 42
Hansen Flood Control Basin,
  Los Angeles 430
Hasbani River 362, 367
health hazard of smog, Los
  Angeles 435
helicopter, cattle mustering in
  Australia 231
herbivores 46
herd size, traditional pastoralism
  190
Hereford cattle, Australia 218
hides 179, 186
High Plains-Ogallala Aquifer
  Regional Resources Study 330
High Plains, Texas, 185, 307,
  314–25; climatic change 53;
  groundwater pumping 456
Hit, Iraq 26
Hofuf, Saudi Arabia 340, 348,
  349
Hoover Dam 25, 142, 238, 267,
  268; Colorado River 411
Hollywood, Los Angeles 402
Homestead Act 1862 138, 139,
  184, 311
homesteading, High Plains 310
Hordeum spantaneum 59
Huleh marshes 362; drainage of
  366
human induced degradation 473
hunter-gatherers 57, 62, 69, 70,
  71; aborigines 74
hyaenas 46

hydraulic gold mining 155
hydrocarbons, Los Angeles 433,
  434
hydro-electricity, Aswan High
  dam 252; dams 86; Volga
  River 396
hydrological conditions, climatic
  change 454
hydrological cycle, complexity of
  454
hydrology, Israel 360–3
hydrophytes 41
hyperion sewage treatment
  works, Los Angeles 418, 419,
  420, 421

Ibex 66
Ibrahimiya Canal, Nile valley
  249
ice advances 47
ice cores 49
ice cover, Arctic Ocean 400, 401
Idjil, Mauretania 200
immigration 341
Imperial Dam 240, 268
Imperial Preference 177
incineration, solid waste 466
indebtedness 452; Great Plains
  314; rural families 461
index of aridity, Martonne 4
India, dryland cities 98;
  population control 460
Indian Ocean, British influence
  334
Indian tribes, Colorado basin
  263; irrigation 263
Indian water rights, Colorado
  basin 275
industrial development 96;
  Eastern Province, Saudi
  Arabia 347; Gulf States 342,
  355; Jubail, Saudi Arabia 348;
  pollution 108
industrial growth 82; Eastern
  Province, Saudi Arabia 354;
  Iran 86
industrial development 458
industrial use of water, Israel
  457
infiltration capacity 39

infrastructure development 89, 156; Eastern Province, Saudi Arabia 348; Gulf States 342; Oman 88; road and rail 154
insect pests, Nile valley 250
insects 44
intensive meat production, feed lots 185
interbasin water transfer, Great Plains 330; impact in Soviet Central Asia 399
interceptor sewer, Tel Aviv 375
international agency, policies 469
international food aid 469
International Monetary Fund 109
Inter-Tropical Front, Sahel 204
introduced species, Australia 236
Iran 62, 69, 86, 89, 472; development plans 86–7; food problems 93; Gulf War 156; land reform 94, 95; population growth 460; qanats 130; rain-fed agriculture 164; Shah 86
Iranian plateau 13, 154
Iraq, Euphrates River 26; land reform 94; rain-fed agriculture 164; Shatt al Arab 351
Irbid, Jordan 168
iron ore 152; Australia 161, 468
iron pyrites 155
irrigated agriculture 457; decline of, Israel 457; development of 126; distribution 135–8; energy use in USA 150; expansion 150; future of 456; increase in yields 151; Tucson 467
irrigated area, countries 137; High Plains of Texas 319; Soviet Central Asia 387; USA 140; Wasatch oasis 293
irrigation 25, 58, 125–52; Avra valley, Arizona 123; boom in Texas 316; Colorado basin 272; economics of operation 131; effects on crop yields 126, 133; efficient use of water 133; Egypt 28; Gulf States 332; High Plains of Texas

314–25; Iraq 21, 26; Isfahan 283; linear programming model 148; Mormons 299; Nile valley 72, 239–63; oases in Saudi Arabia 349; pressure systems 130, 132; surface 130; Tigris-Euphrates 72; Wasatch oasis 299–301; water withdrawal for 136, 137
irrigation canals, unlined in USA 148
irrigation civilisations 125; division of labour 70
irrigation companies, Mormons 300
irrigation development, Egypt 256; Soviet Central Asia 387–93; USA 138
irrigation district, USA 140
irrigation movement, USA 468
irrigation networks, modernisation of in USSR 393; prestige projects 457; Wasatch oasis 299
irrigation projects 42, 457; Soviet Central Asia 390; Sudan 259; USA 455, 457
irrigation pumping, subsidies 471
irrigation systems 69, 130–3
irrigation water, classification 134, 135; Sahara desert 199; sources 128–30
irrigation water rights, Isfahan oasis 286
irrigation water use 455; Blaney and Criddle 5; Colorado River 137; Israel 378, 384; Soviet Central Asia 392, 398; subsidies 144, 149; USA 137, 146, 147
Irtysh River 398; navigation 400
Isfahan, population 288; Safavid rule 283
Isfahan Master Plan 291
Isfahan oasis 278–93; annual precipitation 278; crops 288; floods 288; future water demand 292–3; growth of urban area 290, 291; irrigated area 287, 289; irrigation water

needs 290; land use 285–8;
new projects 292; qanats 286;
Zayandeh River 24
Ismail, Egypt 249
Isna barrage, Egypt, Nile river
249
Israel 10, 13, 254, 457, 468;
immigration 357; precipitation
358, 359; water development
357–85; water resources 467;
water use 262
Istanbul 467
Ivanhoe mine, Australia 160

jackrabbit 44
Jaj River, Iran 72
Jarmo 66
Japan, cultured pearls 334;
market for Australian wheat
178
Jebel Akhdar Mountains, Oman
332
Jebel Aulia dam 250
Jerboa 44
Jericho, Jordan 65; gazelle 66;
irrigation 66, 126
Jerusalem 358; cholera epidemic
373
Jiddah, Saudi Arabia,
desalination plant 352
Johnston Plan, Jordan River 369
Joint Outfall System, California
421
Jonglei canal 254
Jonglei Investigation Team 254
Jordan 12, 13, 360; cereal
production 94; environmental
degradation 170; food imports
460; rain-fed agriculture 164,
168–75
Jordan River 360, 361, 367;
water balance 361
Ju'aymah, Saudi Arabia 349
Jubail, Saudi Arabia 347, 348;
immigrants 341; industry 340
juniper berries 67

Kairouan, Tunisia 277
Kakhovka project, Ukraine 391
Kalahari desert 46, 187

Kalgoorlie, Australia 158, 159,
160; gold 158
Kama River, USSR 395
kangaroo 45, 219; effects of
drought on 232, 235;
Gascoyne Basin, Australia 226
kangaroo rat 44
Kansas 310, 330; irrigation 145;
Ogallala aquifer 323
Kara Kum canal 389
Karshi Steppe project, Soviet
Central Asia 391
Karun River, Iran 288
Kazakhstan 386
Kenya 243
Kerman, Iran 13, 14;
temperatures 14
Khartoum, Sudan 239, 244, 250,
259, 262
Khomeini, Ayatollah 87
Khuzestan, nomadism 181
Kikuyu grass, Australia 219
Konya plain, Turkey 66
Kopet Dagh mountains 389
Kuban River, USSR 391, 392
Kubena Lake, USSR 395
Kuhrang River, Iran, diversion
288
Kuwait 332, 334, 343, 467; oil
production 336; urban growth
340; water needs 350
Kuwait Oil Company,
desalination plant 350

Lacha Lake, USSR 395
Lake Albert 239, 255
Lake Huleh, Israel 372; nature
reserve 366
Lake Kioga 239, 255
lake levels, climatic indicators
454
lake level variations 50
Lake Mead, USA 25; water
storage 268, 270
Lake Nasser, Egypt 28;
evaporation losses 260;
sediment accumulation 253;
seepage losses 260; storage
capacity 252
Lake Onega, USSR 395, 396

Lake Powell, water storage 270
Lake Tana, Ethiopia 239, 243, 244, 255
Lake Tiberias 360, 362, 365, 366, 380; pollution 373; storage capacity 369
Lake Victoria 239, 243, 255
Lake Zeribar, Iran 70
land capability class, Wadi Ziqlab, Jordan 174
land degradation 453, 469, 474; South Africa 188
land reclamation schemes, Egypt 256
land reform 94–6, 461
land settlement policy, use of irrigation 457
land use, Australia 216
Lander Creek, Australia, Walbiri tribe 75
landlords, land reform 96; political power 461
landslides, Los Angeles 425
laser levelling techniques, Colorado basin 273
Las Vegas, Nevada, hydro-electricity 142
lead 156
lead in atmosphere, Los Angeles 445, 446
leases, Alice Springs district 228; Australia 220; Gascoyne Basin, Australia 222; stocking rates in Australia 221
Lebanon 93, 360
Lee Ferry, USA 263, 266, 270, 272
legislation, motor vehicle emissions 466
legumes 66
Lena River, USSR 387, 395
lentils 65
Lewis Smith water diversion scheme 328
Libya 195, 201
lignite, Ankara 466
Lima 464
lion 46

Litani River, Lebanon, diversion of 367
livestock farming 67
livestock losses, drought in the Sahel 205
livestock numbers, Niger 211
lizards 44
locusts, Sahel 205
Logan, Utah 293
Long Beach, USA 403
Los Angeles aqueduct, Owens valley 410
Los Angeles County Air Pollution Control District 432
Los Angeles County Flood Control District 427
Los Angeles plain 403
Los Angeles River 36, 403, 409; climate 406; floods 426, 430; sewage collection 418
Los Angeles, USA 29, 98, 141, 142, 420, 422, 464, 466, 467; conurbation 402–48; crime 402; environmental management 402–48; flood control system 111; smog 466
Los Angeles Water Department 409
low interest rates, for irrigation projects 456
Lubbock, USA 316
lynxes 46

machinery 461
Mackenzie River 327
Maghreb 179
maize 211
major dam projects, irrigation networks 89
Malakal 243, 254
Mali 200, 474
mallee shrubs, expansion of wheat farming into 177
mammals 44
management of air pollution, Los Angeles 439–42
management strategies, pastoral activity 190
Managil irrigation project, Sudan 251

Man and Biosphere Programme, UNESCO 6
Mancos Project, Colorado 268
Manson Project, Washington 143
map of world distribution of arid regions, UNESCO 6, 7, 8
marginal land, commercial pastoralism 185, 461; rain-fed agriculture 164
Marginal Lands Scheme, Australia 177
marine pollution, Gulf 347
maritime trade routes, Gulf 333
market economies 469
markets, commercial pastoralism 183, 186
Marra mountains, Sudan 196
Maryat, Egypt 258
Mashar swamps, 255; evaporation reduction 254
mass movement 33
Master Gas System, Saudi Arabia 347
Mauretania 83
meat 179; commercial pastoralism 182
medical care 458
Mediterranean 472
Mediterranean, check dams 452; terraces 453
Mediterranean Sea 62
Mekorot Water Company, Israel 364
Melbourne, Australia 178
merino sheep, Australia 218; Gascoyne Basin, Australia 223
mesolithic 58
mesophytes 41
Metropolitan Water District, Los Angeles 411
Mexicali Valley, Mexico 275; water demand 276
Mexican Water Treaty 271; water from Colorado River 268
Mexico 275, 327, 403; irrigation 268

Mexico City 464; smog 466; subsidence 114
Middle East 452, 458, 472; check dams 452; food imports 459; mining 468; nomadism 179; oil 458; oil revenues 469; oil rich countries 460; urban growth rates 99
migrants, Gulf States 341; Los Angeles 403; Oman 88
migration routes, nomads 181
migration, to urban areas 461
Milankovitch 48
milk 179; Sahel 207
millepedes 44
millet, Niger 211; Sahel 207
mineral exploitation 152–62
mineral prospecting, Australia 161
mining, boom 157; 161–2; capital investment 157; catalyst to development 162; environmental pollution 158; infrastructure provision 162; Western Australia 158–61
mining boom, Walbiri tribe 77
mining of groundwater, Ogallala aquifer 317
mining of vegetation, Gascoyne Basin, Australia 224
Ministry of Irrigation, Egypt 255
Mississippi River 145; water imports for Texas 322
Missouri River 330
modernisation of agriculture 94
modern societies 82–96
Mollisols 37, 38
molluscs 65
mono basin, USA 410
monoculture, commercial agriculture 165
monsoon, Ethiopia 195
Montana, irrigation 145
Moon Lake Project, Utah 268
moors 198, 199, 200
Mormons 293, 298; irrigation 295; Salt Lake City 278
Mormon settlements 296–9
Morocco 195, 199; cereal production 94

Mount Hermon 362
Mount Newman, Australia, iron ore 161
Mount San Gorgonio 406
Mount Tom Price, Australia, iron ore 161
movement of pastoralists, Sahara 201
mudflows 34; Los Angeles 425
mule deer 45
mules 168
multi-nationals, mining 153
multi-purpose dam schemes, development planning 86; Iran 86
Mureybet 64, 65
Murray Basin, Australia; salinity problems 457
Muscat, Oman, urban growth 340
myxomatosis, Australia 220

Na'aman River 360
Nabateans 71
Nag Hammadi barrage, Nile river 249
Nahrawan Canal 127
Nasser, President 251
Natal, South Africa 187
National Flood Insurance Programme, USA 430
National Sewerage Plan, Israel 375
National Water Carrier 365, 378, 379; Israel 366–71, 375, 377
National Water Commission, Israel 143
National Water Company, Israel 377
National Water Council, Israel 376
Natufian hunter gatherers 64, 65
natural gas 152, 156
natural vegetation, degraded 472
Navajo Project, New Mexico 270
NAWAPA 413
Nebraska 314; irrigation 145; Ogallala 323; water imports 330

Negev 71, 365; groundwater 363; irrigation 367
Nekouabad regulator, Zayandeh river 290
neolithic 71, 72
Nevada 464
Newlands Project, Nevada 141
New Mexico, Ogallala 323; USA 10, 11
New South Wales, wheat production 175
New World 8
Ngalia country, Australia 75; Walbiri tribe 75
nickel, Australia 161
Niger 200, 474; pastoralism 202, 213
Nigeria 199, 207
Niger River 196, 213
Nile basin 238; hydraulic works 242
Nile delta 256
Nile flood 28, 68, 247; fallow period 247; sediment 247
Nile flood plain 67, 69; irrigation 67, 471; urban developments 69
Nile river 21, 83, 109, 196; Aswan High Dam 28; coastal erosion 253; discharge 244, 245; discharge variations 51; flow at Aswan 27; hydrology 243–6; integrated development 262; irrigation system 28; regulation of 255; water supply 262
Nile valley 198; development of basin irrigation 127; flood height 128; irrigation 126
Nile Water Agreement 251, 259
nitrate pollution, Israel 372
nitrates, Egypt 250
nitric oxide, Los Angeles 434
nitrogen dioxide, Los Angeles 435, 446
nomadic population groups 180
nomadism 179; Niger 202; patterns of movement 180; traditional pastoralism 179

nomadism in Sahel, survival of 213
nomads 57, 72; management of, Niger 209; raiding parties 180; Sahel 207
North Africa, relative humidities 16; urban growth rates 99
North America 45
North American Water and Power Alliance NAWAPA 326, 327; water transfer 326, 327
North Tahrir, Egypt 256
Northern Highlands, Jordan 168
Northern Territory 75; evaporation 17
Nubarya Canal, Egypt 258
Nubian Sandstone aquifer 29, 30, 301; Negev 363; Sinai 363
nuclear weapons, dryland testing 164
Nularbor Plain, Australia 15; evaporation 17
nutrients 47

oases, 277–306; major urban centres 277; Sahara 199
oasis life 199
Ob River 387, 393, 398, 400; navigation 400
ocean outfall for sewerage, Los Angeles 416, 420
Oceans Plan, California 423
offshore oilfields 156
Ogallala aquifer 29, 30, 314, 317; outcrop 324; recharge 318; storage volumes 320; water table decline 317, 318
Ogden, Utah 306
oil 152, 156
oil pipelines 156
oil production, Gulf States 335
oil revenues 83, 336; Fifth Plain, Iran 87; Gulf States 334–8; imports in Saudi Arabia 349; Oman 87; Saudi Arabia 337, 460
Oklahoma 330; Ogallala aquifer 323
olives 62

Omaha, Nebraska 310
Oman 88, 89, 332, 343; development plan 87
Onega Gulf, USSR 396
OPEC 338; quadrupling oil revenues 378; oil price rise 87
open-cast mining 152
Orange County Sanitation District 421, 422
Orange Free State, South Africa 187
Orange groves, Los Angeles 402
Oregon 312
origins of agriculture 58
Oroville dam, Feather River 411
Oryx 43, 46
ostrich 46
ostrich feathers, trade 200
Othents 38
overgrazing 39, 452; Alice Springs district 230, 232; commercial pastoralism 190; Escalante, Utah 304; Gascoyne Basin, Western Australia 223; overgrazing, grassland ecosystem 462; nomads 180; South Africa 188
overland telegraph lines, Australia 227, 228
overpumping of aquifers, Israel 364
overstocking, Australia 221; South Africa 189
overwatering, irrigation 457
Owen Falls Dam 239, 255
Owens Valley, California, water for Los Angeles 410, 467
ox drawn plough 68
oxen 168
oxides of nitrogen, Los Angeles 433, 434
ozone, control problems in Los Angeles 442; Los Angeles 435
ozone levels, Los Angeles 439
ozone levels in the atmosphere, Los Angeles 443

paddocks, animal numbers in Australia 233; Gascoyne Basin, Western Australia 226

Pakistan, population growth 460
palaeolithic 58
Palestine, British Mandate 364
Palestine Water Company 364
Palestinian refugees 173
Palo Verde Project, California and Arizona 268
Palos Verdes Hills 36; Los Angeles 403
Pastoral Appraisement Board, Australia 223
pastoral farming 57
pastoral industry, decline of Alice Springs district 230
pastoral leases, Alice Springs district 233; Australia 184, 215; subdivision 217
pastoral modernisation zone, Niger 207
pastoral practices, Australia 217–21
pastoral systems 179–91
pastoralism, oasis based 200; retreat to interior of Australia 217; Sahel 204
pastoralists, Australia 468
Parker dam, Colorado 411
Parker Davies Project, Arizona, California and Nevada 268
pearl fishing, Gulf, 334
peas 67
Pechora River 395, 396; navigation 400; water quality 400
Peking, China 98
Penman, potential evapotranspiration 19
perennial irrigation, Nile valley 247, 248–63
perpetual leases, South Africa 187
Persian Gulf 33; oil pollution 156
Perth, Australia 160
petrochemical plants, Gulf States 338
petroleum, Los Angeles 402
Phoenix, Arizona 118, 241, 276, 464; water demand 276; water needs 272

phosphates 152
photochemical reactions of smog, Los Angeles 432
photochemical smog, Los Angeles 408
pigs 62, 69
Pilbura mine, Australia, iron ore 161
pipelines 455
pistachio nuts 66, 67
planning regulations, Third World Cities 108
pleuroneumonia, Sahel 207
Plio-Pleistocene aquifer, Israel 363, 364
pollution, atmospheric 108; automobiles 108; industrial 108; oil industry 156
pollution of aquifers, Gulf States 347
pollution of beaches, Israel 375
population control 460; India 460; China 460
population growth 89–94 451, 458, 469; arid countries 91; Egypt 254; industrialisation 90; Israel 357; Los Angeles 403; post Second World War 92; Spain 460; strategies to limit 460; urbanisation 90
population growth rates, Niger 203
population pressure, Sahel 213
population projections, Colorado basin 275
Porter-Colone Water Quality Act 1969 423
ports, Gulf 334
potential evapotranspiration, turbulent-transfer approach 19; water balance approach 19
Powell, John Wesley 138, 139, 140, 141, 266
Powell National Forest, Utah 303
precipitation 4, 8–13; Agadez, Niger 205; Australia 9; frontal systems 13; intensity 10; inter annual variability 6; inter-quartile range 10; rain-days

10, 13; snowfall 11; Tahoua,
Niger 205; variability 10
precipitation variations, Los
Angeles 407
Pre-Pottery Neolithic 65
primary productivity 47
productivity gains, agriculture
458
profit motive, commercial
pastoralism 191
Pronghorn 45
Provo, Utah 293, 306
Psamments 38
pumping costs of water, Tucson,
Arizona 118
pumping of fossil water, Gulf
States 345

Qattara depression 30
qanat 74, 128; cross section and
plan 129; discharge 130;
lengths 130; mother well
depths 130; systems 130; water
flow 130
Qatar 343; oil discovery 336;
urban growth rates 340
Qatif 349; Saudi Arabia 340
Qishon River 360
quaternary era 47
Queensland 230; dingo-proof
fences 220

rabbits, Australia 220; Gascoyne
Basin, Western Australia 226;
myxomatosis 220
Rahud river 244
rail freight rates, wheat
production, Australia 177
railway, Alice Springs, Australia
230
rain fed agriculture 58, 67,
163–78, 472; bumper crops
164; commercial 165; crop
failure 163, 164; government
subsidies 168; low yields 163;
Niger 202, 208, 211; solar
energy 164; traditional 165;
wheat 165
rain gauges, recording 10

rain water harvesting 70; Niger
70
rainfall variability 164; Agadez,
Niger 205
ranching 182; Australia 182;
distribution of in Gascoyne
Basin, Western Australia 226;
North America 182; South
Africa 182; South America
182
ranges, commercial pastoralism
183
Ras'al Khafji, Saudi Arabia,
desalination plants 352
Ras Tanurah, Saudi Arabia 349;
oil termimal 347
rates of soil erosion 453
Raudhatain, Kuwait,
groundwater 351
Reclamation Act 1902 141, 142,
144, 266
Reclamation Bureau, USA 141
Reclamation Project Act 1939
143
Reclamation Reform Act 1982
144, 145
Reclamation Service 266
recreational activity 471
recreation, drylands 452
Red Cross 473
Red Indians 183
Red Sea, relative humidities 16
refrigerated rail wagons, Great
Plains 311
refrigeration, introduction of 184
refuse, urban areas 108
relative humidity 15–16; diurnal
and seasonal variations 16
rendzinas 463
reservation doctrine 275
retirement communities, South
West USA 464
revenues from oil, Middle East
469
reverse osmosis 467
Reza Shah, nomad
sedentarisation 181
Rift Valley, Jordan 168

Rinderpest, Sahel 207
river, discharge decrease
  downstream 23
river basin development 238–76
river crossings, Nile river 253
river discharge, variations 51
river diversions, European part
  of USSR 394; Siberian part of
  USSR 397
river regimes, modified by man
  25; snow melt 21
river runoff 21–8; water balance
  approach 21
Riyadh 348
road construction, Oman 88
road infrastructures, Gulf States
  354
road train, Alice Springs District
  230
road system, Gulf States 348,
  349
Rocky Mountain Trench 327
Rocky Mountains 307
rodents 44
Roosevelt Dam, Arizona 141
Roseires, Sudan 244; Dam 250
runoff farms 70; Negev 70
runoff fluctuations Colorado
  River, USA 455
rural areas, population growth
  92
rural indebtedness 462
rural institutions, co-operatives
  95; land reform 95
rural populations 86
rural urban migration 92

Sabkhas 33; Saudi Arabia 353
Sacramento River, California
  411
safe yield of aquifers, Israel 370
Sago palm 58
Sahara 46; temperatures 14;
  traditional pastoralism
  195–215
Saharan trade, disruption 202
Sahel 53, 198, 200, 201, 473;
  desertification 469; nomadism
  180, 204, 214; pastoralism 202;
  rainy season 205

Sakieh, Egypt 248
saline soils, Egypt 250;
  reclamation of 151; Soviet
  Central Asia 392
salinisation 150; irrigation 151
salinity hazard, irrigation waters
  135
salt, trade 200
salt crusts 35, 36
salt desert 35
Salt Lake City, USA 277, 294,
  296, 297, 306, 464;
  communications 298
Salt River, Arizona 241
Salt River Project, Arizona,
  water storage capacity 267
salt weathering, cities 114;
  concrete 114
Samarkand, USSR 277
San Bernadino Mountains 406
San Bernadino National Forest,
  smog damage 436
San Bernadino Valley 406
San Bruno Mountains 111
sand dunes 35; oases 36
San Fernando Valley 406, 409,
  435
San Francisco, USA 111
San Francisco Bay area 410, 411
San Gabriel Mission, Los
  Angeles 408
San Gabriel Mountains 406, 412;
  floods 426
San Gabriel River 403; floods
  426, 430
San Gabriel Valley 406
Sanitation District of Los
  Angeles 421
San Joaquin River, California
  412
San Pedro Bay 36, 403
Santa Anna Mountains 403
Santa Anna River 403, 421;
  floods 430
Santa Anna winds, Los Angeles
  408, 433
Santa Cruz River, Arizona,
  irrigation 266; Tucson 117
Santa Fe Flood Control Basin,
  Los Angeles 430

Santa Monica Bay, California 36, 403, 418
Santa Monica, California 406
Santa Monica Mountains 403
Sassanian period, Diyala plains 127
satellite systems, monitoring environmental change 462
Saudi Arabia 93; aquifer systems 32; Euphrates River 26; oil production 336; oil revenues 460; wheat production 459
savannahs 40
Schofield Project, Utah 268
scorpions 44
scrub desert 40
sea breeze, Los Angeles 406, 433
sea fog, Los Angeles 406
Sea of Azov 396, 399
sea water intrusion into aquifer, Israel 363, 365; Los Angeles 422
Sedalia, USA, railhead 310
sedentarisation of nomads, Iran 181
sedentary agriculture 62; population densities 70
sediment and soil movement, Mediterranean region 453
seed agriculture 57, 58
seed production, Sahel 206
self-sustaining economic growth 474
semi-nomadic groups 198
Senegal 199
Senegal, River 196
Sennar Dam 250, 251
Sepulveda Flood Control Basin, Los Angeles 430
settlement, establishment of 64; Great Plains 310
Seventh Approximation 37
sewage, Los Angeles 415–25; USA cities 107
sewage sludge, Los Angeles 423
sewage systems, disease potential 467
sewage systems, Los Angeles 416

sewage treatment technologies 416–18
Shadouf, Nile valley 248
Shah Abbas Dam, Iran 288, 290; Isfahan 283
Shanidar 64, 65
Shasta Dam, California, Sacramento River 411
Shatt al Arab, Iraq 350
shearing, Australia 218
sheep 46, 62, 66, 67, 69, 72, 198; Australia 217
sheep production, Gascoyne Basin, Australia 226; South Africa 188
sheep station boundaries, Gascoyne Basin, Australia 222
sheet erosion 39
shifting cultivation 69
Shorthorn cattle, Australia 218
shrub savannah 195
Shuwaikh, Kuwait, desalination plant 350
Siberian water diversion projects 398
Sierra Nevada, California, gold mining 155
silver 157
single family dwelling, Los Angeles 402
Sioux Indian 310
Siwa oasis, Egypt 30
skins 186
slavery 204
slope gradients, Wadi Ziqlab, Jordan 172
slope stability, landslides 111
smog, Los Angeles 431; harmful effects in Los Angeles 439
snakes 44
snowfall, drylands 10
sodium adsorption rates, irrigation water 135
sodium (alkali) hazard, irrigation water 135
soil classification 37
soil conservation measures, Great Plains 313; Wadi Ziqlab, Jordan 173
soil conservation techniques 453

soil deforestation 165; erosion
452, 462, 463; gully 463;
Jordan 167
Soil erosion rates, Wadi Ziqlab,
Jordan 171, 172, 173
soil moisture reservoirs 24, 163
soil profile 36
soil salinity, Egypt 256;
irrigation 457; Soviet Central
Asia 401
solar radiation, variations 48
Soldier Settlement Schemes,
Australia 177
Sorbat River 243, 254, 255
sorghum, Niger 211; Sudan 259
South Africa, commercial
pastoralism 186–90; diamonds
458; farm amalgamation 178;
gold 458; gold mines 159; land
degradation 189; land prices
188; mining 468, 469; rainfall
variations 54
South Australia 227, 230, 278,
279; dingo-proof fence 220;
wheat production 175
South Coast Air Basin, Los
Angeles 442
South Coast Air Quality
Management District 439
South Coast Air Quality
Management Plan, California
442
Southern California Air
Pollution Control District 439
Soviet Central Asia, irrigation
386–401; water transfers
386–401
Spain, population growth 460
spatial distribution of air
pollutants, Los Angeles 435
spiders 44
spoil heaps, mining 152
sprinkler irrigation 130, 457;
High Plains of Texas 317;
Israel 366, 383; water use 133
stable runoff 137
Strawberry Valley Project, Utah
267
Steppe desert 195
stocking levels, Australia 219

stock reduction scheme, South
Africa 189
stone armouring 38
strip mining 152
strip planting of crops, Wadi
Ziqlab, Jordan 173
Stuart, John McDonald 228
Subdivision of Agricultural Land
Act 1970 188
subsidies, for irrigation projects
456; food supply 462;
government 471; pastoral
industry in Australia 236;
production of wheat in Saudi
Arabia 459
subsistence agriculture 458;
Egypt 472; Middle East 452;
Mediterranean region 452
succulents 41
Sudan 239, 243, 244, 251, 255,
260; agricultural sector 258;
desertification 258, 259;
famine 196; future irrigation
development 262; irrigation
costs 262; labour force 258;
nomadism 180; population
growth 255; potentially arable
land 259; rapid population
growth 262; slaves 200
Sudd swamp 239, 243
Suez Canal 175, 251
Suez, Egypt, groundwater
quality 115; salt weathering
116
Sukhona River, USSR 395, 396;
flow reduction 400
Sulak River, USSR 391
Sulibiyyah, Kuwait, water well
field 351
sulphates, salt weathering 114
sulphur dioxide in the
atmosphere, Los Angeles 434,
439, 446
Sultan Qaboos 87
Sumer 26, 68
summer crops, Nile valley 247
summer monsoon 244; Australia
16
Sun Belt, USA 464; urbanisation
97, 451

superphosphates, Egypt 250
surface water, Sahel 206
surface water withdrawals,
 Colorado River 272
sustainable development 451,
 461, 473
Sydney, Australia 178
Syrdarya River, USSR 389, 399;
 water problems 392
Syria 64, 360, 472; Euphrates
 River 26, 28

Tabqa Dam, Syria 26
Tahal, Israel 376, 377, 385
Tahoua, Niger 202; French
 occupation 204
Tallow 186
Tamanrasset, Algeria 200
Tamboor, Nile valley 248
Tanninim River, Israel 360, 370
Taoudenni, Mali 200
Taro 58
Tasmania 74, 75
Tchin Tabaraden, Niger 209;
 livestock losses 212
Tehachapi Pass, San Gabriel
 Mountains 412
Tehran, Iran 10, 87, 464, 467;
 air pollution 108
Tel Aviv, Israel 358, 365, 467;
 municipal water supply 365;
 sewage treatment 375; water
 needs 364
telegraph, Australia 160;
 England to Australia 175
telegraph network 184
temperature 13–15; ranges 13
temperature inversion, Los
 Angeles 433
Terek river 391
terraces 452; collapse of 453
Terra Rossa 462
Texas 310, 314, 321, 473;
 groundwater law 325; growing
 season 309; irrigation 145;
 water imports 330
Texas Department of Water
 Resources 318
Texas Water Development
 Board 321

Texas Water Plan 320, 321, 322
thermal processing of sewage,
 Los Angeles 424
Third World cities, water
 provision and sewage disposal
 109
Thornthwaite, potential
 evapotranspiration 19
Tibesti Uplands, Chad 196, 198,
 200
Tigris–Euphrates lowlands 21,
 67, 238; agricultural
 productivity decline 127;
 irrigation 126; saline soils 68;
 snowmelt floods 126; urban
 development 69
Tigris River, drain for irrigation
 waters 127
Tijuana, Mexico, growth rates
 99
Tobolsk, USSR 398, 400
tomatoes, Niger 211
tourist activity, 'honey-pot' sites
 463
trace-mineral deficiencies,
 pastures in Australia 220;
 wheat farming in Australia
 177
tractors, soil erosion 168
trade, camel caravan 200
traditional agricultural methods,
 environmental damage 452
traditional land use 198
traditional pastoralism 179–82
traditional societies 57–82
Trans-Arabia Pipeline 348, 372
transport, commercial
 pastoralism 190; subsidies in
 Australia 178; wheat
 production in Australia 175
Trans-Texas Canal 321
Transvaal, South Africa 187, 189
tree-ring chronologies 51
tribal federations, Gulf 334
trickle irrigation 130, 383, 457;
 saline waters 132
triticum boeticum 59
triticum dicoccoides 59
Tsalmon, Israel 369
Tuaregs 198, 199, 200, 202, 211;

confederations 204; Niger 203; tribute 180
tuberculosis, Alice Springs district 231
Tubu 198
Tucson, Arizona 276, 415, 464, 466, 467; changing attitudes to water use 123; growth rate 99; overpumping of groundwater 117; per capita water use 119; water demand 122, 272, 276; water rating policy 121; water supply 116–25
Tunisia 201
Turkey 69, 93, 472; Euphrates River 26, 28; rain-fed agriculture 164; water surplus 21

Uganda 243
Uinta Mountains 263
Uintah basin 270
Ukraine, irrigation 391
Umm Al Aish, Kuwait, groundwater 351
Umm ar Radhuma aquifer 353
Uncompahge Project, Utah 267
underground water conservation districts 320
undersea aqueduct, California 413
UNEP, desertification 469
United Arab Emirates 332; oil production 336; urban growth rates 340
United Nations 451, 473
United States Geological Survey 140
Universal Soil Loss Equation 39, 40
Upper Colorado Compact 1948 270
Upper Volta 199
uranium, Australia 161, 468
urban agglomerations, dryland nations 98
urban areas, air pollution 466; prosperity 462; attractions 461
urban/industrial water demand 25; Saudi Arabia 354

Urban management policies 99–109
urban system 464
urban waste waters, recovery 468
urbanisation 97–125, 464, 471; building materials 108; economic activity 464; Gulf States 334, 338–45; slope stability 111
US Corps of Engineers, flood control in Los Angeles 427
USA 10, 35, 254, 327, 463, 464, 468, 474; dryland cities 98; groundwater use 146; irrigation 136, 138, 145; lake levels 51; laser levelling 132; rain-days 11; resource use of cities 107; urban growth rates 98; urban water demands 142
USSR 254, 464; dryland cities 98; irrigation 136; stable runoff 137
Ustalfs 38
Utah 299, 306
Utah Valley 295

Varamin Plain, Iran 72; irrigation systems 73
Varzaneh, Iran 284
vegeculture 58
vegetables 72
vegetation 40–3
vegetation changes, Australia 234;· Gascoyne Basin, Australia 227; Niger 213
vegetation, deterioration 463
vertisols 37, 38
Vetchling 67
veterinary programme, Sahel 207
Victoria, Australia, wheat production 175
village, rain-fed agriculture 164
vine 62
Virgin Lands Programme 1954–60; USSR 389
Volga River 387, 389, 391, 393, 395, 396, 399.

Volga–Ural canal 391
Vostok ice core 49
Vozhe lake, USSR 395
Vychegda River, USSR 395

Wadi Ziqlab, Jordan 169–75;
    modernisation of agriculture
    170
Walbiri tribe, Australia 75, 77;
    food sources 78–83; tribal
    divisions 76
wallaby 45; Australia 219
Wall Street Crash 1929 313
Walmalla Territory, Australia,
    Walbiri tribe 75
Wasatch Mountains 293, 306;
    precipitation 296
Wasatch Oasis, animals 293;
    crops 293, 298; population
    growth 300, 301, 306; Salt
    Lake City 293
Wasia aquifer 30
wasps 44
waste water disposal 467
Wastewater Facilities Plan, Los
    Angeles 423
wastewater reclamation, Israel
    373–6; Los Angeles 424
Water and Power Resource
    Service 266
water associations 299
water availability 47; Sudan 259
water balance 17–21; Berkeley,
    California 20; energy balance
    approach 4, 5; Israel 385;
    Penman 4; Thornthwaite 4;
    turbulent transfer approach 4,
    5
Water Code, 1903, Utah 300
Water Commission, Israel 377
water conservation measures,
    High Plains of Texas 317, 322
water consumption, irrigation in
    Israel 457
water costs, California Aqueduct
    413
water demands, Eastern
    Province, Saudi Arabia 354;
    Gulf States 350

water distribution systems,
    Kuwait 351
water, domestic and industrial
    use 466
water erosion 39
water imports, Great Plains
    325–31
Water Law, 1959, Israel 376,
    378
waterlogging of soils 150; Egypt
    250, 256; Soviet Central Asia
    392
water management, Israel 370
water master, Mormons 300
water meters, Los Angeles 409
water pollution, sewage 108
water prices, USA 149
water pricing 455
water quality, California State
    Water Project 422; Colorado
    River Aqueduct 422; decline
    in Israel 371
water rights, Owens Valley 410
water resource development,
    Israel 364–7; multi-purpose
    approach 142
water resources, Amudarya 396;
    Australia 218; Colorado basin
    265; commercial pastoralism
    183; Gulf States 345; Iran 86;
    Los Angeles 414; Soviet
    Central Asia 387; Syrdarya
    396; Tucson 117
water shortages 455, 456; Gulf
    States 355; Israel 378
water storage, Lake Nasser 251;
    reservoirs 25
water supply, Egypt 261; Los
    Angeles 408
water subsidies, Israel 383
water surplus 30; Middle East
    22; water balance 21
water table rise 151; Egypt 250;
    West Nubarya Project, Egypt
    258
water transfer schemes, Soviet
    Central Asia 393–9
water use, Israel 379, 381
water utilisation rates, for crops
    456

water wells, Alice Springs district 230, 232; commercial pastoralism 190; drilling in Australia 236; effect on cattle numbers in Australia 235; overpumping 128
Welton–Mohawk Project 273
West Bank, Jordan, Palestine refugees 173
Western Australia, evaporation 17; farm amalgamation 178; gold 160; wheat production 175, 177
West Nubarya Project, Egypt 256; land reclamation scheme 257
wetlands, Huleh marshes 371
Whittier Narrows Flood Control Basin, Los Angeles 430
wheat 64, 66, 69; wild ancestors 59
wheat exports, Saudi Arabia 459
wheat imports, Middle East 458
wheat production, Australia 175–8; Middle East 93; Niger 211; Saudi Arabia 349, 459
wheat storage facilities, Saudi Arabia 459
White Nile 239, 243, 244, 255
wild ass 64
wild boar 66
wind 16–17; dust 17
wind erosion 39, 165, 168; Jordan 175
wind pumps 184; Great Plains 310; High Plains of Texas 314
Wind River Range 263
Winters Doctrine, Indian water rights 275

woodlice 44
wool, 179; Australia 217; Gascoyne Basin, Australia 226; South Africa 188
World Bank 109

xeralfs 38
xerophytes 41

Yalpari region, Australia, Walbiri tribe 75
yams 58
Yarkon–Negev Project, Israel 369, 376
Yarkon River, Israel 365, 370, 375
Yarkon–Tanninim aquifer 379
Yarmouk River, Jordan 360, 362, 369
Yemen, South 83, 84
Yenesei River, USSR 387, 395, 399, 400
yields, cereals 94
Yukon River 327
Yuma Project, USA 267

Zagros Mountains, Iran 62, 70, 278, 284, 288; nomadism 181; snowmelt 21
Zamankhan, Iran 284
Zawi Chemi 64
Zayandeh River, Iran 23, 24; discharge 285; irrigation districts 286; Isfahan 278; precipitation variations 283; river management 288–93; water quality variations 284
Zifta barrage, Nile River 249
zinc 156